Shape and Energies of Geometric Objects

SERIES ON KNOTS AND EVERYTHING

ISSN: 0219-9769

The Series on Knots and Everything: is a book series polarized around the theory of knots. Volume 1 in the series is Louis H Kauffman's Knots and Physics.

One purpose of this series is to continue the exploration of many of the themes indicated in Volume 1. These themes reach out beyond knot theory into physics, mathematics, logic, linguistics, philosophy, biology and practical experience. All of these outreaches have relations with knot theory when knot theory is regarded as a pivot or meeting place for apparently separate ideas. Knots act as such a pivotal place. We do not fully understand why this is so. The series represents stages in the exploration of this nexus.

Details of the titles in this series to date give a picture of the enterprise.

Published:

Vol. 80: *Shape and Energies of Geometric Objects*
by L. H. Kauffman, L. S. Velimirović, M. S. Najdanović & S. R. Rančić

Vol. 79: *High Dimensional Knotting: An Illustrated Guide*
by D. Roseman

Vol. 78: *Four-Dimensional Paper Constructions After Möbius, Klein and Boy*
by E. Ogasa

Vol. 77: *Quipu: Decorated Permutation Representations of Finite Groups*
by Y. Bae, J. S. Carter & B. Kim

Vol. 76: *Combinatorial Knot Theory*
by R. A. Fenn

Vol. 75: *Scientific Legacy of Professor Zbigniew Oziewicz: Selected Papers from the International Conference "Applied Category Theory Graph-Operad-Logic"*
edited by H. M. C. García, José de Jesús Cruz Guzmán, L. H. Kauffman & H. Makaruk

Vol. 74: *Seeing Four-Dimensional Space and Beyond: Using Knots!*
by E. Ogasa

Vol. 73: *One-Cocycles and Knot Invariants*
by T. Fiedler

Vol. 72: *Laws of Form: A Fiftieth Anniversary*
edited by L. H. Kauffman, F. Cummins, R. Dible, L. Conrad, G. Elisbury, A. Crompton & F. Grote

More information on this series can also be found at http://www.worldscientific.com/series/skae

Series on Knots and Everything — Vol. 80

Shape and Energies of Geometric Objects

Louis H Kauffman
University of Illinois at Chicago, USA

Ljubica S Velimirović
University of Niš, Serbia

Marija S Najdanović
University of Priština in Kosovska Mitrovica, Serbia

Svetozar R Rančić
University of Niš, Serbia

World Scientific

NEW JERSEY • LONDON • SINGAPORE • BEIJING • SHANGHAI • HONG KONG • TAIPEI • CHENNAI • TOKYO

Published by

World Scientific Publishing Co. Pte. Ltd.
5 Toh Tuck Link, Singapore 596224
USA office: 27 Warren Street, Suite 401-402, Hackensack, NJ 07601
UK office: 57 Shelton Street, Covent Garden, London WC2H 9HE

Library of Congress Control Number: 2025032273

British Library Cataloguing-in-Publication Data
A catalogue record for this book is available from the British Library.

Series on Knots and Everything — Vol. 80
SHAPE AND ENERGIES OF GEOMETRIC OBJECTS

ISBN 978-981-98-1945-4 (hardcover)
ISBN 978-981-98-1946-1 (ebook for institutions)
ISBN 978-981-98-1947-8 (ebook for individuals)

For any available supplementary material, please visit
https://www.worldscientific.com/worldscibooks/10.1142/14487#t=suppl

Desk Editors: Nambirajan Karuppiah/Angeline Husni

Typeset by Stallion Press
Email: enquiries@stallionpress.com

This book is dedicated to our families.

Preface

The primary motivation for this book is to analyze application of geometry in shape consideration of different objects. A useful descriptor of the shape of organs, tissues, and cells are the curvature and its functionals. Curvature of the surface that bounds them has a strong influence on cells and is intrinsically linked to the biological processes which occur within them. A deeper understanding of the interactions between shape and biology has fundamental implications for our healthcare, due to its importance in the processes of embryogenesis, organogenesis and the design and development of new materials for tissue engineering. This can be applied to develop bioinspired structures with controlled geometries. Biological actuators provide a rich source of inspiration for the architectural design of their synthetic analogs. We try to offer the mathematical way of consideration of shape.

In the first part, we consider the shape. That is all the geometrical information that is invariant to translations, rotations, and size changes. We consider deformable objects at small, infinitesimal deformations that are of special interest. The shape of geometric objects is determined by their geometric curvature as well as curvature-based functionals which are said to be the energies of curves and surfaces. Geometric objects that are discussed are curves, surfaces, and knots. Different kinds of curvatures and their properties and influence to the shape are discussed.

The second part is devoted to the study of curves and knots using geometric approach and Frenet–Serret equations. Some interesting bending fields and classes of curves are considered. To study energies and shape of curves and knots, we use Willmore energy, total curvature, torsional energy, total torsion, total normalcy, and Möbius energy. We investigate infinitesimal bending of the first and the second order both analytically and using our proper visualization tool.

In the third chapter, we consider surfaces by methods and tools of differential geometry and tensor calculus. As a result of infinitesimal bending of surfaces, all its line elements are rotated with uniquely defined rotational vector field. Corresponding bending field also known as the velocity field is completely determined by kinematic system of equations. We consider the level of variability, i.e. the variation of curvatures (normal, mean, Gaussian, geodesic), Willmore energy, total mean curvature as well as volume of generalized cone under infinitesimal bending. We point to the application of the obtained results in the calculation of the volume of a solid model.

The fourth chapter considers infinitesimal bending of surfaces of revolution using Cohn-Vossen's method, which is the basic for this class of surfaces. Many works at a theory of deformation of surfaces are dedicated to infinitesimal bending problem of surfaces of revolution. The problem of generating the field of infinitesimal bending is in this case reduced to the problem of solving the system of ordinary differential equations. We obtain the non-rigidity condition for toroid with quadrangular meridian. We find the geometrical meaning of this condition and classify toroid surfaces with quadrangular meridian as rigid and non-rigid. Further, we examine the variation of the volume of surfaces of revolution under infinitesimal bending. Infinitesimal bending of higher order is discussed and some examples of deformed toroid are visualized.

The last chapter on knot dynamics is a departure from the main theme of this book. It is a self-contained research paper by Louis H. Kauffman on the construction of examples of unknots and also knots that can be resistant to simplification by energy gradient and force evolution methods in programming models. We expect that our investigations of energy and Willmore energy will eventually shed light on such phenomena.

This book is based on many different works by authors from recent decades, but also on some new results. Many papers relevant to the research are cited in the literature.

This book is part of the COST project CA22153 EuroCurvoBioNet.

Finally, we would like to mention that this book does not have an encyclopedic character and that, although many topics are discussed, for more detailed individual questions about them and a more studious background, we recommend consulting the references listed in the book. Our goal is primarily to approach different geometric objects and their geometric quantities from the same point of view, keeping in mind a flexible shape definition that takes into account near-isometric deformations, such as infinitesimal bending.

About the Authors

Louis Hirsch Kauffman is an American mathematician, mathematical physicist, and professor in the Department of Mathematics, Statistics, and Computer Science at the University of Illinois at Chicago. His research spans topology, knot theory, topological quantum field theory, quantum information theory, and diagrammatic and categorical mathematics. He is best known for introducing and developing the bracket polynomial, the Kauffman polynomial and the theory of virtual knots and links. He is the author of numerous books related to knots and physics; he is the editor in chief and founding editor of the (World Scientific) *Journal of Knot Theory and Its Ramifications*; he is the recipient of two Lester Ford Awards of the Mathematical Association of America; he is a Fellow of the American Mathematical Society; he is the recipient of the Bertalanffy Award for innovative systems thinking and the Warren McCulloch Award and Norbert Wiener Medal of the American Society for Cybernetics.

Ljubica S Velimirović is a mathematician specializing in differential geometry, particularly in the geometric analysis of curves, surfaces, and knots. Her scientific work focuses on the interplay between geometry and physics through the study of shape and energy functionals. She has contributed to the theory of infinitesimal bending, exploring how small deformations affect the structure and energy of geometric objects. Her research often bridges theoretical mathematics with applications in civil engineering, biology, and computer graphics. She has published extensively in international journals and authored textbooks used in higher education.

Marija S Najdanović is an associate professor at the Faculty of Sciences and Mathematics in Kosovska Mitrovica, Serbia, and a senior research associate in the field of mathematical sciences. Her research focuses on differential geometry, particularly on infinitesimal deformations of geometric objects (curves, surfaces, and manifolds) and the application of geometry to natural processes. She has published more than 40 research papers and two textbooks.

Svetozar R Rančić is an associate professor in the Department of Computer Science, Faculty of Mathematics and Science, University of Niš. He holds a PhD in computer science and teaches courses in computer graphics, software design and development, and compiler construction. His research interests include computer graphics, scientific computation, and visualization of curves and surfaces. He has published more than 30 papers in scientific journals. He has been working as a software developer and team leader, participating in many international software projects.

Contents

Chapter 1

On Shape and Energies

1.1 Shape

The **shape** (Old English: gesceap, created thing) of an object located in some space is a geometrical description of the part of that space occupied by the object, as determined by its external boundary — abstracting from location and orientation in space, size, and other properties such as color, content, and material composition.

In geometry, two subsets of Euclidean space have **the same shape** if one can be transformed to the other by a combination of translations, rotations (together also called rigid transformations), and uniform scalings. In other words, the shape of a set of points is all the geometrical information that is invariant to translations, rotations, and size changes. Having the same shape is an equivalence relation, and accordingly a precise mathematical definition of the notion of shape can be given as being an equivalence class of subsets of Euclidean space having the same shape.

Geometric objects that have the same shape are said to be **similar**. If they also have the same scale, they are said to be **congruent**. Therefore, two congruent objects have the same shape as well as the same size, while similar objects may have different size.

A more flexible definition of shape takes into consideration the fact that we often deal with **deformable shapes** in reality (e.g. a person in different postures, a tree bending in the wind or a hand with different finger positions). By allowing also isometric (or near-isometric) deformations, the intrinsic geometry of the object will stay the same, while sub-parts might be located at very different positions in space. This definition uses the fact that, geodesics (curves measured along the surface of the object) stay the same, independent of the isometric embedding. This means that the distance from a finger to a toe of a person measured along the body is always the same, no matter how the body is posed.

1.2 Curvatures

The shape of curves and surfaces as geometric objects in Euclidean 3-dimensional space is determined by their geometric curvature. Intuitively, the curvature gives the answer to the question how much a curve deviates from being a straight line, or a surface deviates from being a plane. In the case of a circle, smaller ones bend more sharply, the velocity vectors change rapidly and hence have higher curvature. Straight lines themselves have zero curvature. Generally, for the curve $C : \mathbf{r} = \mathbf{r}(s)$ $(\mathbf{r} : \mathcal{I} \mapsto \mathcal{R}^3)$, $\mathcal{I} \subseteq \mathcal{R}$, the **curvature** is defined as

$$k(s) = \|\mathbf{r}''(s)\| = \|\mathbf{t}'(s)\|,$$

where s is the arc length and $\mathbf{t}$ is the unit tangent vector of the curve. In the case of space curves, apart the curvature, we need the information whether the curve deviates from the plane i.e. how sharply it is twisting out of the plane. The answer to this question gives the **torsion** (second curvature). In fact, the torsion measures the speed of rotation of the binormal vector at the given point. It is found from the equation

$$\mathbf{n}_2' = -\tau \mathbf{n}_1,$$

where $\mathbf{n}_1$ and $\mathbf{n}_2$ are unit normal and binormal vector of the curve, respectively. The larger the torsion is, the faster the binormal vector rotates around the axis given by the tangent vector. Obviously, each plane curve (with non-vanishing curvature) has zero torsion at all points.

The curvature and the torsion determine the shape of the curve. Namely, for a given smooth functions $k(s) > 0$ and $\tau(s)$, there exists a biregular curve such that its curvature and torsion are given by $k(s)$ and $\tau(s)$, respectively. Moreover, any other curve satisfying the same conditions differs from the first one by a rigid motion. This is a well known result in differential geometry.

While for curves we differentiated the tangent vector in order to obtain the curvature, in the case of surface S, we will differentiate the surface normal ν in order to estimate how it changes from point to point. For that purpose we define the **shape operator** $\underline{S}$ applied to a tangent vector $\mathbf{v}_p$ as the negative of the derivative of a surface normal $\boldsymbol{\nu}$ in the direction $\mathbf{v}_p$, i.e.

$$\underline{S}(\mathbf{v}_p) = -D_{\mathbf{v}}\boldsymbol{\nu},$$

(see [Gray (1998)]). Therefore, the shape operator at a point $p \in S$ is a linear transformation that measures how S bends in different directions. It is easy to see that the shape operator of a plane is identically zero at all points of the plane. For a nonplanar surface, the surface normal $\boldsymbol{\nu}$ will twist and turn from point to point, and $\underline{S}$ will be nonzero.

Although the shape operator gives us a lot of information about the shape by measuring the bending of the surface in different directions, it is much easier to handle numbers that summarize the information that shape operator may give. In this regard, we introduce the following terms.

The **normal curvature** $\bar{K}$ in the direction $\mathbf{v}_p$ is

$$\bar{K}(\mathbf{v}_p) = \frac{\underline{S}(\mathbf{v}_p) \cdot \mathbf{v}_p}{\|\mathbf{v}_p\|^2}.$$

It measures how a surface bends in the direction $\mathbf{v}_p$. Geometrically, the absolute value of the normal curvature is equal to the curvature of the plane curve formed by the intersection of the surface with the plane passing through $\mathbf{v}_p$ meeting the surface perpendicularly.

The eigenvalues of the shape operator at $p \in S$ turn out to be the maximum and minimum of the normal curvature at p. These are the **principal curvatures** k_1 and k_2.

The **Gaussian curvature** K at p is

$$K(p) = \det(\underline{S}(p)).$$

It is related to the principal curvatures by

$$K = k_1 k_2.$$

The Gaussian curvature permits us to distinguish four kinds of points on a surface: elliptic (if $K(p) > 0$ or, equivalently, k_1 and k_2 have the same sign), hyperbolic (if $K(p) < 0$ or, equivalently, k_1 and k_2 have opposite signs), parabolic (if $K(p) = 0$, but $\underline{S}(p) \neq 0$ or, equivalently, exactly one of k_1 and k_2 is zero) and planar (if $K(p) = 0$ and $\underline{S}(p) = 0$ or, equivalently, $k_1 = k_2 = 0$). In this way, the local shape of the surface, in a small neighborhood of the point in question, is determined. Thus, at an elliptic point all the curves on the surface through that point curve in the same direction, either towards the normal to the surface or away from it. At a hyperbolic point, some curves on the surface through that point curve towards the normal to the surface and others away from it. At a parabolic point, all curves on the surface through that point curve towards the normal to S or away from it except for one that does not curve at all with respect to the normal to S. At last, at a planer point curves on the surface through that point do not curve at all with respect to the normal to S ([Fernández (2005)]). A surface with zero Gaussian curvature at all points is the **developable surface**.

The **mean curvature** H at p is

$$H(p) = \frac{1}{2} tr(\underline{S}(p)).$$

Applying the principal curvatures, we have

$$H = \frac{k_1 + k_2}{2}.$$

A surface with zero mean curvature is the **minimal surface** — the surface that locally minimizes its area.

The Gaussian and mean curvatures are the most important curvatures in surface theory.

The Gaussian curvature belongs to the **intrinsic geometry** of the surface. This means that the Gaussian curvature is invariant under **isometric** (distance-preserving) transformations. Isometrical surfaces have the same Gaussian curvature. The properties from the intrinsic geometry can be expressed only in terms of the first fundamental form of the surface. Apart the Gaussian curvature, so-called the **geodesic curvature** $\underset{g}{K}$ has the same property. The geodesic curvature is the rate of change of the angle that the tangent to the curve makes with parallel direction along the curve. In the case of the plane, the parallel direction is fixed and the geodesic curvature reduces to the usual curvature.

A curve on the surface is said to be the **geodesic** if its geodesic curvature is zero at all points. Equivalently, it is a curve representing the shortest path between two points on the surface. In the plane, the geodesics are straight lines. On the sphere, the geodesics are great circles.

1.3 Energies

In the study of geometric properties of curves and surfaces, it is useful to observe curvature-based functionals which are said to be the **energies** of curves and surfaces. The energies are frequently used in many areas like mathematics, physics, chemistry, biology, thin plate and shell engineering, etc. In physics, mechanical engineering and biology they are used to model the behavior of bending surfaces: car bodies, roofs of buildings, polymer vesicles, and biological membranes. In geometric modeling, these functionals are used for surface fairing, feature detection and defining variational surfaces; in computer animation, curvature-based functionals are used for simulating motion of deformable objects, most commonly, cloth animation [Zorin (2005)].

1.3.1 *Energies of Curves*

We are going to list some of the major energies of the curve.

The **Helfrich energy** of the curve $C : \mathbf{r} = \mathbf{r}(s)$ $(\mathbf{r} : \mathcal{I} \mapsto \mathcal{R}^3)$, $\mathcal{I} \subseteq \mathcal{R}$ is given by

$$\mathcal{E}b_\lambda(C) = \frac{1}{2}\int_{\mathcal{I}} (k - c_0)^2 \, ds + \lambda \mathcal{L}(C), \tag{1.1}$$

where k and s denote the curvature and the arc length of C, respectively, and $\mathcal{L}(C) = \int_{\mathcal{I}} ds$ is the length of C. The map $c_0 : \mathcal{I} \mapsto \mathcal{R}$ is called spontaneous curvature. The constant $\lambda \in \mathcal{R}$ is taken to be positive, so that the growth in length of a curve is penalized. The above functional is motivated by the modeling of cell membranes [Helfrich (1973)].

The **Willmore energy** is the special case of the Helfrich energy where $c_0 = 0$ and $\lambda = 0$, and is given by

$$\mathcal{W}(C) = \frac{1}{2}\int_{\mathcal{I}} k^2 \, ds. \tag{1.2}$$

It can also be historically motivated by the so-called Euler-Bernoulli model of elastic rods [Truesdell (1983)]. Willmore energy penalizes bending models the stiffness of a polymer, and it has been used to model the elastic properties of DNA [Capovilla *et al.* (2002)].

The Helfrich and Willmore energies are mathematically very interesting. Both functionals have been extensively investigated analytically and numerically in recent years and the literature is by now rather vast.

The **total curvature** of the curve C is defined by the following equation:

$$\mathcal{K}(C) = \int_{\mathcal{I}} k(s)\, ds. \tag{1.3}$$

The total curvature of a curve C is a quantity which measures the total turning of the tangent vector. It was studied by W. Fenchel [Fenchel (1929)]. In 1929, he proved that the total curvature of a closed curve must be at least 2π, with equality if and only if the curve is planar and convex. K. Borsuk in [Borsuk (1947)] has extended this result to n dimensional space. Many papers are dedicated to the study of the total curvature.

The **torsional energy** is defined as the half of the integral of the squared torsion

$$\mathcal{E}_{\tau}(C) = \frac{1}{2} \int_{\mathcal{I}} \tau^2(s)\, ds \tag{1.4}$$

Together with the Willmore energy, it represents an important characteristic of elastic filaments.

The **total torsion** is another global property of a curve C. It is defined by the integral

$$\mathcal{T}(C) = \int_{\mathcal{I}} \tau(s)\, ds, \tag{1.5}$$

where τ is the torsion of C. Geometrically, the total torsion is a measure of the binormal indicatrix. It is well known that for any real number p, there is a closed curve C such that its total torsion is equal to p. Also, a classical result in differential geometry assures that the total torsion of a closed spherical curve in the three-dimensional space vanishes. Besides, if a surface is such that the total torsion vanishes for all closed curves, it is part of a sphere or a plane. There are many investigations about the total torsion (see, for instance, [Pansonato and Costa (2008); Qin and Li (2002)]).

The **total normalcy** is a geometric quantity which measures the normal indicatrix of a curve. It is defined by the equation

$$\omega(C) = \int_I \sqrt{k^2(s) + \tau^2(s)}\, ds. \tag{1.6}$$

The **Möbius energy** of the curve C is

$$\mathcal{E}(C) = \iint \left(\frac{1}{\|\mathbf{r}(s) - \mathbf{r}(t)\|^2} - \frac{1}{l(s,t)^2} \right) \|\dot{\mathbf{r}}(s)\| \, \|\dot{\mathbf{r}}(t)\| \, ds dt,$$

where $l(s,t)$ denotes the minimum distance along the curve C between points $\mathbf{r}(s)$ and $\mathbf{r}(t)$. This functional has had its impact on the modeling of macromolecules such as DNA and proteins.

1.3.2 *Energies of Knots*

Energies of curves are especially important in the theory of knots since they allow understanding the details of a knot shape.

By the term **knot**, we mean a closed, self-avoiding curve in a three-dimensional space. From the point of view of a physicist, knots are tied on an infinitely thin, perfectly flexible and infinitely strong filament, the ends of which have been joined. From the topological point of view, we have more general definition. According to that definition, we consider a knot as a equivalence class of closed curves, whereby a curve C' is equivalent to C if it can be continuously deformed into C without crossing itself.

Equivalent knots are considered the same in the topological sense and determine a special class of knots named **knot type**. It is therefore possible to think of a knot as a curve with small but positive thickness which allows us to present it as a tube. In classic knot theory, mathematicians are often interested in knot classes and how to distinguish between them. Contrary to this approach, geometric knot theory deals with the specific shape of knots and how to find or compute particularly nice representatives of a given knot class. In this sense, knot theory studies relations between the geometry of space curves and the knot types they represent. In order to determine the "quality" of the knot, several geometric curvature-based functionals or energies are considered. An optimal representative is then said to be a minimizer of energy among all curves of a given knot class. Therefore, in geometrical sense, we could observe small deformations of knots as the family of different curves. For connection between knots and physics, we refer to [Kauffman (2001)].

The first knot energy was propose by Jun O'Hara in [O'Hara (1991)], which is invariant under Möbius transformations. This is why it is also known as Möbius energy. The Möbius energy of knots has been studied in [Freedman *et al.* (1994); Rawdon and Simon (2002)] etc. A well-known Fary-Milnor theorem speaks of the total curvature of knots. According to this theorem, the total curvature of any knot is greater than 4π [Fáry (1949); Milnor (1950)]. For different knot energies, we refer to [Blatt and Reiter (2014); O'Hara (1991); Simon (1996)].

The knowledge of curvatures and curvature based functionals of a knot is essential, since it allows one to understand the details of its shape. If, for instance, in a certain part of a knot its curvature and torsion are both constant, one can conclude that this part of the knot has a helical shape [Baranska *et al.* (2008)].

We are going to demonstrate on an example how the curvature and the torsion affect on the knots. Figure 1.1 shows us the trefoil knot

$$\mathbf{r}(u) = (4\cos(2u) + 2\cos(u), 4\sin(2u) - 2\sin(u), \sin(3u)),$$

colored in the function of the curvature k, the torsion τ, the squared curvature k^2, and the normalcy $\sqrt{k^2 + \tau^2}$.

Graphics of the curvature, the torsion, the squared curvature and the normalcy, respectively, of the trefoil knot are presented in Fig. 1.2.

1.3.3 *Energies of Surfaces*

When it comes to surface energy, especially important and most commonly used is so-called the **Willmore energy**. For the oriented surface $S : \mathbf{r} = \mathbf{r}(u, v)$, $\mathbf{r} : \mathcal{D} \to \mathcal{R}^3$, $\mathcal{D} \subseteq \mathcal{R}^2$, the Willmore energy is

$$\mathcal{W}(S) = \iint_S (H^2 - K)\, dA,$$

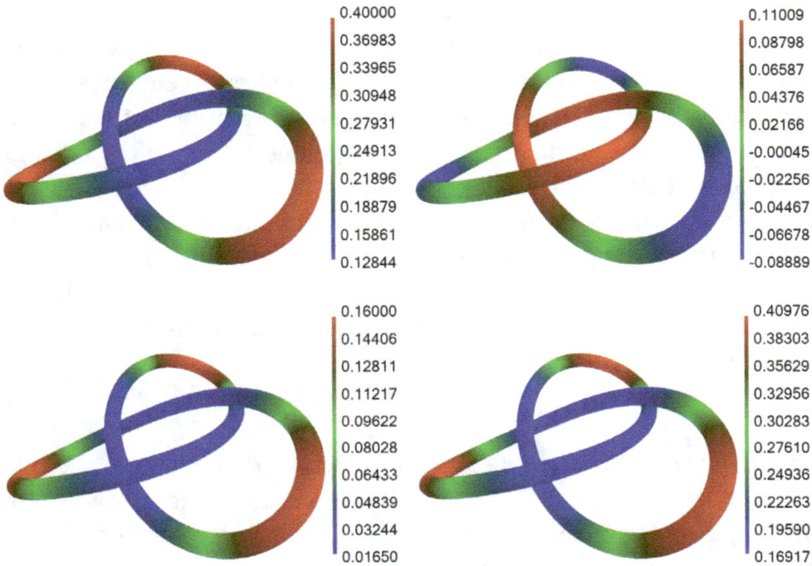

Fig. 1.1. The trefoil knot colored in the function of the curvature, the torsion, the squared curvature and the normalcy, respectively.

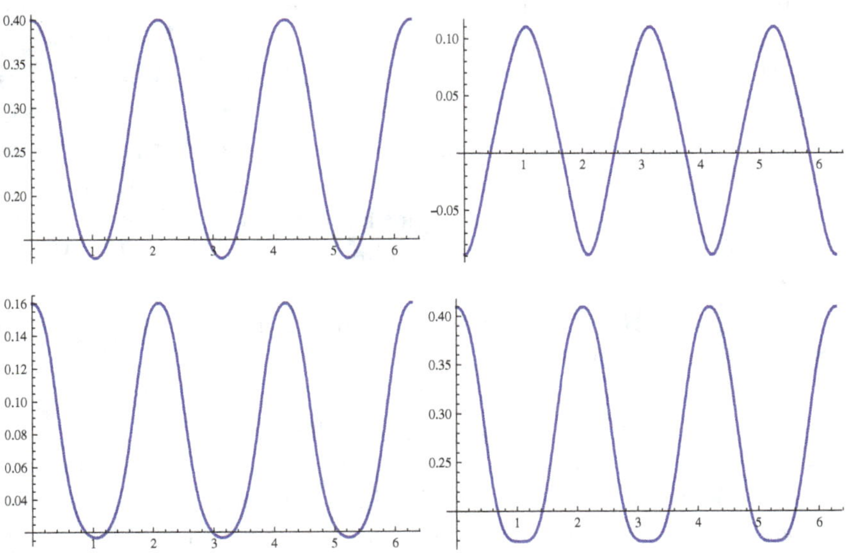

Fig. 1.2. Graphics of the curvature, the torsion, the squared curvature and the normalcy, respectively, of the trefoil knot.

where H and K are the mean and the Gaussian curvature of S, respectively, dA is an element of the surface area. The Willmore energy is a quantitative measure of how much a given surface deviates from a round sphere. This energy plays an important role in digital geometry processing, geometric modeling, theory of membranes, theory of shell structures etc.

In biology, the Willmore energy is a special case of so-called **elastic bending energy**, also known as the **Helfrich energy**, which was studied by Helfrich [Helfrich (1973)] and which describes the properties of equilibrium shape of membranes. Precisely,

$$\mathcal{E}_b(S) = \iint_S [\alpha + \beta(H - c_0)^2 - \gamma K]\, dA, \tag{1.7}$$

where α represents the surface tension, which demonstrate the interaction effects between the vesicle material and the ambient fluid material; H is the mean curvature of the membrane surface, K is the Gaussian curvature; β and γ are elastic constants which are determined by the interaction and properties of the materials that form the membrane; c_0 is the spontaneous curvature that describes the asymmetry effect of the membrane or its environment. For a symmetric membrane, $c_0 = 0$.

With a constant α, the first term in (1.7) can be neglected as it remains constant for vesicles with a given surface area. By taking $c_0 = 0$, $\beta = \gamma$, elastic bending energy reduced to the Willmore energy. Willmore energy has been widely studied, for instance in [Barrett *et al.* (2016); Bernard and Rivière (2014); Gruber *et al.* (2019); Marques and Neves (2014); Velimirović *et al.* (2010b); Weiner (1978)].

Another important curvature-based functional is the **total mean curvature** defined by the equation

$$\mathcal{H}(S) = \iint_S H\, dA,$$

where H is the mean curvature. The total mean curvature has been extensively studied in many papers, for instance [Alexandrov (2009); Almgren and Rivin (1992); Chen (2014); Velimirović and Ćirić (2011)].

The surface integral of the Gaussian curvature K over some region T of a surface is called the **integral curvature** or **total Gaussian**

curvature, i.e.

$$\mathcal{K}(T) = \iint_T K\, dA.$$

The integral curvature has a simple geometrical interpretation. When we draw through a point O lines parallel to the surface normals at the points of a region T on a surface S bounded by an arc C, then they intersect the unit sphere in a region T_1 bounded by an arc C_1 which is the spherical image of C. It is easy to prove that the integral curvature of a region on the surface is equal to the area of its spherical image. This property was noted by Gauss.

The integral curvature is considered in one of the most important theorems of differential geometry, known as **Gauss-Bonnet theorem**. It states that if the Gaussian curvature K of a surface is continuous in a simply connected region A bounded by a closed curve ∂A composed of n smooth arcs making at the vertices exterior angles $\beta_1, \ldots, \beta_n$, then:

$$\oint_{\partial A} \underset{g}{K}\, ds + \iint_A K\, dA + \sum_{i=1}^{n} \beta_i = 2\pi,$$

where $\underset{g}{K}$ represents the geodesic curvature of the arcs. One of the consequences of this theorem says that the integral curvature of a surface which is topologically equivalent to the sphere is 4π. Furthermore, the integral curvature of a surface which is topologically equivalent to a torus is zero.

1.4 Deformations

The study of deformations dates from the ancient times and stems from purely practical problems. In physics, a deformation means a change of a shape of body under the influence of the external or internal forces. In the case when the body returns to its original shape after the termination of the effect of forces, it said to be elastic, otherwise, if it is deformed permanently, we say it is plastic. Testing of deformations of different materials is of great importance in civil engineering. Measurement of deformations of high buildings allows us to provide an adequate level of safety and security from potential damage and disaster. Deformations are in close connection with thin elastic shell and has a huge application from the mechanical point of view. In biology, the notion of deformation has also found his place. We are talking about the elasticity of the cell membrane in connection with the fluidity that allows the proteins to move along the membrane.

In geometry, the problem of deformations is covered by so-called the **surface bending theory**. The surface bending theory considers the bending of surfaces, i.e. the isometrical deformations, as well as, the infinitesimal bending of surfaces and presents one of the main parts of **global differential geometry**.

Under bending, surface is included in continuous family of isometrical surfaces, so that any curve on the surface preserves its arc length. The angles are also preserved.

Two surfaces are **trivially isometrical** if we get one from another by rigid motion or by plane symmetry (or by finite number of such transformations). A surface is **uniquely defined** if there are only trivially isometrical surfaces. Each uniquely defined surface is rigid in a sense of isometric bending (as there are not isometrical surfaces bent from initial).

On the other hand, **infinitesimal bending** of surfaces is not an isometric deformation, or roughly speaking it is an isometric deformation with appropriate precision. Arc length is stationary under infinitesimal bending with a given precision. In the case of infinitesimal bending, the surface is deformed so that in the initial moment of a deformation, the arc length on the surface is stationary, i.e. initial velocity of its change is zero. Precisely, infinitesimal bending is determined with the equation

$$ds_\epsilon^2 - ds^2 = o(\epsilon^m), \quad \epsilon \geq 0, \quad \epsilon \to 0,$$

$m \geq 1$, which means that the difference of the squares of the line elements of deformed and initial surface is an infinitesimal of the order higher then ϵ^m [Efimov (1948)]. The basic task of flexibility and rigidity (isometrical bending) problems is to find as many surfaces, that represent the class of uniquely defined, rigid, surfaces, as well as, to find those that represent the class of bendable, flexible, surfaces.

Under infinitesimal bending of surfaces, various geometric magnitudes succumb to the effects of a given deformation. Some magnitudes remain unchanged, i.e. stationary (with a given precision), while others suffer some changes. Besides, the arc length which is stationary by definition, some other magnitudes are also not changed, for instance, the coefficients of the first fundamental form, the determinant of the first and the second fundamental form, Cristoffel's symbols, Gaussian and geodesic curvature, etc. However, coefficients of the second fundamental form are changing under infinitesimal bending, as well as many other geometric magnitudes. The level of variability of a geometric magnitude is measured by so-called the **variation** which is zero in the case of stationary magnitudes, or the magnitudes which are the same for both the initial and deformed surface (with a given precision), and different from zero in the case of changeable geometric magnitudes.

Bending of surfaces could be picturesque perceived as bending, twisting, buckling of a sheet of paper without stretching, creasing and creating of sharp edges. If we draw lines on the paper and then bend the paper, the length of the lines will remain the same, as well as the angle between two lines and the area of the region bounded by the closed curved line. However, some properties will be changed under such a "motion" of the paper. A straight line will not be straight, therefore the curvature of a line is changed.

The following main tasks are imposed ahead of the infinitesimal bending theory: (1) to find as many as possible representatives of rigid surfaces, as well as flexible or bendable surfaces; (2) to check the variation of different geometric magnitudes, (3) to connect obtained results with other sciences in order to find possible applications.

A concept of infinitesimal bending dealt first with infinitesimal bending of surfaces and then with the same problem in the theory of curves and manifolds. However, for the sake of simplicity, in this book we will first study the infinitesimal bending of curves.

Infinitesimal bending of curves and surfaces is studied, for instance, in [Alexandrov (2009); Belov (1968); Belova *et al.* (2021); Broscăteanu *et al.* (2024); Cohn-Vossen (1959); Connelly (1980); Efimov (1948); Fomenko (1967); Gözütok *et al.* (2020); Ivanova-Karatopraklieva and Sabitov (1995); Jimez (2018); Maksimivić *et al.* (2021); Najdanović and Velimirović (2017a); Perlova (1969); Pogorelov (1969); Vekua (1959); Velimirović (2001); Velimirović and Rančić (2010)]. For deformations on manifolds, we refer to [Hinterleitner *et al.* (2008); Mikeš *et al.* (2019); Najdanović (2017); Najdanović *et al.* (2022); Rýparová and Mikeš (2019); Velimirović *et al.* (2007)].

1.5 Visualization

Infinitesimal bending of curves and surfaces has influence on their geometric shapes and geometric magnitudes. It is interesting to view new deformed shapes together with some colorings aimed for visually representing local values of some geometric magnitude. In this book, first four chapters, except the last one, consider infinitesimally bent curves, knots represented as curves as well as surfaces using explicitly defined functions. Over time, for research purposes, we have developed software tools: for calculating and visualizing curves called CurvBend, as well as software for calculating and visualizing surfaces called SurfBand. Almost all of the figures in the first four chapters were created using our software.

CurvBend is developed having in mind curves and knotted curves lying on a torus, defined by a parametric equation. Our aim is to calculate and visualize influence of infinitesimal bending fields on both deformed shape and geometrical magnitudes like curvature, torsion, Willmore energy, ... We will visualize the shape of curves and knotted curves before and after infinitesimal bending fields deform the curves. We calculate appropriate magnitudes of curves for basic and deformed ones. Although in a few examples, we use color for paint curve shapes in most examples, we give figures containing the presented curves, where we use colors to indicate the local values of magnitudes at different points of the curves. For every example, we calculated a range of appropriate values at points and then mapped that range to a color scale. Points on curves are colored according the value mapped to an appropriate color. We also calculated numerically geometrical magnitude for basic and deformed curves and knotted curves.

There are several approaches for calculating and visualizing curves. It is possible to hardcode the curve definition as an appropriate function in a programming code in some programming language, and then manually, on paper, calculate derivations and all necessary mathematical expressions and hardcode them in a program. In this way, for any changes in curve definition, the infinitesimal bending fields, or any mathematical expression, one should again manually recalculate and change the program. Another approach suggests using some software package that is capable of carrying out symbolic, numeric, and visualizing activities, like *Mathematica* or *Matlab*. We decided to develop our own software that can perform the needed calculations and is capable of carrying out some symbolic and numeric operations, with additional result visualization capabilities.

We intended to obtain flexibility for defining curves and bending fields by allowing the software user to enter strings representing them. Then, the string representations are divided lexically, and then analyzed semantically, and finally, we built the appropriate expression tree for the entered function. An expression tree is built of nodes: one node for each operation, mentioned elementary function, or part of an expression (variables and constants). With some more info of used variables, we built the so-called internal form for every function we had to deal with: curve parametric definitions and the bending fields of the first and second order. We used the obtained expression tree for calculating function values easily and effectively by passing values for variables, and every node in the expression tree can conduct the appropriate calculation, and the result of the calculation is passed up to the parent node and finally to the tree root with the final result. The expression tree is also a kind of object-oriented tree that is capable of two important things: giving a new expression tree that represents a derivative over some variables and giving the string function representation of itself. With these abilities, we can automatize forming necessary functions for unit tangent, binormal, and necessary symbolical calculations. Developed software do not have the ability of symbolical integration, but we solved such tasks using numerical integration.

Such an approach further allowed us to convert a string representation to our internal representation based on the expression tree and vice versa. A software user can enter strings that define curves and bending fields (in parametric representation) that include arbitrary elementary functions. After that, on the user's demand (on a button click), the software can perform all the other necessary activities: parsing, building expression trees, calculating function values, creating a 3D model of curves, and representing colored curves.

The visualization of an obtained curve is based on a polygonal line that connects points on a curve. At every such point, we should calculate the following functions: the curve; the first, second, and third derivative; and also the local values of appropriate geometrical magnitude. This is necessary to obtain the transformed shape of the curve and for curve coloring. For the purpose of calculating integrals numerically, we also perform the same calculations at every subdivision point. For the sake of efficiency, we transform strings to their internal function form once, and then evaluate it as many times as we need, according to the principle "parse once, evaluate many times". Our software tool is written in *Microsoft Visual C++* (*VS 2022*). The 3D modeling and visualization with semitransparent objects are carried out in *OpenGL (1.5)*. For better visibility, the curves have a tube around them.

SurfBend is our visualization tool devoted to visual representation infinitesimally bent surfaces and rotational surfaces. Those rotational surfaces were obtained by revolution of a meridian in the shape of polygon. It is also able to show circles formed by apices of polygon and its infinitesimally deformed shape, as well as, visualize surfaces created during deformation. Our goals are to create an easy to use tool for:

- polygon and deformation definition. The ability to numerically check flexibility condition and obtain symbolic definition of bent surface is used. Our tool can transform symbolic definition of surface into a numerically efficient internal structure and use it for necessary calculation during surface visualization;
- visual presentation and possibility of quick basic and 3D calculations. It is very useful to interactively examine bent surfaces and the influence of infinitesimal bending fields on them. Obtained 3D object can be rotated by mouse move, colored, and viewed in fill or wire mode with adjustable semi transparency.

SurfBend is developed in Object Oriented language C++. It uses explicitly defined functions with n independent variables over a set of elementary functions. Each elementary function is wrapped by an appropriate class, and a class hierarchy is introduced to support building a tree structure as an expression tree of the function. Every class in the hierarchy has overridden abstract members:

```
double evaluate (double * arguments);
Function * derive(int argNumber);
```

The function `evaluate` is designed to calculate the value of the wrapped elementary function and as an argument it takes an array with double values of n parameters for which we want to calculate the function value. The function `derive` is designed to build a new tree structure according to the derivation rules for the elementary functions, widened with the rules for composite functions, having as argument the ordinal of the ith independent variable on which the partial derivation is desired. The return value is also of the `Function * type`, so this enables us to build the tree structure for an arbitrary order partial derivation, and calculate values of the obtained function by calling the evaluate member function. This ability is valuable in light calculation on the surface, also for needed geometrical magnitudes of the surface.

Efficiency considerations force us to give solution based on principle: parse once evaluate many times. After parsing, we build an internal tree structure — an expression tree of the function as described in the Object Oriented design pattern Composite [Gamma *et al.* (1995)]. We also used the famous OO patterns Singleton and Abstract Factory to produce function objects and build trees and evaluate functions. Visualization of bent surfaces is obtained using OpenGL. Rising control for the interactive level has been achieved using MFC [Shepherd and Kruglinski (2003)]. We are able to rotate a 3D object and see it from different angles and points of view. It is possible to use scroll bars, to easily, interactively adjust important parameters for bending calculations like ε, the number of points, ... Apices of meridian form circles and during deformation describe surfaces and we are able to give visualization of such surfaces in fill or wire mode with ability to adjust semi transparency of hidden lines. 3D model of surface is drawn on fill or wire mode with the ability to adjust the semi transparency of hidden lines. The tool offers easy, interactively, a click away adjustment, unlike special function call for calculation and visualization as usual in general mathematical visualization package. This function should be called again with changed parameters to make a new picture of 3D object.

Chapter 2

Curves and Knots

2.1 Curves

We begin with some basic definitions about curves that will be used throughout the book, according to [Carmo (1976); Schlichtkrull (2011); Struik (1961)]. We assume the reader is relatively familiar with some key concepts of differential geometry, so some details are omitted.

Definition 2.1. A **parameterized curve** C is a map $\mathbf{r} : \mathcal{J} \to \mathcal{R}^3$ of class $C^k\,(k \geq 1)$, where $\mathcal{J}$ is an open interval in $\mathcal{R}$. A curve is said to be **regular** if $\dot{\mathbf{r}}(u) \neq 0$ for all $u \in \mathcal{J}$. (Here dot denotes a derivative with respect to u.)

Usually we denote the parameterized curve as follows:

$$C : \mathbf{r} = \mathbf{r}(u), \quad u \in \mathcal{J} \subseteq \mathcal{R}. \tag{2.1}$$

Unless otherwise stated, we will assume all curves are regular. Also, we will not always specify the class of differentiability and will assume that k is a sufficiently large number as required by the consideration in question.

Further, we can consider a curve on a closed interval $[a, b]$ as the restriction of a curve defined on an open interval $\mathcal{J} \supset [a, b]$ in the sense of Definition 2.1.

Definition 2.2. A parameterized curve $C : \mathbf{r} = \mathbf{r}(u)$, $u \in [a, b]$ is **closed** if

$$\begin{aligned} \mathbf{r}(a) &= \mathbf{r}(b) \\ \dot{\mathbf{r}}(a) &= \dot{\mathbf{r}}(b) \\ \ddot{\mathbf{r}}(a) &= \ddot{\mathbf{r}}(b) \\ &\vdots \end{aligned} \tag{2.2}$$

Definition 2.3. A parameterized curve $C : \mathbf{r} = \mathbf{r}(u)$, $u \in [a, b]$ is **simple** if $\mathbf{r}$ is injective on $[a, b)$.

Throughout the book, we are mainly concerned with simple curves.

Further, for a given $u \in \mathcal{J}$, we define the **arc length** of the curve from u_0 to u as

$$s(u) = \int_{u_0}^{u} \|\dot{\mathbf{r}}(u)\|\, du. \tag{2.3}$$

As $\|\dot{\mathbf{r}}(u)\| > 0$, the inequality $u_2 < u_3$ implies $s(u_2) < s(u_3)$, therefore the function $s = s(u)$ is monotonically increasing and there exists inverse function $u = u(s)$. Putting that into Eq. (2.1) we obtain $\mathbf{r} = \mathbf{r}[u(s)]$ and write

$$C : \mathbf{r} = \mathbf{r}(s) = \mathbf{r}[u(s)], \quad s \in \mathcal{I}. \tag{2.4}$$

We say the curve is **parameterized by arc length**. Also, $ds = \|\dot{\mathbf{r}}(u)\|\, du$, so if s grows when u grows we have $ds = \|\dot{\mathbf{r}}(u)\, du\| = \|d\mathbf{r}\|$, i.e. $\|\frac{d\mathbf{r}}{ds}\| = 1$. If we denote derivative by arc length with $'$, we have $\|\mathbf{r}'(s)\| = 1$. The vector $\mathbf{t}(s) = \mathbf{r}'(s)$ is called the **unit tangent vector**.

The quantity $\|\mathbf{t}'(s)\| = \|\mathbf{r}''(s)\|$ measures the rate of change of the angle that neighboring tangents make with the tangent at s. Said differently, $\|\mathbf{r}''(s)\|$ quantifies how quickly the curve pulls away from the tangent at s (within a neighborhood of s). This leads to a definition.

Definition 2.4. Let $\mathbf{r} = \mathbf{r}(s)$, $s \in \mathcal{I}$ be parameterized by arc length. The quantity $k(s) = \|\mathbf{r}''(s)\|$ is the **curvature (I curvature, flexion)** of the curve at s.

In what follows, the inner and cross product of vectors in $\mathcal{R}^3$ are denoted by $\cdot$ and $\times$, respectively.

If $k(s) \neq 0$, we define the **unit normal vector** $\mathbf{n}_1$ by $\mathbf{r}''(s) = k(s)\mathbf{n}_1(s)$. This vector is in fact normal to $\mathbf{t}(s)$. Indeed, from $\mathbf{t}\cdot\mathbf{t} = 1$, we obtain by differentiation $2\mathbf{t}\cdot\mathbf{t}' = 0$, and finally $\mathbf{t}\perp\mathbf{t}'$, i.e. $\mathbf{t}\perp\mathbf{n}_1$. If $\mathbf{t}'(s) = \mathbf{r}''(s) = \mathbf{0}$, and thus $k(s) = 0$, for some $s \in \mathcal{I}$, we call s a **singular point**. Such a case should be considered separately. Our attentions are focused on curves parameterized by arc length without singular points. The plane determined by $\mathbf{t}$ and $\mathbf{n}_1$ is called the **osculating plane** at s. The quantity $R(s) = \frac{1}{k(s)}$ is called the **radius of curvature**.

Now let $\mathbf{n}_2(s) = \mathbf{t}(s) \times \mathbf{n}_1(s)$, so $\mathbf{n}_2$ is normal to the osculating plane. Since $\mathbf{t}$ and $\mathbf{n}_1$ are of unit length, so is $\mathbf{n}_2$. This vector is the **unit binormal vector** at s. The vector $\mathbf{b}'(s)$ measures how quickly the curve pulls away from the osculating plane of s (within a neighborhood of s). By differentiating the equations $\mathbf{n}_2(s)\cdot\mathbf{n}_2(s) = 1$ and $\mathbf{n}_2(s) = \mathbf{t}(s) \times \mathbf{n}_1(s)$ we obtain $\mathbf{n}_2'(s)\cdot\mathbf{n}_2(s) = 0$ and $\mathbf{n}_2'(s) = \mathbf{t}(s) \times \mathbf{n}_1'(s)$, respectively. Thus, we see that $\mathbf{n}_2'$ is normal to both $\mathbf{n}_2$ and $\mathbf{t}$. Therefore, $\mathbf{n}_2'$ must be collinear to $\mathbf{n}_1$. This leads to a definition.

Definition 2.5. Let $\mathbf{r} = \mathbf{r}(s)$ be a curve parameterized by arc length without singular points. The quantity $\tau(s)$ defined by $\mathbf{n}_2'(s) = -\tau(s)\mathbf{n}_1(s)$ is the **torsion (II curvature)** of the curve at s.

Notice that for each s, we have an orthonormal basis $\{\mathbf{t}, \mathbf{n}_1, \mathbf{n}_2\}$ called **Frenet–Serret frame**. The following three equations

$$\begin{aligned} \mathbf{t}' &= k\mathbf{n}_1, \\ \mathbf{n}_1' &= -k\mathbf{t} + \tau\mathbf{n}_2, \\ \mathbf{n}_2' &= -\tau\mathbf{n}_1, \end{aligned} \tag{2.5}$$

are known as **Frenet–Serret equations**. These equations are quite important and extremely useful.

For curves not parameterized by arc length, the vectors of the orthonormal basis can be directly obtain by the next formulas:

$$\mathbf{t}(u) = \frac{\dot{\mathbf{r}}}{\|\dot{\mathbf{r}}\|}, \quad \mathbf{n}_1(u) = \frac{(\dot{\mathbf{r}}\cdot\dot{\mathbf{r}})\ddot{\mathbf{r}} - (\dot{\mathbf{r}}\cdot\ddot{\mathbf{r}})\dot{\mathbf{r}}}{\|\dot{\mathbf{r}}\|\|\dot{\mathbf{r}}\times\ddot{\mathbf{r}}\|}, \quad \mathbf{n}_2(u) = \frac{\dot{\mathbf{r}}\times\ddot{\mathbf{r}}}{\|\dot{\mathbf{r}}\times\ddot{\mathbf{r}}\|}. \tag{2.6}$$

The curvature and the torsion are now

$$k(u) = \frac{\|\dot{\mathbf{r}} \times \ddot{\mathbf{r}}\|}{\|\dot{\mathbf{r}}\|^3}, \quad \tau(u) = \frac{[\dot{\mathbf{r}}, \ddot{\mathbf{r}}, \dddot{\mathbf{r}}]}{\|\dot{\mathbf{r}} \times \ddot{\mathbf{r}}\|^2}, \tag{2.7}$$

and Frenet–Serret equations are

$$\begin{aligned} \dot{\mathbf{t}}(u) &= \|\dot{\mathbf{r}}(u)\| k(u)\mathbf{n}_1(u) \\ \dot{\mathbf{n}}_1(u) &= \|\dot{\mathbf{r}}(u)\|(-k(u)\mathbf{t}(u) + \tau(u)\mathbf{n}_2(u)) \\ \dot{\mathbf{n}}_2(u) &= -\|\dot{\mathbf{r}}(u)\|\tau(u)\mathbf{n}_1(u). \end{aligned} \tag{2.8}$$

2.2 Infinitesimal Bending of Curves

Let us consider a curve

$$C : \mathbf{r} = \mathbf{r}(u), \quad u \in \mathcal{J} \subseteq \mathcal{R}, \tag{2.9}$$

included in a family of the curves

$$C_\epsilon : \tilde{\mathbf{r}}(u, \epsilon) = \mathbf{r}_\epsilon(u) = \mathbf{r}(u) + \epsilon \overset{(1)}{\mathbf{z}}(u) + \epsilon^2 \overset{(2)}{\mathbf{z}}(u) + \cdots + \epsilon^m \overset{(m)}{\mathbf{z}}(u), \quad m \geq 1, \tag{2.10}$$

where $\epsilon \geq 0$, $\epsilon \to 0$ and we get C for $\epsilon = 0$ ($C = C_0$). The fields $\overset{(j)}{\mathbf{z}}(u) \in C^k$, $k \geq 1$, $j = 1, \ldots m$, are vector functions defined in the points of C.

Definition 2.6 ([Efimov (1948)]). A family of curves C_ϵ is an **infinitesimal bending of the order** m **of the curve** C if

$$ds_\epsilon^2 - ds^2 = o(\epsilon^m). \tag{2.11}$$

The field $\overset{(j)}{\mathbf{z}} = \overset{(j)}{\mathbf{z}}(u)$ is the **infinitesimal bending field of the order** j, $j = 1, \ldots m$, of the curve C.

Based on the definition of the infinitesimal bending of the order m, the next relation is valid

$$d\mathbf{r}_\epsilon^2 - d\mathbf{r}^2 = (d\mathbf{r}_\epsilon - d\mathbf{r}) \cdot (d\mathbf{r}_\epsilon + d\mathbf{r}) = o(\epsilon^m).$$

From here we have

$$d\left(\sum_{j=1}^{m}\epsilon^j\overset{(j)}{\mathbf{z}}\right)\cdot d\left(2\mathbf{r}+\sum_{j=1}^{m}\epsilon^j\overset{(j)}{\mathbf{z}}\right)=o(\epsilon^m),$$

or more precise

$$\left(\epsilon d\overset{(1)}{\mathbf{z}}+\epsilon^2 d\overset{(2)}{\mathbf{z}}+\cdots+\epsilon^m d\overset{(m)}{\mathbf{z}}\right)\cdot\left(2d\mathbf{r}\right.$$
$$\left.+\epsilon d\overset{(1)}{\mathbf{z}}+\epsilon^2 d\overset{(2)}{\mathbf{z}}+\cdots+\epsilon^m d\overset{(m)}{\mathbf{z}}\right)=o(\epsilon^m).$$

The necessary and sufficient condition for the left side to be infinitesimal value with respect to ϵ^m is to be

$$d\mathbf{r}\cdot d\overset{(1)}{\mathbf{z}}=0,$$

$$2d\mathbf{r}\cdot d\overset{(2)}{\mathbf{z}}+(d\overset{(1)}{\mathbf{z}})^2=0,$$

$$2d\mathbf{r}\cdot d\overset{(3)}{\mathbf{z}}+2d\overset{(1)}{\mathbf{z}}\cdot d\overset{(2)}{\mathbf{z}}=0,$$

....

$$2d\mathbf{r}\cdot d\overset{(m)}{\mathbf{z}}+\sum_{l=1}^{m-1}d\overset{(l)}{\mathbf{z}}\cdot d\overset{(m-l)}{\mathbf{z}}=0,$$

or, shorter

$$d\mathbf{r}\cdot d\overset{(1)}{\mathbf{z}}=0,\quad 2d\mathbf{r}\cdot d\overset{(j)}{\mathbf{z}}+\sum_{l=1}^{j-1}d\overset{(l)}{\mathbf{z}}\cdot d\overset{(j-l)}{\mathbf{z}}=0,\ \text{for } j=2,\ldots,m. \tag{2.12}$$

According to that, the next theorem states.

Theorem 2.1 ([Efimov (1948)]). *The necessary and sufficient condition for the curves* C_ϵ, (2.10), *to be infinitesimal bending of the m-th order of the curve* C, (2.9), *is to be valid* (2.12).

If the infinitesimal bending is reduced to the rigid motion of the curve, without internal deformations, we say it is the **trivial** infinitesimal bending. The corresponding bending field is also called trivial.

2.2.1 *Infinitesimal Bending of the First Order*

For the curve

$$C : \mathbf{r} = \mathbf{r}(u), \quad u \in \mathcal{J} \subseteq \mathcal{R}, \tag{2.13}$$

infinitesimal bending of the first order, or shorter, infinitesimal bending, is a family of the curves

$$C_\epsilon : \tilde{\mathbf{r}}(u, \epsilon) = \mathbf{r}_\epsilon(u) = \mathbf{r}(u) + \epsilon \mathbf{z}(u), \tag{2.14}$$

where $\mathbf{z}(u)$ is the infinitesimal bending field (of the first order). The necessary and sufficient condition for infinitesimal bending reduces to

$$d\mathbf{r} \cdot d\mathbf{z} = 0, \tag{2.15}$$

Since

$$d\mathbf{r} = \dot{\mathbf{r}}(u)du, \quad d\mathbf{z} = \dot{\mathbf{z}}(u)du,$$

according to (2.15), infinitesimal bending field of the curve C satisfies

$$\dot{\mathbf{r}}(u) \cdot \dot{\mathbf{z}}(u) = 0. \tag{2.16}$$

According to [Vekua (1959)], we have the next theorem.

Theorem 2.2. *Under infinitesimal bending of the curve C, the line element undertakes a non-negative addition, which is the infinitesimal value of order at least 2 with respect to ϵ, i.e.*

$$ds_\epsilon - ds = o(\varepsilon) \geq 0. \tag{2.17}$$

Proof. Based on (2.16), we have

$$ds_\epsilon = \|\dot{\mathbf{r}}_\epsilon(u)\|\, du = \|\dot{\mathbf{r}}(u) + \epsilon\dot{\mathbf{z}}(u)\|\, du = (\|\dot{\mathbf{r}}(u)\|^2 + \epsilon^2\|\dot{\mathbf{z}}(u)\|^2)^{\frac{1}{2}}\, du$$

$$= \|\dot{\mathbf{r}}(u)\| \left(1 + \epsilon^2 \frac{\|\dot{\mathbf{z}}(u)\|^2}{\|\dot{\mathbf{r}}(u)\|^2}\right)^{\frac{1}{2}} du = ds \left(1 + \epsilon^2 \frac{\|\dot{\mathbf{z}}(u)\|^2}{\|\dot{\mathbf{r}}(u)\|^2}\right)^{\frac{1}{2}}$$

After using the Maclaurin formula we get

$$ds_\epsilon = ds \left(1 + \epsilon^2 \frac{\|\dot{\mathbf{z}}(u)\|^2}{2\|\dot{\mathbf{r}}(u)\|^2} - \epsilon^4 \frac{\|\dot{\mathbf{z}}(u)\|^4}{8\|\dot{\mathbf{r}}(u)\|^4} + \dots\right)$$

i.e.

$$ds_\epsilon - ds = \epsilon^2 \frac{\|\dot{\mathbf{z}}(u)\|^2}{2\|\dot{\mathbf{r}}(u)\|^2} ds - \dots,$$

which leads to (2.17). □

The next theorem is related to determination of the infinitesimal bending field of the curve C.

Theorem 2.3 ([Velimirović (2001)]). *The infinitesimal bending field for the curve C, (2.13), is*

$$\mathbf{z}(u) = \int [p(u)\mathbf{n}_1(u) + q(u)\mathbf{n}_2(u)]\, du, \tag{2.18}$$

where $p(u)$ and $q(u)$ are arbitrary integrable functions, and vectors $\mathbf{n}_1(u)$ and $\mathbf{n}_2(u)$ are respectively unit principal normal and binormal vector fields of the curve C.

Proof. According to (2.16), we have

$$\dot{\mathbf{r}} \cdot \dot{\mathbf{z}} = 0, \quad \text{i.e. } \dot{\mathbf{r}} \perp \dot{\mathbf{z}}, \tag{2.19}$$

for nontrivial bending field. Based on that we conclude that $\dot{\mathbf{z}}$ lies in the normal plane of the curve C, i.e.

$$\dot{\mathbf{z}}(u) = p(u)\mathbf{n}_1(u) + q(u)\mathbf{n}_2(u), \tag{2.20}$$

where $p(u)$ and $q(u)$ are arbitrary integrable functions. Integrating (2.20), we obtain (2.18). □

According to (2.6), infinitesimal bending field can be written in the form

$$\mathbf{z}(u) = \int [p(u)\frac{(\dot{\mathbf{r}} \cdot \dot{\mathbf{r}})\ddot{\mathbf{r}} - (\dot{\mathbf{r}} \cdot \ddot{\mathbf{r}})\dot{\mathbf{r}}}{\|\dot{\mathbf{r}}\|\|\dot{\mathbf{r}} \times \ddot{\mathbf{r}}\|} + q(u)\frac{\dot{\mathbf{r}} \times \ddot{\mathbf{r}}}{\|\dot{\mathbf{r}} \times \ddot{\mathbf{r}}\|}]du$$

where $p(u)$ and $q(u)$ are arbitrary integrable functions, or in the form

$$\mathbf{z}(u) = \int [P_1(u)\dot{\mathbf{r}} + P_2(u)\ddot{\mathbf{r}} + Q(u)(\dot{\mathbf{r}} \times \ddot{\mathbf{r}})]du$$

where $P_i(u)$, $i = 1, 2$, i $Q(u)$ are arbitrary integrable functions, too.

A curve parameterized by arc length. Suppose that the curve

$$C : \mathbf{r} = \mathbf{r}(s) = \mathbf{r}[u(s)], \quad s \in \mathcal{I} \subseteq \mathcal{R}, \tag{2.21}$$

is parameterized by the arc length s. Let us consider an infinitesimal bending of the curve C,

$$C_\epsilon : \tilde{\mathbf{r}}(s, \epsilon) = \mathbf{r}_\epsilon(s) = \mathbf{r}(s) + \epsilon\mathbf{z}(s). \tag{2.22}$$

As the vector field $\mathbf{z}$ is defined in the points of the curve C, it can be presented in the form

$$\mathbf{z} = z\mathbf{t} + z_1\mathbf{n}_1 + z_2\mathbf{n}_2, \tag{2.23}$$

where $z\mathbf{t}$ is tangential and $z_1\mathbf{n}_1 + z_2\mathbf{n}_2$ is normal component, z, z_1, z_2 are the functions of s.

According to (2.15), the necessary and sufficient condition for the field $\mathbf{z}$ to be infinitesimal bending field of the curve C is

$$\mathbf{r}' \cdot \mathbf{z}' = 0, \tag{2.24}$$

i.e. $\mathbf{t} \cdot \mathbf{z}' = 0$, where $'$ denotes a derivative with respect to arc length s. Substituting Eq. (2.23) into the previous equation and using Frenet–Serret equations (2.5), we obtain the next theorem.

Theorem 2.4 ([Najdanović (2015)]). *The necessary and sufficient condition for the field* $\mathbf{z}$ (2.23) *to be an infinitesimal bending field of the curve* C, (2.21), *is*

$$z' - kz_1 = 0, \tag{2.25}$$

where k *is the curvature of* C.

Example 2.1. Let us find an infinitesimal bending field of the helix parameterized by the arc length

$$C : \mathbf{r} = \left(\sqrt{2}\cos\frac{s}{2}, \ \sqrt{2}\sin\frac{s}{2}, \ \frac{\sqrt{2}s}{2}\right), \quad s \in \mathcal{I}. \tag{2.26}$$

The vector fields of the orthonormal basis are

$$\begin{aligned}
\mathbf{t} &= \mathbf{r}' = \left(-\frac{\sqrt{2}}{2}\sin\frac{s}{2},\ \frac{\sqrt{2}}{2}\cos\frac{s}{2},\ \frac{\sqrt{2}}{2}\right),\\
\mathbf{n}_1 &= \frac{\mathbf{r}''}{\|\mathbf{r}''\|} = \left(-\cos\frac{s}{2},\ -\sin\frac{s}{2}, 0\right),\\
\mathbf{n}_2 &= \mathbf{t}\times\mathbf{n}_1 = \left(\frac{\sqrt{2}}{2}\sin\frac{s}{2},\ -\frac{\sqrt{2}}{2}\cos\frac{s}{2},\ \frac{\sqrt{2}}{2}\right).
\end{aligned} \tag{2.27}$$

The curvature and the torsion are respectively

$$k = \|\mathbf{r}''\| = \frac{\sqrt{2}}{4}, \qquad \tau = \frac{\sqrt{2}}{4}. \tag{2.28}$$

According to Theorem 2.3, for $p = q = 1$ we obtain infinitesimal bending field for C in the form

$$\mathbf{z} = \left(-2\sin\frac{s}{2} - \sqrt{2}\cos\frac{s}{2},\ 2\cos\frac{s}{2} - \sqrt{2}\sin\frac{s}{2},\ \frac{\sqrt{2}s}{2}\right).$$

Let us decompose vector field $\mathbf{z}$ via the orthonormal basis, as in Eq. (2.23). We obtain

$$\mathbf{z} = \frac{s+2\sqrt{2}}{2}\mathbf{t} + \sqrt{2}\mathbf{n}_1 + \frac{s-2\sqrt{2}}{2}\mathbf{n}_2. \tag{2.29}$$

2.2.1.1 *Rotational Field*

There is another vector field, related to the field $\mathbf{z}$, which is characterized by infinitesimal bending of the first order.

Theorem 2.5. *Let $\mathbf{z}$ be an infinitesimal bending field of the curve $C : \mathbf{r} = \mathbf{r}(u)$. Then, there exists a vector field $\mathbf{y} = \mathbf{y}(u)$, unique up to the tangent component, so that*

$$d\mathbf{z} = \mathbf{y} \times d\mathbf{r}. \tag{2.30}$$

Proof. If $\mathbf{z}$ is an infinitesimal bending field, than $\dot{\mathbf{r}} \cdot \dot{\mathbf{z}} = 0$. It follows immediately that there is a vector field $\mathbf{y}$ so that

$$\dot{\mathbf{z}} = \mathbf{y} \times \dot{\mathbf{r}}, \tag{2.31}$$

i.e. Eq. (2.30) is valid. Let us prove the uniqueness of the field $\mathbf{y}$. If $\mathbf{y}$ and $\bar{\mathbf{y}}$ satisfy the conditions of the theorem, then

$$\dot{\mathbf{z}} = \mathbf{y} \times \dot{\mathbf{r}} = \bar{\mathbf{y}} \times \dot{\mathbf{r}} \quad \Rightarrow \quad (\mathbf{y} - \bar{\mathbf{y}}) \times \dot{\mathbf{r}} = \mathbf{0}.$$

From here we conclude that either $\mathbf{y} = \bar{\mathbf{y}}$ or $\mathbf{y} - \bar{\mathbf{y}}$ and $\dot{\mathbf{r}}$ are collinear, i.e. $\mathbf{y} - \bar{\mathbf{y}} = \lambda \dot{\mathbf{r}} \Rightarrow \mathbf{y} = \bar{\mathbf{y}} + \lambda \dot{\mathbf{r}}$. This means that the fields $\mathbf{y}$ and $\bar{\mathbf{y}}$ can only be distinguished by the tangent component. □

Definition 2.7. The field $\mathbf{y}$ determined by (2.30) is the **rotational field** of the curve C under infinitesimal bending with the bending field $\mathbf{z}$.

In addition to the field $\mathbf{y}(u)$, we consider the field $\mathbf{s}(u)$ defined with the equation

$$\mathbf{s} = \mathbf{z} - \mathbf{y} \times \mathbf{r}, \quad \text{i.e. } \mathbf{z} = \mathbf{y} \times \mathbf{r} + \mathbf{s}. \tag{2.32}$$

Definition 2.8. The field $\mathbf{s}$ defined by Eq. (2.32) is the **translation field** under infinitesimal bending generating by $\mathbf{z}$.

If we consider the curve C parameterized by the arc length s, we have

$$\frac{d\mathbf{z}}{du} = \mathbf{y} \times \frac{d\mathbf{r}}{du} \Rightarrow \frac{d\mathbf{z}}{ds}\frac{ds}{du} = \mathbf{y} \times \frac{d\mathbf{r}}{ds}\frac{ds}{du},$$

and

$$\mathbf{z}' = \mathbf{y} \times \mathbf{r}' = \mathbf{y} \times \mathbf{t}. \tag{2.33}$$

Let us look at the decomposition of the vector fields $\mathbf{z}$ and $\mathbf{y}$ over the vectors of Frenet–Serret frame:

$$\mathbf{z} = z\mathbf{t} + z_1\mathbf{n}_1 + z_2\mathbf{n}_2, \quad \mathbf{y} = y\mathbf{t} + y_1\mathbf{n}_1 + y_2\mathbf{n}_2.$$

Substituting into Eq. (2.33), using the Frenet–Serret equations and the condition (2.25), we get

$$\begin{aligned} y_1 &= -\tau z_1 - z_2', \\ y_2 &= kz + z_1' - \tau z_2. \end{aligned} \tag{2.34}$$

These equations uniquely determine the normal components y_1 and y_2 of the rotational vector field, while the tangent component y is not determined.

2.2.2 *Infinitesimal Bending of the Second Order*

Let

$$C_\epsilon : \tilde{\mathbf{r}}(u,\epsilon) = \mathbf{r}_\epsilon(u) = \mathbf{r}(u) + \epsilon \overset{(1)}{\mathbf{z}}(u) + \epsilon^2 \overset{(2)}{\mathbf{z}}(u) \tag{2.35}$$

be the second-order infinitesimal bending of the curve $C : \mathbf{r} = \mathbf{r}(u)$. Then, the system (2.12) is reduced to

$$d\mathbf{r} \cdot d\overset{(1)}{\mathbf{z}} = 0, \quad 2d\mathbf{r} \cdot d\overset{(2)}{\mathbf{z}} + d\overset{(1)}{\mathbf{z}} \cdot d\overset{(1)}{\mathbf{z}} = 0, \tag{2.36}$$

which is equivalent to the next system of differential equations:

$$\dot{\mathbf{r}} \cdot \overset{(1)}{\dot{\mathbf{z}}} = 0, \quad 2\dot{\mathbf{r}} \cdot \overset{(2)}{\dot{\mathbf{z}}} + \overset{(1)}{\dot{\mathbf{z}}} \cdot \overset{(1)}{\dot{\mathbf{z}}} = 0. \tag{2.37}$$

Theorem 2.6 ([Najdanović and Velimirović (2017a)]). *Under infinitesimal bending of the second order of the curve C, the line element satisfies the next equation*

$$ds_\epsilon - ds = o(\varepsilon^2). \tag{2.38}$$

Proof. Based on Eqs. (2.37), we obtain the following:

$$\begin{aligned} ds_\epsilon &= \|\dot{\mathbf{r}}_\epsilon(u)\|\, du = \|\dot{\mathbf{r}}(u) + \epsilon \overset{(1)}{\dot{\mathbf{z}}}(u) + \epsilon^2 \overset{(2)}{\dot{\mathbf{z}}}(u)\|\, du \\ &= (\|\dot{\mathbf{r}}(u)\|^2 + 2\epsilon^3 \overset{(1)}{\dot{\mathbf{z}}}(u) \cdot \overset{(2)}{\dot{\mathbf{z}}}(u) + \epsilon^4 \|\overset{(2)}{\dot{\mathbf{z}}}(u)\|^2)^{\frac{1}{2}}\, du \\ &= \|\dot{\mathbf{r}}(u)\| \left(1 + \frac{2\epsilon^3 \overset{(1)}{\dot{\mathbf{z}}}(u) \cdot \overset{(2)}{\dot{\mathbf{z}}}(u) + \epsilon^4 \|\overset{(2)}{\dot{\mathbf{z}}}(u)\|^2}{\|\dot{\mathbf{r}}(u)\|^2}\right)^{\frac{1}{2}} du \end{aligned} \tag{2.39}$$

Further, we get

$$ds_\epsilon = ds\left(1 + \frac{2\epsilon^3 \overset{(1)}{\dot{\mathbf{z}}}(u)\cdot \overset{(2)}{\dot{\mathbf{z}}}(u) + \epsilon^4 \|\overset{(2)}{\dot{\mathbf{z}}}(u)\|^2}{2\|\dot{\mathbf{r}}(u)\|^2}\right.$$

$$\left. - \frac{(2\epsilon^3 \overset{(1)}{\dot{\mathbf{z}}}(u)\cdot \overset{(2)}{\dot{\mathbf{z}}}(u) + \epsilon^4 \|\overset{(2)}{\dot{\mathbf{z}}}(u)\|^2)^2}{8\|\dot{\mathbf{r}}(u)\|^4} + \cdots\right)$$

i.e.

$$ds_\epsilon = ds + \epsilon^3 \frac{\overset{(1)}{\dot{\mathbf{z}}}(u)\cdot \overset{(2)}{\dot{\mathbf{z}}}(u)}{\|\dot{\mathbf{r}}(u)\|^2} ds + \epsilon^4 \frac{\|\overset{(2)}{\dot{\mathbf{z}}}(u)\|^2}{2\|\dot{\mathbf{r}}(u)\|^2} ds - \epsilon^6 \cdots + \epsilon^7 \cdots + \cdots .$$

□

Theorem 2.7 ([Najdanović and Velimirović (2017a)]). *Infinitesimal bending fields of the first and the second order of the curve* C *are respectively*

$$\overset{(1)}{\mathbf{z}} = \int [p(u)\mathbf{n}_1 + q(u)\mathbf{n}_2]\, du, \tag{2.40}$$

$$\overset{(2)}{\mathbf{z}} = \int \left[-\frac{p^2(u)+q^2(u)}{2\|\dot{\mathbf{r}}\|}\mathbf{t} + r(u)\mathbf{n}_1 + g(u)\mathbf{n}_2\right] du, \tag{2.41}$$

where $p(u)$, $q(u)$, $r(u)$, $g(u)$ *are arbitrary integrable functions and vectors* $\mathbf{t}$, $\mathbf{n}_1$, $\mathbf{n}_2$ *are unit tangent, principal normal and binormal vector fields, respectively, of the curve* C.

Proof. According to the first equation in (2.37), we conclude that $\overset{(1)}{\dot{\mathbf{z}}}$ lies in the normal plane of the curve C, i.e.

$$\overset{(1)}{\dot{\mathbf{z}}} = p(u)\mathbf{n}_1 + q(u)\mathbf{n}_2, \tag{2.42}$$

where $p(u)$ and $q(u)$ are arbitrary integrable functions. Integrating the previous equation, we obtain Eq. (2.40). If we put Eq. (2.42) into the second equation in (2.37), we obtain

$$2\dot{\mathbf{r}} \cdot \overset{(2)}{\dot{\mathbf{z}}} + p^2(u) + q^2(u) = 0,$$

wherefrom, we have

$$\mathbf{t} \cdot \overset{(2)}{\dot{\mathbf{z}}} = -\frac{p^2(u) + q^2(u)}{2\|\dot{\mathbf{r}}\|}.$$

Therefore,

$$\overset{(2)}{\dot{\mathbf{z}}} = -\frac{p^2(u) + q^2(u)}{2\|\dot{\mathbf{r}}\|}\mathbf{t} + r(u)\mathbf{n}_1 + g(u)\mathbf{n}_2,$$

where $r(u)$ and $g(u)$ are arbitrary integrable functions. □

A curve parameterized by arc length. Let us consider an infinitesimal bending of the second order of the curve $C : \mathbf{r} = \mathbf{r}(s) = \mathbf{r}[u(s)]$, $s \in \mathcal{I} \subseteq \mathcal{R}$ parameterized by the arc length s:

$$C_\epsilon : \tilde{\mathbf{r}}(s, \epsilon) = \mathbf{r}_\epsilon(s) = \mathbf{r}(s) + \epsilon \overset{(1)}{\mathbf{z}}(s) + \epsilon^2 \overset{(2)}{\mathbf{z}}(s). \tag{2.43}$$

Since the vector fields $\overset{(1)}{\mathbf{z}}$ and $\overset{(2)}{\mathbf{z}}$ are defined in the points of the curve C, they can be presented in the form

$$\overset{(j)}{\mathbf{z}} = \overset{(j)}{z}\mathbf{t} + \overset{(j)}{z}_1\mathbf{n}_1 + \overset{(j)}{z}_2\mathbf{n}_2, \quad j = 1, 2, \tag{2.44}$$

where $\overset{(j)}{z}\mathbf{t}$ is a tangential and $\overset{(j)}{z}_1\mathbf{n}_1 + \overset{(j)}{z}_2\mathbf{n}_2$ is a normal component, $\overset{(j)}{z}, \overset{(j)}{z}_1, \overset{(j)}{z}_2$ are the functions of s.

In this case, the system of differential equations (2.37) gets the form

$$\mathbf{r}' \cdot \overset{(1)}{\mathbf{z}}{}' = 0, \quad 2\mathbf{r}' \cdot \overset{(2)}{\mathbf{z}}{}' + \overset{(1)}{\mathbf{z}}{}' \cdot \overset{(1)}{\mathbf{z}}{}' = 0. \tag{2.45}$$

Using the equations (2.45), the expressions (2.44) for the infinitesimal bending fields of the first and the second order respectively, as well as Frenet–Serret equations (2.5), we prove the next theorem.

Theorem 2.8 ([Najdanović and Velimirović (2017a)]). *Necessary and sufficient conditions for the fields* $\overset{(j)}{\mathbf{z}}$, $j = 1, 2$, *to be infinitesimal bending fields of the corresponding order of the curve* C *are*

$$\begin{aligned} &\overset{(1)}{z}{}' - k\overset{(1)}{z_1} = 0, \\ &\overset{(2)}{z}{}' - k\overset{(2)}{z_1} = -\frac{1}{2}\{[k\overset{(1)}{z} + \overset{(1)}{z_1}{}' - \tau\overset{(1)}{z_2}]^2 + [\tau\overset{(1)}{z_1} + \overset{(1)}{z_2}{}']^2\} \end{aligned} \tag{2.46}$$

where k *is the curvature and* τ *is the torsion of* C.

2.2.3 *Some Special Classes of Curves*

In this section, we are going to discuss some special classes of curves under infinitesimal bending of the first order.

2.2.3.1 *Plane Curves*

We will consider closed plane curve at polar coordinates

$$K\,:\,\rho = \rho(\theta), \quad \theta \in [0, 2\pi]. \tag{2.47}$$

Under infinitesimal bending, this curve is included in a family of curves

$$K_\epsilon\,:\,\rho_\epsilon = \rho_\epsilon(\theta), \quad \theta \in [0, 2\pi], \quad (\epsilon \geq 0,\, \epsilon \to 0). \tag{2.48}$$

Equations of these curves in vector form are:

$$K\,:\,\mathbf{r} = \mathbf{r}(\theta), \quad \theta \in [0, 2\pi], \tag{2.49}$$

$$K_\epsilon\,:\,\mathbf{r}_\epsilon = \mathbf{r}(\theta) + \varepsilon\,\mathbf{z}(\theta), \quad \theta \in [0, 2\pi], \tag{2.50}$$

where $\mathbf{z}(\theta)$ is a field of infinitesimal bending.

Theorem 2.9 ([Velimirović (2001)]). *Infinitesimal bending field that plane curve*

$$K\,:\,\rho = \rho(\theta) \tag{2.51}$$

under infinitesimal bending includes in a family of planes curves

$$K_\epsilon\,:\,\rho_\epsilon = \rho_\epsilon(\theta) \tag{2.52}$$

is

$$\mathbf{z}(\theta) = \rho(\theta)\sin\theta\mathbf{i} - \rho(\theta)\cos\theta\mathbf{j}, \tag{2.53}$$

where $\mathbf{i}$ *and* $\mathbf{j}$ *are unit vectors in the direction of Cartesian axes* Ox *and* Oy*, respectively.*

Proof. The necessary and sufficient condition for the field $\mathbf{z}(\theta)$ to be infinitesimal bending field of the curve $\mathbf{r}(\theta)$ is

$$\dot{\mathbf{r}}(\theta) \cdot \dot{\mathbf{z}}(\theta) = 0. \tag{2.54}$$

From here, we conclude that $\dot{\mathbf{z}} \perp \dot{\mathbf{r}}$, i.e. that $\dot{\mathbf{z}}$ is in the normal plane of the curve, i.e.

$$\dot{\mathbf{z}}(\theta) = p(\theta)\mathbf{n}_1 + q(\theta)\mathbf{n}_2, \tag{2.55}$$

where $p(\theta)$ and $q(\theta)$ are arbitrary functions. In order to stay in the plane of the curve, we choose

$$q(\theta) = 0. \tag{2.56}$$

So, we have

$$\dot{\mathbf{z}}(\theta) = p(\theta)\mathbf{n}_1(\theta),$$

i.e.

$$\mathbf{z}(\theta) = \int p(\theta)\mathbf{n}_1(\theta)\, d\theta + c. \tag{2.57}$$

We will take $c = 0$. As the equation of the curve K in vector form is

$$K\,:\,\mathbf{r}(\theta) = \rho(\theta)\cos\theta\,\mathbf{i} + \rho(\theta)\sin\theta\,\mathbf{j}, \tag{2.58}$$

we calculate infinitesimal bending field of this curve according to (2.57).

Substituting $\mathbf{n}_1$ with respect to (2.6), we have

$$\mathbf{z} = \int p(\theta)\frac{(\rho\ddot{\rho} - \rho^2 - 2\dot{\rho}^2)[(\rho\cos\theta + \dot{\rho}\sin\theta)\mathbf{i} + (\rho\sin\theta - \dot{\rho}\cos\theta)\mathbf{j}]}{\sqrt{\rho^2 + \dot{\rho}^2}|\rho\ddot{\rho} - \rho^2 - 2\dot{\rho}^2|}d\theta.$$

As

$$\rho(\theta)\cos\theta + \dot{\rho}(\theta)\sin\theta = (\rho(\theta)\sin\theta)^{\cdot},$$
$$\rho(\theta)\sin\theta - \dot{\rho}(\theta)\cos\theta = -(\rho(\theta)\cos\theta)^{\cdot},$$

we choose

$$p(\theta) = ||\dot{\mathbf{r}}|| = \sqrt{\rho(\theta)^2 + \dot{\rho}(\theta)^2},$$

that gives

$$\mathbf{z}(\theta) = \int [(\rho(\theta)\sin\theta)^{\cdot}\mathbf{i} - (\rho(\theta)\cos\theta)^{\cdot}\mathbf{j}]\, d\theta = \rho(\theta)\sin\theta\mathbf{i} - \rho(\theta)\cos\theta\mathbf{j}. \tag{2.59}$$

□

Corollary 2.1 ([Velimirović (2001)]). *If the plane curve*

$$K : \rho = \rho(\theta), \quad \theta \in [0, 2\pi],$$

under infinitesimal bending stays at original plane, the equation of deformed curve will be

$$K_\epsilon : \rho_\epsilon(\theta) = \rho(\theta)\sqrt{1+\epsilon^2}. \tag{2.60}$$

Proof. According to the previous theorem, we have

$$\mathbf{r}_\epsilon = \mathbf{r} + \epsilon\mathbf{z} = \rho(\theta)(\cos\theta + \epsilon\sin\theta)\mathbf{i} + \rho(\theta)(\sin\theta - \epsilon\cos\theta)\mathbf{j},$$

i.e. we obtain (2.60). □

2.2.3.2 *Spherical Curves*

We are asking the question: Is it possible to infinitesimally bend a curve on the sphere (i.e. spherical curve), so that all bent curves belong to the same sphere? The next theorem answers this question.

Theorem 2.10 ([Velimirović *et al.* (2010a)]). *Let $C : \mathbf{r} = \mathbf{r}(t)$, $t \in \mathcal{J}$ be a curve on the sphere $S : ||\mathbf{r} - \mathbf{q}|| = r$, with the center $\mathbf{q} \in \mathcal{R}^3$ and the radius $r > 0$. There is no nontrivial vector field $\mathbf{z}(t)$ so that the family of the curves*

$$C_\epsilon : \mathbf{r}_\epsilon = \tilde{\mathbf{r}}(t,\epsilon) = \mathbf{r}(t) + \epsilon\mathbf{z}(t), \quad (\epsilon \geq 0,\, \epsilon \to 0), \quad \tilde{\mathbf{r}}(t,0) = \mathbf{r}(t), \tag{2.61}$$

belongs to the sphere S.

Proof. Suppose that the family of the curves (2.61) belongs to the sphere S. It means that it is valid:

$$||\mathbf{r}_\epsilon - \mathbf{q}|| = r, \quad \forall \epsilon \geq 0.$$

Squaring of this equation, we get $||\mathbf{r}_\epsilon - \mathbf{q}||^2 = r^2$, i.e. $||\mathbf{r}(t) + \epsilon \mathbf{z}(t) - \mathbf{q}||^2 = r^2$, wherefrom, we have

$$||\mathbf{r}(t)||^2 + \epsilon^2||\mathbf{z}(t)||^2 + ||\mathbf{q}||^2 - 2\epsilon \mathbf{q} \cdot \mathbf{z}(t) - 2\mathbf{q} \cdot \mathbf{r}(t) + 2\epsilon \mathbf{z}(t) \cdot \mathbf{r}(t) = r^2. \tag{2.62}$$

On the other hand, $||\mathbf{r}(t) - \mathbf{q}||^2 = r^2$, i.e.

$$||\mathbf{r}(t)||^2 - 2\mathbf{q} \cdot \mathbf{r}(t) + ||\mathbf{q}||^2 = r^2. \tag{2.63}$$

Substituting (2.63) into (2.62), we get for $\epsilon \neq 0$,

$$2\mathbf{z}(t) \cdot \mathbf{r}(t) - 2\mathbf{q} \cdot \mathbf{z}(t) + \epsilon||\mathbf{z}(t)||^2 = 0. \tag{2.64}$$

The condition (2.64) must be valid for each $\epsilon > 0$. It is valid when

$$\mathbf{z}(t) \cdot (\mathbf{r}(t) - \mathbf{q}) = 0 \tag{2.65}$$

and

$$||\mathbf{z}(t)||^2 = 0. \tag{2.66}$$

But from (2.66), we obtain $\mathbf{z}(t) = \mathbf{0} = (0, 0, 0)$, and the theorem is proved. □

Corollary 2.2 ([Velimirović *et al.* (2010a)]). *There is no infinitesimal bending field of a spherical curve which that curve leaves on the same sphere.*

2.2.3.3 *Curves on Ruled Surfaces*

Theorem 2.11 ([Najdanović and Velimirović (2018)]). *Let a ruled surface S be given by*

$$S : \mathbf{r} = \mathbf{r}(u, v) = \boldsymbol{\rho}(u) + v\,\mathbf{e}(u) \quad (u \in \mathcal{J} \subseteq \mathcal{R}, v \in \mathcal{R}, \|\mathbf{e}(u)\| = 1), \tag{2.67}$$

with a directrix $\boldsymbol{\rho} = \boldsymbol{\rho}(u)$ *and generatrices in the direction of the vector* $\mathbf{e}(u)$, *and let a curve*

$$C : \mathbf{r} = \mathbf{r}(t) = \mathbf{r}(u(t), v(t)) \tag{2.68}$$

be on the surface S. *Then, the infinitesimal bending field which given curve leaves on the surface* S *is*

$$\mathbf{z}(t) = c\,\mathbf{e}(u(t))\, e^{-\int \frac{\dot{u}^2(\boldsymbol{\rho}_u \cdot \mathbf{e}_u + v\|\mathbf{e}_u\|^2)}{\dot{u}\boldsymbol{\rho}_u \cdot \mathbf{e} + \dot{v}}\, dt}, \tag{2.69}$$

where $\dot{u}\boldsymbol{\rho}_u \cdot \mathbf{e} + \dot{v} \neq 0$ *and* c *is a constant.*

Proof. Let

$$C : \mathbf{r} = \mathbf{r}(t) = \boldsymbol{\rho}(u(t)) + v(t)\,\mathbf{e}(u(t))$$

be a curve on the surface S and

$$C_\epsilon : \mathbf{r}_\epsilon(t) = \boldsymbol{\rho}(u(t)) + v(t)\,\mathbf{e}(u(t)) + \epsilon\,\mathbf{z}(t)$$

be an infinitesimal bending of the curve C determined with the field $\mathbf{z}$. As the family of curves C_ϵ, $\epsilon \geq 0$, belongs to the surface S, the field $\mathbf{z}$ is in the form

$$\mathbf{z}(t) = \mathbf{e}(u(t))\, z_1(t), \tag{2.70}$$

where $z_1(t)$ is a real continuous differentiable function.

Having in mind that $\mathbf{z}$ is an infinitesimal bending field, the condition (2.16) must be satisfied, i.e.

$$(\boldsymbol{\rho}_u\, \dot{u} + \dot{v}\,\mathbf{e} + v\,\mathbf{e}_u\, \dot{u}) \cdot (\mathbf{e}_u\, \dot{u}\, z_1 + \mathbf{e}\, \dot{z}_1) = 0. \tag{2.71}$$

Since $\|\mathbf{e}\| = 1$, it must be valid $\mathbf{e} \cdot \dot{\mathbf{e}} = 0$ ($\mathbf{e} \perp \dot{\mathbf{e}} \Leftrightarrow \dot{\mathbf{e}} \cdot \mathbf{e} + \mathbf{e} \cdot \dot{\mathbf{e}} = 0 \Leftrightarrow (\mathbf{e} \cdot \mathbf{e})^{\cdot} = 0 \Leftrightarrow (\|\mathbf{e}\|^2)^{\cdot} = 0 \Leftrightarrow \|\mathbf{e}\| = const$). Using this fact, we obtain homogenous linear differential equation

$$z_1 \dot{u}^2(\boldsymbol{\rho}_u \cdot \mathbf{e}_u + v\|\mathbf{e}_u\|^2) + \dot{z}_1(\dot{u}\boldsymbol{\rho}_u \cdot \mathbf{e} + \dot{v}) = 0, \tag{2.72}$$

whose solution is

$$z_1(t) = ce^{-\int \frac{\dot{u}^2(\boldsymbol{\rho}_u \cdot \mathbf{e}_u + v\|\mathbf{e}_u\|^2)}{\dot{u}\boldsymbol{\rho}_u \cdot \mathbf{e} + \dot{v}}\, dt}, \tag{2.73}$$

c is a constant and $\dot{u}\boldsymbol{\rho}_u \cdot \mathbf{e} + \dot{v} \neq 0$. Putting (2.73) into (2.70), we obtain (2.69). □

The ruled surfaces are not the only surfaces on which it is possible to bend curves infinitesimally. We will show that in the following example.

Example 2.2. Let S be the rotation surface in $\mathcal{R}^3$ of class C^∞, given as the graph of the function

$$F(x,y) = \begin{cases} e^{\frac{-1}{(r^2-1)^2}}, & r > 1 \\ 0, & 0 \le r \le 1 \end{cases} \qquad x^2 + y^2 = r^2.$$

Note that S is not a ruled surface (in fact it is not ruled in any open neighborhood of any point of the unit circle $(\cos\theta, \sin\theta, 0)$). Consider the curve $C \subset S$ given by

$$C : \mathbf{r}(t) = (t, o, f(t)),$$

where

$$f(t) = \begin{cases} e^{\frac{-1}{(t^2-1)^2}}, & |t| > 1 \\ 0, & |t| \le 1 \end{cases}$$

Let

$$g(t) = \begin{cases} 0, & |t| \ge 1 \\ e^{\frac{-1}{(t^2-1)^2}}, & |t| < 1 \end{cases}$$

Then, the family of curves given by

$$C_\epsilon : \mathbf{r}_\epsilon = (t, \epsilon g(t), f(t))$$

belongs to the surface S and presents infinitesimal bending of the curve C determined with the field $\mathbf{z} = (0, g(t), 0)$.

Infinitesimal Bending of Curves on the Cylinder A cylinder is an example of the ruled surfaces. Let us find an infinitesimal bending field of a curve on the cylinder, that leaves the given curve on the cylinder after bending.

Theorem 2.12 ([Najdanović and Velimirović (2018)]). *Let the cylinder be given by the equation $S : x^2+y^2 = a^2$. Let $C : \mathbf{r} = \mathbf{r}(t)$, $t \in (\alpha, \beta)$ be a curve on the cylinder S and $\mathbf{z}(t)$ be a vector field of class*

C^1 which given curve includes in the family of the curves $C_\epsilon : \mathbf{r}_\epsilon = \mathbf{r}(t) + \epsilon \mathbf{z}(t)$, $\epsilon \geq 0$, $\epsilon \to 0$, on the cylinder S, under infinitesimal bending.

(a) *If the curve C is in the plane $z = const$, then infinitesimal bending field is $\mathbf{z}(t) = z_3(t)\mathbf{k}$, where $\mathbf{k} = (0, 0, 1)$ and $z_3(t)$ is an arbitrary real function of class C^1.*

(b) *Otherwise, infinitesimal bending field is a constant vector $\mathbf{z} = c\mathbf{k}$, where c is a real constant. Bending is rigid and reduces to the translation along z-axis.*

Proof. The parametric equation of the cylinder of height h and of semi-axis a is

$$\mathbf{r}(u, v) = (a \cos u, a \sin u, v), \quad u \in [0, 2\pi], \quad v \in [0, h]. \tag{2.74}$$

Let

$$\mathbf{r}(t) = \mathbf{r}(u(t), v(t)) = (a \cos u(t), a \sin u(t), v(t)), \quad t \in (\alpha, \beta) \tag{2.75}$$

be the curve which lies on the cylinder (2.74). Infinitesimal bending will be

$$\begin{aligned} \mathbf{r}_\epsilon(t) &= \mathbf{r}(t) + \epsilon \mathbf{z}(t) \\ &= (a \cos u(t) + \epsilon z_1(t), a \sin u(t) + \epsilon z_2(t), v(t) + \epsilon z_3(t)), \end{aligned} \tag{2.76}$$

where $\mathbf{z}(t) = (z_1(t), z_2(t), z_3(t))$ and $z_1(t), z_2(t), z_3(t)$ are real continuous differentiable functions. Since the curves $\mathbf{r}_\epsilon$ are on the surface (2.74), it must be valid

$$(a \cos u(t) + \epsilon z_1(t))^2 + (a \sin u(t) + \epsilon z_2(t))^2 = a^2,$$

i.e. after recombination

$$2a \left(\cos u(t) z_1(t) + \sin u(t) z_2(t)\right) + \epsilon \left(z_1^2(t) + z_2^2(t)\right) = 0,$$

for each $\epsilon > 0$. Therefore, it must be valid

$$\cos u(t) z_1(t) + \sin u(t) z_2(t) = 0 \tag{2.77}$$

and

$$z_1^2(t) + z_2^2(t) = 0. \tag{2.78}$$

From (2.78), we have $z_1(t) = z_2(t) = 0$, so $\mathbf{z}(t) = (0, 0, z_3(t))$. Since $\mathbf{z}$ is infinitesimal bending field, it holds (2.16). Therefore,

$$(-a \sin u(t)\dot{u}(t), a \cos u(t)\dot{u}(t), \dot{v}(t)) \cdot (0, 0, \dot{z}_3(t)) = 0,$$

i.e.

$$\dot{v}(t)\dot{z}_3(t) = 0. \tag{2.79}$$

From here, it is $\dot{v}(t) = 0$ or $\dot{z}_3(t) = 0$. We distinguish two cases:

- If $\dot{v}(t) = 0$, then we have $v(t) = const$, i.e. the curve $\mathbf{r}(t)$ is in the plane $z = const$. Infinitesimal bending field in this case is $\mathbf{z}(t) = (0, 0, z_3(t)) = z_3(t)\mathbf{k}$.
- If $\dot{z}_3(t) = 0$, i.e. $z_3(t) = const$, then we obtain infinitesimal bending field of an arbitrary curve on the cylinder in the form $\mathbf{z} = (0, 0, c) = c\mathbf{k}$, c is a real constant.

□

Example 2.3. Let us consider some curves on the cylinder $S : \mathbf{r}(u, v) = (3 \cos u, 3 \sin u, v)$ and their infinitesimal bending.

For the circle $C_1 : \mathbf{r}(t) = (3 \cos t, 3 \sin t, 0)$, which is in the plane $z = 0$, we can take infinitesimal bending field $\mathbf{z}(t) = t\mathbf{k}$. In Fig. 2.1, we can see the curve C_1 and bent curves for $\epsilon = 0.2, 0.4, 0.6, 0.8, 1$. Obviously, the curve C_1 clefts under that infinitesimal bending and is included in a family of helices on the cylinder.

For the same circle $C_1 : \mathbf{r}(t) = (3 \cos t, 3 \sin t, 0)$, we can take $\mathbf{z}(t) = t(2\pi - t)\mathbf{k}$. As it is valid $\mathbf{r}_\epsilon(t = 0) = \mathbf{r}_\epsilon(t = 2\pi)$, the curve remains closed (see Fig. 2.2; bent curves are obtained for $\epsilon = 0.1, 0.2, 0.3, 0.4, 0.5$).

If we take the curve $C_2 : \mathbf{r}(t) = (3 \cos t^3, 3 \sin t^3, t^3)$ and infinitesimal bending field $\mathbf{z} = 4\mathbf{k}$, we get only rigid motion, i.e. translation along the z-axis (Fig. 2.3).

Infinitesimal Bending of Curves on Hyperbolic Paraboloid

A hyperbolic paraboloid is an example of doubly ruled surfaces. Below, we will find infinitesimal bending field of an arbitrary curve belonging to this surface.

Theorem 2.13 ([Najdanović and Velimirović (2018)]). *Let a hyperbolic paraboloid be given by the equation* $S : \mathbf{r}(u, v) = (u, v, uv)$

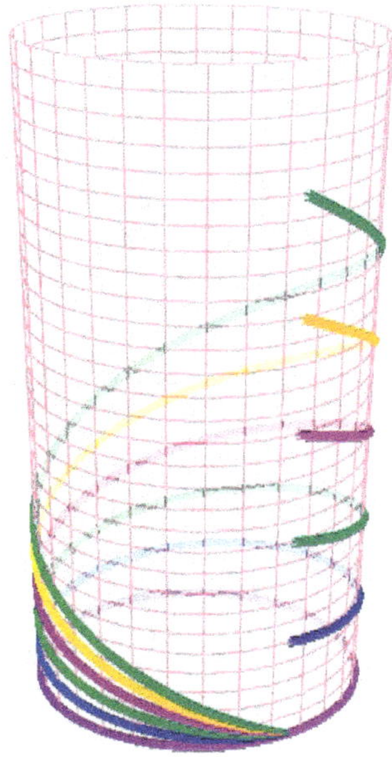

Fig. 2.1. Infinitesimal bending of the circle C_1 on the cylinder.

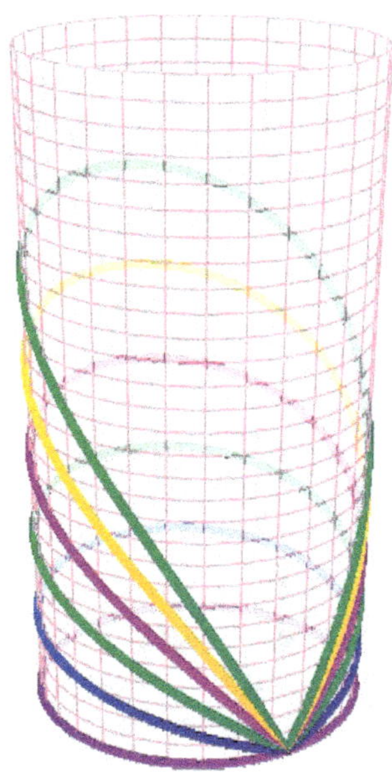

Fig. 2.2. Infinitesimal bending of the circle C_1 on the cylinder.

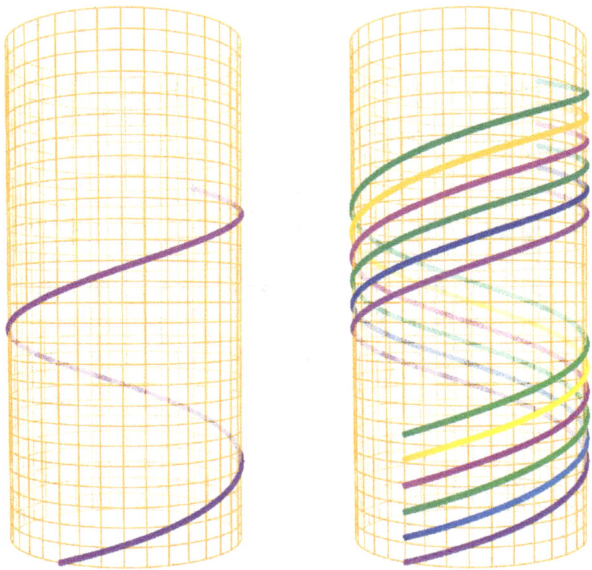

Fig. 2.3. Infinitesimal bending of the curve C_2 on the cylinder.

and a curve on it by $C : \mathbf{r}(t) = \mathbf{r}(u(t), v(t))$. *Let* $\mathbf{z}(t)$ *be a vector field of class* C^1 *which given curve under infinitesimal bending includes in the family of the curves* $C_\epsilon : \mathbf{r}_\epsilon = \mathbf{r}(t) + \epsilon\mathbf{z}(t)$, $\epsilon \geq 0$, $\epsilon \to 0$, *on the hyperbolic paraboloid* S. *Then, the equations*

$$\mathbf{z}(t) = ce^{-\int \frac{(uv)^{\cdot}\dot{u}}{\dot{v}+u(uv)^{\cdot}}dt}(0, 1, u(t)), \tag{2.80}$$

$\dot{v} + u(uv)^{\cdot} \neq 0$, *and*

$$\mathbf{z}(t) = ce^{-\int \frac{(uv)^{\cdot}\dot{v}}{\dot{u}+v(uv)^{\cdot}}dt}(1, 0, v(t)), \tag{2.81}$$

$\dot{u} + v(uv)^{\cdot} \neq 0$, *determine the field* $\mathbf{z}(t)$. c *is an arbitrary constant.*

Proof. Let

$$\begin{aligned}\mathbf{r}_\epsilon(t) &= \mathbf{r}(t) + \epsilon\mathbf{z}(t) \\ &= (u(t) + \epsilon z_1(t), v(t) + \epsilon z_2(t), u(t)v(t) + \epsilon z_3(t))\end{aligned} \tag{2.82}$$

be an infinitesimal bending of C determined by the field $\mathbf{z}(t) = (z_1(t), z_2(t), z_3(t))$, z_1, z_2, z_3 are real continuous differentiable functions. Since the curves (2.82) are on the surface S, the following condition is satisfied

$$(u(t) + \epsilon z_1(t))(v(t) + \epsilon z_2(t)) = u(t)v(t) + \epsilon z_3(t), \quad \forall \epsilon \geq 0.$$

By dividing with $\epsilon \neq 0$, we obtain

$$z_1(t)v(t) + z_2(t)u(t) - z_3(t) + \epsilon z_1(t)z_2(t) = 0. \tag{2.83}$$

Since the condition (2.83) must be valid for each $\epsilon > 0$, it must be

$$z_1(t)v(t) + z_2(t)u(t) - z_3(t) = 0 \tag{2.84}$$

and

$$z_1(t)z_2(t) = 0. \tag{2.85}$$

From (2.85) we get $z_1(t) = 0$ or $z_2(t) = 0$. We distinguish two cases:

- Let $z_1(t) = 0$ be valid. Then, Eq. (2.84) reduces to $z_2(t)u(t) = z_3(t)$. From here, we have

$$\mathbf{z}(t) = (0, z_2(t), u(t)z_2(t)). \tag{2.86}$$

Since $\mathbf{z}(t)$ is the infinitesimal bending field, Eq. (2.16) is valid. By substituting $\dot{\mathbf{r}} = (\dot{u}, \dot{v}, \dot{u}v + u\dot{v})$ and $\dot{\mathbf{z}} = (0, \dot{z}_2, \dot{u}z_2 + u\dot{z}_2)$, in Eq. (2.16) we obtain

$$(\dot{v} + (uv)^{\cdot}u)\dot{z}_2 + (uv)^{\cdot}\dot{u}z_2 = 0. \tag{2.87}$$

Solving the homogenous linear differential equation (2.87) by z_2, we obtain

$$z_2(t) = ce^{-\int \frac{(uv)^{\cdot}\dot{u}}{\dot{v}+u(uv)^{\cdot}}dt}, \dot{v} + u(uv)^{\cdot} \neq 0. \tag{2.88}$$

From here and from (2.86), we get (2.80).
- We solve the case $z_2(t) = 0$ similarly and get (2.81).

It is easy to show that the fields (2.80) and (2.81) present infinitesimal bending fields of the curve C which that curve leave on the hyperbolic paraboloid S, after bending. □

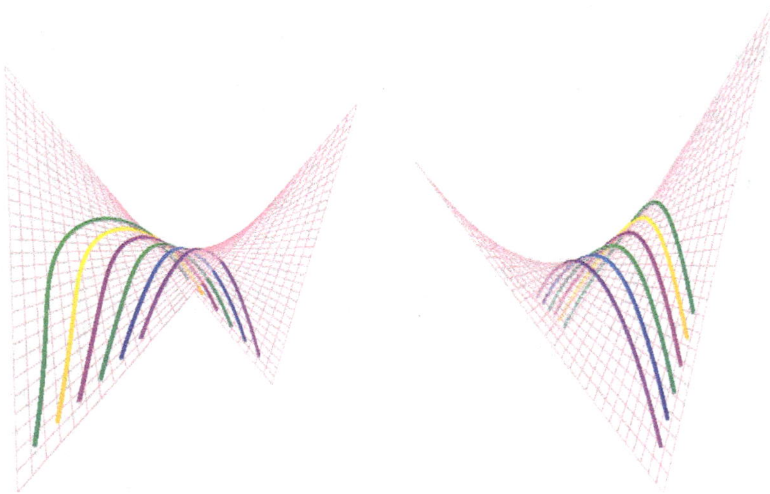

Fig. 2.4. Infinitesimal bending of the curve C_3 on the hyperbolic paraboloid.

Example 2.4. Let the curve $C_3 : \mathbf{r}(t) = (t, t, t^2)$, $t \in (a, b) \subseteq \mathcal{R}$, be given on the surface $S : \mathbf{r}(u, v) = (u, v, uv)$. According to Theorem 2.13, after necessary calculations, we obtain that infinitesimal bending fields of the curve C_3 which given curve leave on the surface S, for $c = 1$, have the following form

$$\mathbf{z}_1(t) = \left(0, \frac{1}{\sqrt{1+2t^2}}, \frac{t}{\sqrt{1+2t^2}}\right)$$

and

$$\mathbf{z}_2(t) = \left(\frac{1}{\sqrt{1+2t^2}}, 0, \frac{t}{\sqrt{1+2t^2}}\right).$$

The resulting deformations are shown in Fig. 2.4. We can see the curve C_3 and bent curves $C_{3\epsilon}$ for $\epsilon = 0.25, 0.5, 0.75, 1$ under the fields $\mathbf{z}_1$ and $\mathbf{z}_2$.

Example 2.5. For the curve $C_4 : \mathbf{r}(t) = (t, t^2, t^3)$, the corresponding infinitesimal bending fields are

$$\mathbf{z}_1(t) = \left(0, \frac{1}{\sqrt{2+3t^2}}, \quad \frac{t}{\sqrt{2+3t^2}}\right)$$

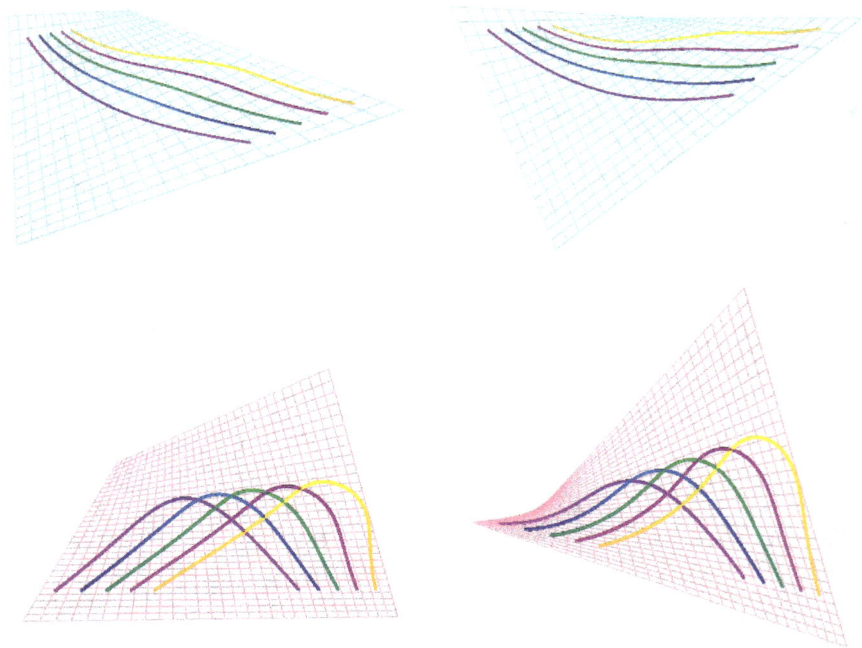

Fig. 2.5. Infinitesimal bending of the curve C_4 on the hyperbolic paraboloid under fields $\mathbf{z}_1$ and $\mathbf{z}_2$.

and

$$\mathbf{z}_2(t) = \left(\frac{1}{\sqrt{1+3t^4}}, \quad 0, \quad \frac{t^2}{\sqrt{1+3t^4}} \right).$$

Graphical presentation of the family of bent curves $C_{4\epsilon}$ under the fields $\mathbf{z}_1$ and $\mathbf{z}_2$ is given in Fig. 2.5.

2.3 Knots

A knot is a closed, self-avoiding curve. From the topological point of view, we have more general definition. In order to give a topological definition, let us introduce some basic terms.

A **homeomorphism** between two topological spaces is continuous bijection $h : \mathcal{X} \to \mathcal{Y}$ whose inverse is also continuous. If such a map exists, then $\mathcal{X}$ and $\mathcal{Y}$ are **homeomorphic** or **topology equivalent**.

Let be given a closed curve C in $\mathcal{R}^3$. The **homeomorphic** curves C, C' are called **isotopic** if there exists a continuous family of curves C_t depending on t, $(0 \leq t \leq 1)$, such that C_t is homeomorphic to C, $C_0 = C$ and $C_1 = C'$. We say that C is a **knot** if it is homeomorphic to a circle, but is not isotopic to a circle.

According to this definition, another knot C' is equivalent to C if it can be continuously deformed into C without crossing itself during the process. Equivalent knots are considered the same in the topological sense and determine an equivalence class of knots named **knot type**. It is therefore possible to think of a knot as a curve with small but positive thickness which allows us to present it as a tube. But in geometrical sense, we can observe particular representative of a knot type as a closed self-avoiding curve in 3-dimensional space.

The simplest knot is the **unknot** also known as the **trivial knot** which can be deformed to a geometric circle in $\mathcal{R}^3$. Two other rather easy knots are the **trefoil** and the **figure-eight knots**.

Usually knots are represented and communicated by their projections into a plane. For a given knot, almost all directions yields projections with only finitely many points of singularity, all of which are all double points (see [Simon (1996)]). The number of double points in a particular projection of a knot is called the **crossing number** of the projection. The crossing number of a knot type is the minimum, over all knots realizing that type, of the crossing numbers of the individual knots. Thus, the unknot has crossing number zero, the trefoil knot three and the figure-eight knot four. There are no other knots with a crossing number this low, and just two knot types have crossing number five. The first one is one of an infinite class called **torus knots** because they can be drawn on the surface of a standard torus. The other five crossing knot type is also one of an infinite class, called **twist knots**, where the "clasp" at the top remains the same, but the number of "twists" at the bottom varies. The number of knot types with a particular crossing number increases rapidly as the crossing number increases.

2.3.1 *Infinitesimal Bending of Knots*

In geometrical sense, small deformations of a knot (a particular curve from the corresponding knot type) can be considered as the family

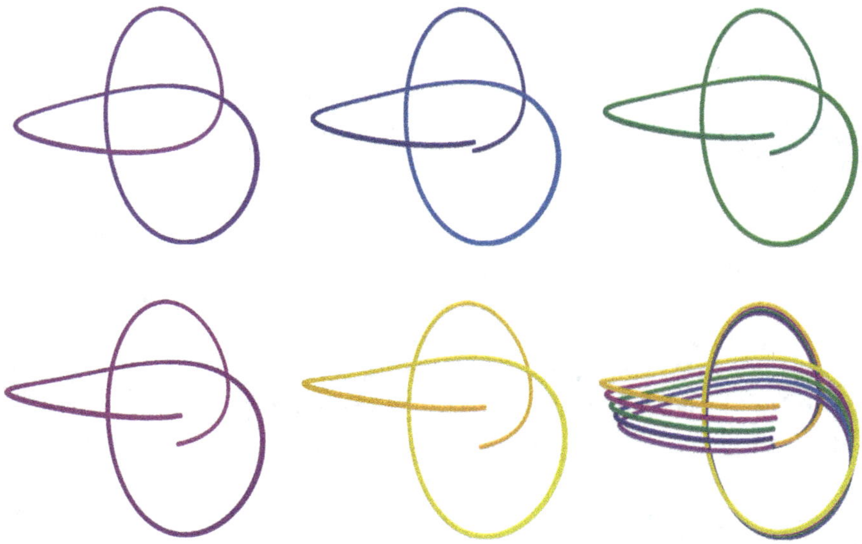

Fig. 2.6. The trefoil knot and its infinitesimal bending for $\epsilon = 0.0005, 0.001, 0.0015, 0.002$.

of different curves. We are going to demonstrate this fact through examples.

Let us consider the trefoil knot

$$\mathbf{r}(u) = (4\cos(2u) + 2\cos(u), 4\sin(2u) - 2\sin(u), \sin(3u)).$$

The field

$$\mathbf{z} = \int \frac{d\mathbf{r}}{du} \times \frac{d^2\mathbf{r}}{du^2}\,du = \left(60\cos(u) - 6\cos(2u) - \frac{3}{2}\cos(4u) + \frac{12}{5}\cos(5u),\ -60\sin(u) - 6\sin(2u) + \frac{3}{2}\sin(4u) + \frac{12}{5}\sin(5u),\ 124u - \frac{16}{3}\sin(3u)\right),$$

$u \in [0, 2\pi]$, is corresponding infinitesimal bending field. This field is obtained, according to Theorem 2.3, for $p(u) = 0$ and $q(u) = \|\frac{d\mathbf{r}}{du} \times \frac{d^2\mathbf{r}}{du^2}\|$. Obviously, $\mathbf{z}(0) \neq \mathbf{z}(2\pi)$, and the knot "'is torn" under this infinitesimal bending, see Fig. 2.6.

Fig. 2.7. The trefoil knot and its infinitesimal bending for ϵ = 0.00025, 0.0005, 0.00075, 0.001.

If we want to get a family of closed curves under infinitesimal bending of a knot, we must specify the condition $\mathbf{z}(0) = \mathbf{z}(2\pi)$ for the infinitesimal bending field.

For $p(u) = \|(\frac{d\mathbf{r}}{du} \times \frac{d^2\mathbf{r}}{du^2}) \times \frac{d\mathbf{r}}{du}\|$, $q(u) = 0$, we obtain

$$\begin{aligned}
\mathbf{z} &= \int \left(\frac{d\mathbf{r}}{du} \times \frac{d^2\mathbf{r}}{du^2}\right) \times \frac{d\mathbf{r}}{du}\, du \\
&= \Bigg(303\sin(u) - 540\sin(2u) - \frac{53}{2}\sin(4u) + \frac{46}{5}\sin(5u) + \frac{9}{7}\sin(7u) \\
&\quad + \frac{9}{4}\sin(8u), 303\cos(u) + 540\cos(2u) - \frac{53}{2}\cos(4u) - \frac{46}{5}\cos(5u) \\
&\quad + \frac{9}{7}\cos(7u) - \frac{9}{4}\cos(8u), -12(-17\cos(3u) + \cos(6u))\Bigg).
\end{aligned}$$

All bent curves are also closed, see Fig. 2.7.

Because of topological properties of a knot, it can be represented as a tube around a curve. It looks like a rope, but without examination any physical characteristic of the rope. Some examples are given below.

Fig. 2.8. The trefoil knot: basic and infinitesimally bent with $\epsilon = 0.3$.

Fig. 2.9. The trefoil knot: infinitesimally bent with $\epsilon = 0.6$. and all knots together.

A trefoil knot, given by scalar parametric equations: $x = \sin(u) + 2\sin(2u)$, $y = \cos(y) - 2\cos(2u)$, $z = -\sin(3u)$, is presented in Figs. 2.8 and 2.9, together with infinitesimal bending determined by the bending field (2.18) for $p(u) = \cos(3u)$ and $q(u) = \sin(3u)$.

A figure eight knot is given by the parametric equations: $x = (2 + \cos(2u)) \cdot \cos(3u)$, $y = (2 + \cos(2u)) \cdot \sin(3u)$, $z = \sin(4u)$. The basic and infinitesimally bent figure eight knot are given in Figs. 2.10 and 2.11.

The bending fields are defined by: $p(u) = \cos(6u)$ and $q(u) = \sin(6u)$.

2.4 Surfaces Defined by Bending Curves

Let us remind of the first order infinitesimal bending for a curve

$$C_\epsilon : \tilde{\mathbf{r}}(u, \epsilon) = \mathbf{r}_\epsilon(u) = \mathbf{r}(u) + \epsilon \mathbf{z}(u), \tag{2.89}$$

where $\mathbf{z}(u)$ is an infinitesimal bending field (of the first order).

Fig. 2.10. Figure eight knot: basic and infinitesimally bent with $\epsilon = 1.4$.

Fig. 2.11. Figure eight knot: basic and infinitesimally bent with $\epsilon = 1.4$ together.

We can consider the definition of the infinitesimal bending of the curve as a vector parametric equation of a surface defined by two free variables: one of them is free variable u which define the curve and another is bending parameter ϵ.

There is a connection between infinitesimal bending of curves and ruled surfaces. Namely, let us consider a ruled surface

$$S: \boldsymbol{\rho}(u,v) = \mathbf{r}(u) + v\mathbf{e}(u), \quad u \in \mathcal{J} \subseteq \mathcal{R}, \quad v \in \mathcal{R}, \quad \|\mathbf{e}(u)\| = 1,$$

with the directrix $C: \mathbf{r} = \mathbf{r}(u)$ and the generatrices in the direction of $\mathbf{e}(u)$. If the directrix C is also the striction line of the ruled surface S (see [Gray (1998)]), then the condition

$$\dot{\mathbf{r}}(u) \cdot \dot{\mathbf{e}}(u) = 0$$

Fig. 2.12. Surface on trefoil knot: basic and infinitesimally bent with $\epsilon = 2.0$.

is valid. Therefore, $\mathbf{e}(u)$ is the infinitesimal bending field and the family of bent curves (infinitesimal bending)

$$C_\epsilon : \tilde{\mathbf{r}}(u, \epsilon) = \mathbf{r}_\epsilon(u) = \mathbf{r}(u) + \epsilon\mathbf{e}(u)$$

belongs to the ruled surface S. (For characterization with some examples, see [Gözütok *et al.* (2020)].)

Note that each curve of the family C_ϵ is "parallel" with C, i.e. the segment of each generatrix between C and C_ϵ is of the same length:

$$\|\tilde{\mathbf{r}}(u_1, \epsilon) - \mathbf{r}(u_1)\| = \|\epsilon\mathbf{e}(u_1)\| = \epsilon = \|\epsilon\mathbf{e}(u_2)\| = \|\tilde{\mathbf{r}}(u_2, \epsilon) - \mathbf{r}(u_2)\|,$$

since $\|\mathbf{e}(u_1)\| = \|\mathbf{e}(u_2)\| = 1$. For $\epsilon \in [0, R]$, the corresponding knot infinitesimal bending determine a ribbon of width R.

While being bent, curve is deformed and moved through the space. In the following examples, we will present some of such surfaces.

The first example is a surface obtained by bending trefoil knot given by $\mathbf{r}(u) = (\sin(u) + 2\sin(2u), \cos(u) - 2\cos(2u), -\sin(3u))$, bending field is defined by $p(u) = \cos(3u)$ and $q(u) = \sin(6u)$, see Fig. 2.12.

Fig. 2.13. p3q2 knot: basic and infinitesimally bent with $\epsilon = 0.6, 1.2, 1.8$.

A second example is a surface obtained by bending a kind of torus knot, topologically equivalent to the trefoil knot, given by scalar parametric equations: $x = (\cos(2u) + 2) \cdot \cos(3u)$, $y = (\cos(2u) + 2) \cdot \sin(3u)$, $z = -\sin(2u)$.

Bending field is defined by $p(u) = \cos(2u)$ and $q(u) = \sin(2u)$. For the bent curves, see Fig. 2.13, obtained surface is on Fig. 2.14.

2.5 Curves on Surfaces with a Given Precision

Let

$$C : \mathbf{r} = \mathbf{r}(t) = (x(t), y(t), z(t)), \quad t \in \mathcal{J} \subseteq \mathbb{R},$$

be the curve on the surface S given by the implicit equation

$$S : F(x, y, z) = 0.$$

Fig. 2.14. Surface on p3q2 knot: basic and infinitesimally bent with $\epsilon = 0.6, 1.2, 1.8$.

So, it is valid

$$F(x(t), y(t), z(t)) = 0.$$

Suppose that

$$C_\epsilon : \mathbf{r}_\epsilon(t) = (x_\epsilon(t), y_\epsilon(t), z_\epsilon(t))$$

is an infinitesimal bending (of the first order) of the curve C and we get C for $\epsilon = 0$. In Section 2.2.3, we considered the infinitesimal bending C_ϵ of the curve C on the given surface S. Now we pose the problem to determinate an infinitesimal bending of C so that all bent curves C_ϵ are on the surface S with a given precision, i.e. so that the following condition is valid:

$$F(x_\epsilon(t), y_\epsilon(t), z_\epsilon(t)) = o(\epsilon). \tag{2.90}$$

We will consider some classes of curves based on papers [Najdanović *et al.* (2021a, 2024); Rančić *et al.* (2019)].

2.5.1 *Spherical Curves*

Let us recall the theorem on spherical curves, which gives certain relations between the curvature and torsion of a space curve that lies on a sphere.

Theorem 2.14 ([Gray (1998)]). *Let $C : \mathbf{r} = \mathbf{r}(s)$, $s \in [0, L]$ be a curve parameterized by arc length on the sphere $S : \|\mathbf{r} - \mathbf{q}\|^2 = r^2$. Then:*

(1) $k \geq \frac{1}{r}$;
(2) *the curvature k and torsion τ of C are related by*

$$\tau^2\left(r^2 - \frac{1}{k^2}\right) = \left(\frac{k'}{k^2}\right)^2; \tag{2.91}$$

(3) $k'(t_0) = 0$ *implies* $\tau(t_0) = 0$ *or* $k(t_0) = \frac{1}{r}$;
(4) $\tau\left(\frac{k''}{k^2} - \frac{2k'^2}{k^3}\right) = \frac{\tau' k'}{k^2} + \frac{\tau^3}{k}$;
(5) $k'(t_0) = k''(t_0) = 0$ *implies* $\tau(t_0) = 0$;
(6) $k \equiv \frac{1}{a}$ *if and only if C is a circle of radius a, necessarily contained in some plane.*

Let

$$C : \mathbf{r} = \mathbf{r}(t) = \mathbf{r}(u(t), v(t)), \quad t \in \mathcal{J} \subseteq \mathcal{R}, \tag{2.92}$$

be a curve on the sphere

$$S : \|\mathbf{r} - \mathbf{q}\|^2 = r^2 \tag{2.93}$$

with a center $\mathbf{q}$ and a radius r, t is an arbitrary parameter.

Suppose that $\mathbf{z}(t)$ is an infinitesimal bending field which the curve C leaves on S with a given precision, i.e. let the next condition be valid:

$$\|\mathbf{r}_\epsilon - \mathbf{q}\|^2 = \|\mathbf{r} + \epsilon \mathbf{z} - \mathbf{q}\|^2 = r^2 + o(\epsilon),$$

whereby $o(\epsilon)$ is an infinitesimal value of at least the second order with respect to ϵ. The previous equation is equivalent to the following one

$$\|\mathbf{r} - \mathbf{q}\|^2 + 2\epsilon(\mathbf{r} - \mathbf{q}) \cdot \mathbf{z} + \epsilon^2 \|\mathbf{z}\|^2 = r^2 + o(\epsilon).$$

Using Eq. (2.93), we obtain $2\epsilon(\mathbf{r} - \mathbf{q}) \cdot \mathbf{z} + \epsilon^2\|\mathbf{z}\|^2 = o(\epsilon)$. Therefore, the left side of the last equation should be an infinitesimal value of at least the second order with respect to ϵ, which is valid if and only if

$$(\mathbf{r} - \mathbf{q}) \cdot \mathbf{z} = 0 \tag{2.94}$$

holds. This is the necessary and sufficient condition for the bent curves to be approximately on the sphere. From this condition, we conclude $\mathbf{z} \perp \mathbf{r} - \mathbf{q}$, i.e. $\mathbf{z}$ lies in the tangent plane of the sphere S, spanned by $\{\mathbf{r}_u, \mathbf{r}_v\}$. Therefore, $\mathbf{z}(t)$ can be represented as

$$\mathbf{z}(t) = a(t)\mathbf{r}_u + b(t)\mathbf{r}_v, \tag{2.95}$$

where $a(t)$ and $b(t)$ are arbitrary real continuous differentiable functions. The necessary and sufficient condition for the infinitesimal bending

$$\dot{\mathbf{r}} \cdot \dot{\mathbf{z}} = 0, \tag{2.96}$$

reduces to the equation

$$\begin{aligned} &\dot{a}(t)\mathbf{r}_u \cdot (\mathbf{r}_u \dot{u} + \mathbf{r}_v \dot{v}) + a(t)(\mathbf{r}_{uu}\dot{u} + \mathbf{r}_{uv}\dot{v}) \cdot (\mathbf{r}_u \dot{u} + \mathbf{r}_v \dot{v}) \\ &\quad + \dot{b}(t)\mathbf{r}_v \cdot (\mathbf{r}_u \dot{u} + \mathbf{r}_v \dot{v}) + b(t)(\mathbf{r}_{vu}\dot{u} + \mathbf{r}_{vv}\dot{v}) \cdot (\mathbf{r}_u \dot{u} + \mathbf{r}_v \dot{v}) = 0. \end{aligned}$$

Taking one of the functions $a(t)$ or $b(t)$ arbitrarily, we get a linear differential equation by the other function. For instance, let us choose the function $b(t)$ arbitrarily. Then, we get the equation by $a(t)$

$$\dot{a}(t)\mathbf{r}_u \cdot (\mathbf{r}_u \dot{u} + \mathbf{r}_v \dot{v}) + a(t)(\mathbf{r}_{uu}\dot{u} + \mathbf{r}_{uv}\dot{v}) \cdot (\mathbf{r}_u \dot{u} + \mathbf{r}_v \dot{v}) + \varphi(t) = 0, \tag{2.97}$$

whereby

$$\varphi(t) = \dot{b}(t)\mathbf{r}_v \cdot (\mathbf{r}_u \dot{u} + \mathbf{r}_v \dot{v}) + b(t)(\mathbf{r}_{vu}\dot{u} + \mathbf{r}_{vv}\dot{v}) \cdot (\mathbf{r}_u \dot{u} + \mathbf{r}_v \dot{v}). \tag{2.98}$$

In the case when the condition $\mathbf{r}_u \cdot (\mathbf{r}_u \dot{u} + \mathbf{r}_v \dot{v}) \neq 0$ holds, Eq. (2.97) reduces to the following form

$$\dot{a}(t) + a(t)\frac{(\mathbf{r}_{uu}\dot{u} + \mathbf{r}_{uv}\dot{v}) \cdot (\mathbf{r}_u \dot{u} + \mathbf{r}_v \dot{v})}{\mathbf{r}_u \cdot (\mathbf{r}_u \dot{u} + \mathbf{r}_v \dot{v})} = -\frac{\varphi(t)}{\mathbf{r}_u \cdot (\mathbf{r}_u \dot{u} + \mathbf{r}_v \dot{v})}$$

whose solution is

$$\begin{aligned} a(t) = e^{-\int \frac{(\mathbf{r}_{uu}\dot{u} + \mathbf{r}_{uv}\dot{v}) \cdot (\mathbf{r}_u \dot{u} + \mathbf{r}_v \dot{v})}{\mathbf{r}_u \cdot (\mathbf{r}_u \dot{u} + \mathbf{r}_v \dot{v})}\, dt} \Bigg(c - \int \frac{\varphi(t)}{\mathbf{r}_u \cdot (\mathbf{r}_u \dot{u} + \mathbf{r}_v \dot{v})} \\ \cdot e^{\int \frac{(\mathbf{r}_{uu}\dot{u} + \mathbf{r}_{uv}\dot{v}) \cdot (\mathbf{r}_u \dot{u} + \mathbf{r}_v \dot{v})}{\mathbf{r}_u \cdot (\mathbf{r}_u \dot{u} + \mathbf{r}_v \dot{v})}\, dt} dt \Bigg). \end{aligned} \tag{2.99}$$

Based on the previous considerations, we obtained an explicit form of the infinitesimal bending field for spherical curves.

Theorem 2.15 ([Najdanović *et al.* (2024)]). *A vector field* $\mathbf{z}(t) = a(t)\mathbf{r}_u + b(t)\mathbf{r}_v$, *where* $b(t)$ *is an arbitrary real continuous differentiable function and* $a(t)$ *is a function given by Eq.* (2.99), *is an infinitesimal bending field of the spherical curve* (2.92) *so that all bent curves are on the sphere* (2.93) *with a given precision.*

Example 2.6. Let us consider a circle C given by inner equations $u = 0$, $v = t$ on the sphere

$$S: \mathbf{r}(u, v) = (x_0 + r\cos u \sin v, y_0 + r\sin u \sin v, z_0 + r\cos v),$$

therefore,

$$C: \mathbf{r}(t) = \mathbf{r}(u(t), v(t)) = (x_0 + r\sin t, y_0, z_0 + r\cos t).$$

According to Eq. (2.95), an infinitesimal bending field is

$$\mathbf{z}(t) = a(t)\mathbf{r}_u + b(t)\mathbf{r}_v = (rb(t)\cos t, ra(t)\sin t, -rb(t)\sin t),$$

In accordance with the condition $\dot{\mathbf{r}} \cdot \dot{\mathbf{z}} = 0$, we obtain $\dot{b}(t) = 0$. Therefore, $b(t) = c = const$ and $a(t)$ is an arbitrary real continuous differentiable function. Finally, infinitesimal bending field has the form:

$$\mathbf{z}(t) = (cr\cos t, a(t)r\sin t, -cr\sin t). \tag{2.100}$$

By a simple check, we conclude that the conditions $\dot{\mathbf{r}} \cdot \dot{\mathbf{z}} = 0$ and $\|\mathbf{r}_\epsilon - \mathbf{q}\|^2 = r^2 + o(\epsilon)$ are valid for this field.

In the case of the curve

$$C: \mathbf{r} = \mathbf{r}(s) \tag{2.101}$$

parameterized by the arc length s, lying on the sphere

$$S: \|\mathbf{r} - \mathbf{q}\|^2 = r^2, \tag{2.102}$$

we have the representation of infinitesimal bending given by the following theorems.

Theorem 2.16 ([Najdanović *et al.* (2024)]). *Infinitesimal bending of the spherical curve C which is on the sphere S with a given precision is*

$$\mathbf{r}_\epsilon = \mathbf{q} + \epsilon z\mathbf{t} - \frac{1}{k}(1 - \epsilon z')\mathbf{n}_1 + \left(\frac{k'}{k^2\tau} + \epsilon\frac{\tau}{k'}z'\right)\mathbf{n}_2, \tag{2.103}$$

where k is the curvature, $k' \neq 0$, τ is the torsion of C, z is an arbitrary real continuous differentiable function, ϵ is an infinitesimal.

Proof. By differentiating Eq. (2.102), we obtain

$$(\mathbf{r} - \mathbf{q}) \cdot \mathbf{t} = 0. \tag{2.104}$$

By differentiating again and using Frenet–Serret equation $\mathbf{t}' = k\mathbf{n}_1$, we obtain

$$1 + (\mathbf{r} - \mathbf{q}) \cdot k\mathbf{n}_1 = 0. \tag{2.105}$$

From Theorem 2.14, we conclude $k \geq \frac{1}{r} > 0$, so we have

$$(\mathbf{r} - \mathbf{q}) \cdot \mathbf{n}_1 = -\frac{1}{k}. \tag{2.106}$$

By differentiating Eq. (2.105) and using Frenet–Serret equation $\mathbf{n}_1' = -k\mathbf{t} + \tau\mathbf{n}_2$, as well as Eqs. (2.104) and (2.105), we obtain

$$-\frac{1}{k}k' + (\mathbf{r} - \mathbf{q}) \cdot k\tau\mathbf{b} = 0.$$

Since $k' \neq 0$ by assumption, we have $\tau \neq 0$ due to Eq. (2.91), so we obtain

$$(\mathbf{r} - \mathbf{q}) \cdot \mathbf{n}_2 = \frac{1}{k^2\tau}k'. \tag{2.107}$$

Finally, based on the decomposition

$$\mathbf{r} - \mathbf{q} = ((\mathbf{r} - \mathbf{q}) \cdot \mathbf{t})\mathbf{t} + ((\mathbf{r} - \mathbf{q}) \cdot \mathbf{n}_1)\mathbf{n}_1 + ((\mathbf{r} - \mathbf{q}) \cdot \mathbf{n}_2)\mathbf{n}_2$$

we obtain

$$\mathbf{r} - \mathbf{q} = -\frac{1}{k}\mathbf{n}_1 + \frac{1}{k^2\tau}k'\mathbf{n}_2. \tag{2.108}$$

We conclude that the vector $\mathbf{r} - \mathbf{q}$ is in the normal plane of the curve C.

Further, let us suppose that $\mathbf{z}$ is the infinitesimal bending field which given spherical curve C leaves approximately on the sphere S. Let the decomposition of $\mathbf{z}$ using the Frenet–Serret frame $\{\mathbf{t},\, \mathbf{n}_1,\, \mathbf{n}_2\}$ be given by

$$\mathbf{z}(s) = z(s)\mathbf{t} + z_1(s)\mathbf{n}_1 + z_2(s)\mathbf{n}_2, \tag{2.109}$$

where $z(s)$, $z_1(s)$ and $z_2(s)$ are real continuous differentiable functions. The field $\mathbf{z}$ satisfies the following two conditions

$$\begin{aligned} &\text{(i)} \quad \mathbf{z}'(s) \cdot \mathbf{r}'(s) = 0, \\ &\text{(ii)} \quad (\mathbf{r} - \mathbf{q}) \cdot \mathbf{z} = 0. \end{aligned}$$

By differentiating Eq. (2.109) and using Frenet–Serret equations, we obtain

$$\mathbf{z}'(s) = (z' - kz_2)\mathbf{t} + (kz + z_1' - \tau z_2)\mathbf{n}_1 + (\tau z_1 + z_2')\mathbf{n}_2.$$

Comparing with the condition (i) we get $z' - \kappa z_1 = 0$, wherefrom, we have

$$z_1 = \frac{1}{k} z'. \tag{2.110}$$

Further, putting Eqs. (2.108) and (2.109) into Eq. (ii), we obtain

$$-\frac{1}{k} z_1 + \frac{1}{k^2 \tau} k' z_2 = 0. \tag{2.111}$$

From here and Eq. (2.110), we express z_2:

$$z_2 = \frac{k\tau}{k'} z_1 = \frac{\tau}{k'} z'. \tag{2.112}$$

According to Eqs. (2.109), (2.110), and (2.112), we obtain

$$\mathbf{z} = z\mathbf{t} + \frac{1}{k} z' \mathbf{n}_1 + \frac{\tau}{k'} z' \mathbf{n}_2. \tag{2.113}$$

Finally, the proof of the theorem is completed after using Eqs. (2.108) and (2.113) in the expression $\mathbf{r}_\epsilon = \mathbf{r} + \epsilon \mathbf{z}$. □

Theorem 2.17 ([Najdanović *et al.* (2024)]). *Let* $\mathbf{r} = \mathbf{r}(s)$ *be a curve and* $\mathbf{z} = z\mathbf{t} + z_1\mathbf{n}_1 + z_2\mathbf{n}_2$ *be a vector field and let the following conditions be valid:*

$$R^2 + \left(\frac{1}{\tau}R'\right)^2 = 1, \tag{2.114}$$

$$Rz_1 + \frac{1}{\tau}R'z_2 = 0, \tag{2.115}$$

where $R = \frac{1}{k}$ *is the radius of the curvature;* k, τ *and* k' *are non zero (except only isolated points). Then, the family of curves* $\mathbf{r}_\epsilon = \mathbf{r} + \epsilon\mathbf{z}$ *is on the unit sphere with a given precision.*

Proof. Let $\boldsymbol{\rho} = \boldsymbol{\rho}(s)$ be given by

$$\boldsymbol{\rho} = \mathbf{r}_\epsilon + R\mathbf{n}_1 + \frac{1}{\tau}R'\mathbf{n}_2 - \epsilon(z\mathbf{t} + z_1\mathbf{n}_1 + z_2\mathbf{n}_2).$$

By differentiating and using Frenet–Serret equations, we obtain

$$\boldsymbol{\rho}' = \left(R\tau + \left(\frac{1}{\tau}R'\right)'\right)\mathbf{n}_2. \tag{2.116}$$

By differentiating Eq. (2.114), we get

$$\frac{1}{\tau}R'\left(\frac{1}{\tau}R'\right)' = -RR',$$

which, together with Eq. (2.116), gives $\frac{1}{\tau}R'\boldsymbol{\rho}' = 0$. Thus, $\boldsymbol{\rho}' = 0$ whenever R' (equivalently k') is non zero. Since $\boldsymbol{\rho}'$ is continuous, $\boldsymbol{\rho}'(s) = 0$ for all s. Therefore, there exists a constant vector $\mathbf{q} \in \mathcal{R}^3$ such that $\boldsymbol{\rho}(s) \equiv \mathbf{q}$. Hence it is

$$\mathbf{r}_\epsilon - \mathbf{q} = -R\mathbf{n}_1 - \frac{1}{\tau}R'\mathbf{n}_2 + \epsilon(z\mathbf{t} + z_1\mathbf{n}_1 + z_2\mathbf{n}_2),$$

and from here

$$\|\mathbf{r}_\epsilon - \mathbf{q}\|^2 = R^2 + \left(\frac{1}{\tau}R'\right)^2 - 2\epsilon\left(Rz_1 + \frac{1}{\tau}R'z_2\right) + o(t).$$

After using (2.114) and (2.115), we obtain $\|\mathbf{r}_\epsilon - \mathbf{q}\|^2 = 1 + o(\epsilon)$. So, $\mathbf{r}_\epsilon$ lies with a given precision on a unit sphere centered at $\mathbf{q}$. □

Remark 2.1. Equation (2.114) presents the characterization of spherical curves and is equivalent with Eq. (2.91) for $r = 1$.

2.5.2 *Curves on Cylinder*

Let the cylinder be given by the implicit equation

$$S : F(x, y, z) \equiv x^2 + y^2 - a^2 = 0, \tag{2.117}$$

$a > 0$, or by the vector parametric equation

$$S : \mathbf{r}(u, v) = (a \cos u, a \sin u, v), \ \ u \in [0, 2\pi], \ \ v \in [0, h]. \tag{2.118}$$

Let

$$C : \mathbf{r}(t) = \mathbf{r}(u(t), v(t)) = (a \cos u(t), a \sin u(t), v(t)), \ t \in \mathcal{J}, \tag{2.119}$$

be the curve on the cylinder S. Suppose that

$$C_\epsilon : \mathbf{r}_\epsilon(t) = (a \cos u(t) + \epsilon z_1(t), a \sin u(t) + \epsilon z_2(t), v(t) + \epsilon z_3(t)) \tag{2.120}$$

is an infinitesimal bending of C, where $z_1(t)$, $z_2(t)$, $z_3(t)$ are real continuous differentiable functions. In order to stay on the cylinder S with a given precision, it is necessary to apply the condition (2.90), so we have

$$(a \cos u(t) + \epsilon z_1(t))^2 + (a \sin u(t) + \epsilon z_2(t))^2 - a^2 = o(\epsilon).$$

From the last equation, we obtain the condition

$$\cos u(t) z_1(t) + \sin u(t) z_2(t) = 0 \tag{2.121}$$

which allows the bent curves to stay on the cylinder S with a given precision. For $\cos u(t) \neq 0$, we can express z_1 as a function of z_2:

$$z_1(t) = -\tan u(t) z_2(t). \tag{2.122}$$

Therefore, we are looking for the infinitesimal bending field in the following form

$$\mathbf{z}(t) = (-\tan u(t) z_2(t), z_2(t), z_3(t)). \tag{2.123}$$

For the field (2.123) to be an infinitesimal bending field, it is necessary that the condition $\dot{\mathbf{r}} \cdot \dot{\mathbf{z}} = 0$ is valid. Since $\dot{\mathbf{r}}(t) =$

$(-a \sin u(t)\, \dot{u}(t), a \cos u(t)\, \dot{u}(t), \dot{v}(t))$ and $\dot{\mathbf{z}}(t) = \Big(-\frac{\dot{u}(t)}{\cos^2 u(t)}\, z_2(t)$ $- \tan u(t) \dot{z}_2(t), \dot{z}_2(t), \dot{z}_3(t)\Big)$, we obtain

$$a\, \dot{u}(t)^2 \frac{\sin u(t)}{\cos^2 u(t)} z_2(t) + \frac{a\, \dot{u}(t)}{\cos u(t)} \dot{z}_2(t) + \dot{v}(t) \dot{z}_3(t) = 0. \tag{2.124}$$

This is the relationship between z_2 and z_3. Let us choose arbitrarily z_3 and solve the linear differential equation by z_2. The solution is

$$z_2(t) = e^{-\int \dot{u}(t) \tan u(t)\, dt} \left[c - \frac{1}{a} \int \frac{\dot{v}(t) \cos u(t)}{\dot{u}(t)} \dot{z}_3(t) e^{\int \dot{u}(t) \tan u(t)\, dt} dt \right], \tag{2.125}$$

$u(t) \neq const$, $\cos u(t) \neq 0$ and c is a constant. If $\dot{v}(t) = 0 \Rightarrow v(t) = const$, the curve C is a circle on the cylinder. In that case, we choose z_3 arbitrarily and determine z_2 from

$$z_2(t) = c e^{-\int \dot{u}(t) \tan u(t)\, dt}.$$

If $u(t) = const$, then Eq. (2.124) reduces to $\dot{v}(t) \dot{z}_3(t) = 0$ wherefrom, we have $\dot{z}_3(t) = 0 \Rightarrow z_3(t) = const$ and we choose z_2 arbitrarily.

Based on the previous considerations, the following theorems hold.

Theorem 2.18 ([Najdanović *et al.* (2021a)]). *The field* $\mathbf{z}(t) = (z_1(t), z_2(t), z_3(t))$ *whose components* z_1 *and* z_2 *satisfy the condition* (2.121) *includes the curve* (2.119) *into the family of deformed curves on the cylinder* (2.117) *with a given precision.* □

Theorem 2.19 ([Najdanović *et al.* (2021a)]). *The field* $\mathbf{z}(t)$ *given by* (2.123) *where* $z_3(t)$ *is arbitrary real continuous differentiable function, and* $z_2(t)$ *is given in* (2.125), *is infinitesimal bending field of the curve* (2.119) *so that all bent curves are on the cylinder* (2.117) *with a given precision.* □

Example 2.7. Let be $u(t) = t$, $v(t) = 0$. Then, the curve C is a circle $\mathbf{r}(t) = (a\cos t, a\sin t, 0)$. We have

$$z_2(t) = ce^{-\int \tan t\, dt} = \bar{c}\cos t,$$

$$z_1(t) = -c\tan t\cos t = -\bar{c}\sin t,$$

z_3-arbitrarily, c, $\bar{c}$ are constants. So, the infinitesimal bending field is $\mathbf{z}(t) = (-\bar{c}\sin t, \bar{c}\cos t, z_3(t))$. By a simple check, we conclude that the conditions $\dot{\mathbf{r}}\cdot\dot{\mathbf{z}} = 0$ and $(a\cos t - \epsilon\bar{c}\sin t)^2 + (a\sin t + \epsilon\bar{c}\cos t)^2 - a^2 = \epsilon^2\bar{c}^2 = o(\epsilon)$ are satisfied.

2.5.3 *Curves on Hyperbolic Paraboloid*

Let the curve $C : \mathbf{r} = \mathbf{r}(t)$ be on the hyperbolic paraboloid

$$S : F(x, y, z) \equiv xy - z = 0 \tag{2.126}$$

with the vector parametric equation

$$S : \mathbf{r}(u, v) = (u, v, uv).$$

Of course, the following equation holds

$$C : \mathbf{r} = \mathbf{r}(t) = \mathbf{r}(u(t), v(t)) = (u(t), v(t), u(t)v(t)). \tag{2.127}$$

Let $\mathbf{z}(t) = (z_1(t), z_2(t), z_3(t))$ be an infinitesimal bending field under which all bent curves are on the surface (2.126) with a given precision. As the infinitesimal bending is in the following form

$$\mathbf{r}_\epsilon(t) = \mathbf{r}(t) + \epsilon\mathbf{z}(t) \equiv (x_\epsilon(t), y_\epsilon(t), z_\epsilon(t)),$$

the condition (2.90) reduces to

$$(u(t) + \epsilon z_1(t))(v(t) + \epsilon z_2(t)) - u(t)v(t) - \epsilon z_3(t) = o(\epsilon)$$

i.e.

$$z_1(t)v(t) + z_2(t)u(t) - z_3(t) = 0. \tag{2.128}$$

This is the necessary and sufficient condition which allows that all bent curves are approximately on S. From Eq. (2.128), we obtain

$$z_3(t) = z_1(t)v(t) + z_2(t)u(t). \tag{2.129}$$

Therefore,

$$\mathbf{z}(t) = (z_1(t), z_2(t), z_1(t)v(t) + z_2(t)u(t)) \tag{2.130}$$

is the required field. It is also necessary to apply the condition $\dot{\mathbf{r}} \cdot \dot{\mathbf{z}} = 0$. Since $\dot{\mathbf{r}} = (\dot{u}, \dot{v}, \dot{u}v + u\dot{v})$, and $\dot{\mathbf{z}} = (\dot{z}_1, \dot{z}_2, \dot{z}_1 v + z_1\dot{v} + \dot{z}_2 u + z_2\dot{u})$, we obtain

$$[\dot{v} + (uv)^{\cdot} u]\dot{z}_2 + (uv)^{\cdot}\dot{u}z_2 + \varphi = 0, \tag{2.131}$$

where

$$\varphi = [\dot{u} + (uv)^{\cdot} v]\dot{z}_1 + (uv)^{\cdot}\dot{v}z_1.$$

If we arbitrarily choose z_1, we obtain the function z_2 by solving the linear differential Equation (2.131), and z_3 from Eq. (2.129). Equation (2.131) reduces to

$$\dot{z}_2 + \frac{(uv)^{\cdot}\dot{u}}{\dot{v} + (uv)^{\cdot} u} z_2 = -\frac{\varphi}{\dot{v} + (uv)^{\cdot} u}, \quad \dot{v} + (uv)^{\cdot} u \neq 0$$

whose solution is

$$z_2 = e^{-\int \frac{(uv)^{\cdot}\dot{u}}{\dot{v}+(uv)^{\cdot} u} dt} \left[c - \int \frac{\varphi}{\dot{v} + (uv)^{\cdot} u} e^{\int \frac{(uv)^{\cdot}\dot{u}}{\dot{v}+(uv)^{\cdot} u} dt} dt \right], \tag{2.132}$$

c is a constant.

In this way, we have proved the following theorems.

Theorem 2.20 ([Najdanović *et al.* (2021a)]). *The field* $\mathbf{z}(t) = (z_1(t), z_2(t), z_3(t))$ *whose components* z_1, z_2 *and* z_3 *satisfy the condition* (2.128) *includes the curve* (2.127) *into the family of deformed curves on the hyperbolic paraboloid* (2.126) *with a given precision.*

Theorem 2.21 ([Najdanović *et al.* (2021a)]). *The field* $\mathbf{z}(t)$ *given by* (2.130) *where* $z_1(t)$ *is arbitrary real continuous differentiable function, and* $z_2(t)$ *is given in* (2.132), *is infinitesimal bending field of the curve* (2.127) *so that all bent curves are on the hyperbolic paraboloid* (2.126) *with a given precision.*

Example 2.8. Let the curve $C : \mathbf{r}(t) = (t, t, t^2)$, $t \in \mathcal{J} \subseteq \mathcal{R}$, be given on the hyperbolic paraboloid $S : \mathbf{r}(u, v) = (u, v, uv)$. Let us find the field $\mathbf{z}$ according to the previous theorem.

Let be $z_1(t) = t$. Since $u(t) = t$, $v(t) = t$, $\dot{u} = 1$, $\dot{v} = 1$, $\dot{z}_1 = 1$, we obtain $\varphi = 1 + 4t^2$. Also,

$$z_2 = e^{-\int \frac{2t}{1+2t^2}dt}\left[c - \int \frac{1+4t^2}{1+2t^2} e^{\int \frac{2t}{1+2t^2}dt} dt\right] = \frac{c}{\sqrt{1+2t^2}} - t.$$

Further, we have

$$z_3 = z_1 v + z_2 u = \frac{ct}{\sqrt{1+2t^2}}.$$

Therefore, the corresponding infinitesimal bending field is

$$\mathbf{z}(t) = \left(t, \frac{c}{\sqrt{1+2t^2}} - t, \frac{ct}{\sqrt{1+2t^2}}\right), \tag{2.133}$$

so

$$\mathbf{r}_\epsilon = \left(t + \epsilon t, t + \epsilon\left(\frac{c}{\sqrt{1+2t^2}} - t\right), \quad t^2 + \epsilon\frac{ct}{\sqrt{1+2t^2}}\right).$$

It is easy to check that the following is true: $\dot{\mathbf{r}} = (1, 1, 2t)$, $\dot{\mathbf{z}} = \left(1, \frac{-2ct}{\sqrt{(1+2t^2)^3}} - 1, \frac{c}{\sqrt{(1+2t^2)^3}}\right)$, $\dot{\mathbf{r}} \cdot \dot{\mathbf{z}} = 0$,

$$(t + \epsilon t)\left(t + \epsilon\left(\frac{c}{\sqrt{1+2t^2}} - t\right)\right) - t^2 - \epsilon\frac{ct}{\sqrt{1+2t^2}}$$
$$= \epsilon^2 t\left(\frac{c}{\sqrt{1+2t^2}} - t\right) = o(\epsilon).$$

2.5.4 *Curves on Helicoid*

Let $C : \mathbf{r} = \mathbf{r}(t)$ be the curve on the helicoid

$$S : F(x, y, z) \equiv \frac{y}{x} - \tan\frac{z}{c} = 0, \tag{2.134}$$

c-constant, with the vector parametric equation

$$\mathbf{r}(u, v) = (u\cos v, u\sin v, cv).$$

Then, the equation of the curve C has the form

$$C : \mathbf{r}(t) = \mathbf{r}(u(t), v(t)) = (u(t)\cos v(t), u(t)\sin v(t), cv(t)). \tag{2.135}$$

Let $\mathbf{z}(t) = (z_1(t), z_2(t), z_3(t))$ be an infinitesimal bending field which given curve leaves on the helicoid S, where $z_1(t)$, $z_2(t)$, $z_3(t)$ are real continuous differentiable functions. Then,

$$\mathbf{r}_\epsilon(t) = (u(t)\cos v(t) + \epsilon z_1(t), u(t)\sin v(t) + \epsilon z_2(t), cv(t) + \epsilon z_3(t))$$

will be the corresponding infinitesimal bending and the condition

$$\frac{u(t)\sin v(t) + \epsilon z_2(t)}{u(t)\cos v(t) + \epsilon z_1(t)} - \tan\frac{cv(t) + \epsilon z_3(t)}{c} = o(\epsilon)$$

must be satisfied. From the last equation, we have

$$\frac{u(t)\sin v(t) + \epsilon z_2(t)}{u(t)\cos v(t) + \epsilon z_1(t)} - \frac{\tan v(t) + \tan(\epsilon\frac{z_3(t)}{c})}{1 - \tan v(t)\tan(\epsilon\frac{z_3(t)}{c})} = o(\epsilon),$$

or, after using Maclaurin series for $y = \tan x$:

$$\frac{u(t)\sin v(t) + \epsilon z_2(t)}{u(t)\cos v(t) + \epsilon z_1(t)} - \frac{\tan v(t) + \epsilon\frac{z_3(t)}{c} + o(\epsilon^2)}{1 - \tan v(t)(\epsilon\frac{z_3(t)}{c} + o(\epsilon^2))} = o(\epsilon).$$

From here, we obtain

$$z_2(t) - u(t)\frac{\sin^2 v(t)}{\cos v(t)}\frac{z_3(t)}{c} - z_1(t)\tan v(t) - u(t)\cos v(t)\frac{z_3(t)}{c} = 0,$$

i.e.

$$z_2(t) = z_1(t)\tan v(t) + \frac{z_3(t)u(t)}{c\cos v(t)}. \tag{2.136}$$

Thus, the field $\mathbf{z}(t)$ has the following form

$$\mathbf{z}(t) = \left(z_1(t), z_1(t)\tan v(t) + \frac{z_3(t)u(t)}{c\cos v(t)}, z_3(t)\right), \tag{2.137}$$

$\cos v(t) \neq 0$, and the functions $z_1(t)$ and $z_2(t)$ are obtained from the condition

$$\dot{\mathbf{r}} \cdot \dot{\mathbf{z}} = 0,$$

where

$$\dot{\mathbf{r}} = (\dot{u}(t)\cos v(t) - u(t)\dot{v}(t)\sin v(t), \quad \dot{u}(t)\sin v(t) + u(t)\dot{v}(t)\cos v(t), c\dot{v}(t)), \quad \dot{\mathbf{z}} = \left(\dot{z}_1(t), \dot{z}_1(t)\tan v(t) + \frac{z_1(t)\dot{v}(t)}{\cos^2 v(t)} + \frac{(\dot{z}_3(t)u(t) + z_3(t)\dot{u}(t))\cos v(t) + z_3(t)u(t)\dot{v}(t)\sin v(t)}{c\cos^2 v(t)}, \dot{z}_3(t)\right).$$

For simplicity, let us consider a helix which is obtained for $u = 1$ and $v = t$. Therefore, the vector parametric equation of a helix is

$$\mathbf{r}(t) = (\cos t, \sin t, ct). \tag{2.138}$$

The field $\mathbf{z}$ reduces to

$$\mathbf{z} = \left(z_1(t),\ z_1(t)\tan t + \frac{z_3(t)}{c\cos t},\ z_3(t)\right), \tag{2.139}$$

$\cos t \neq 0$. Since $\dot{\mathbf{r}} = (-\sin t, \cos t, c)$, $\dot{\mathbf{z}} = (\dot{z}_1(t), \dot{z}_1(t)\tan t + \frac{z_1(t)}{\cos^2 t} + \frac{\dot{z}_3(t)\cos t + z_3(t)\sin t}{c\cos^2 t}, \dot{z}_3(t))$, using the condition $\dot{\mathbf{r}} \cdot \dot{\mathbf{z}} = 0$, we get the equation

$$\dot{z}_3 + \frac{1}{1+c^2} z_3(t)\tan t = -\frac{c}{1+c^2}\frac{z_1(t)}{\cos t}. \tag{2.140}$$

This is the relationship between z_1 and z_3. We choose one of these functions arbitrarily, and we get the other by solving Eq. (2.140). Thus, if z_1 is arbitrary real continuous differentiable function, we obtain

$$z_3(t) = (\cos t)^{\frac{1}{1+c^2}} \left[c_1 - \frac{c}{1+c^2}\int \frac{z_1(t)}{(\cos t)^{\frac{2+c^2}{1+c^2}}}\, dt\right], \tag{2.141}$$

where c_1 is a constant.

Theorem 2.22 ([Najdanović *et al.* (2021a)]). *The field* $\mathbf{z}(t) = (z_1(t), z_2(t), z_3(t))$ *whose components* z_1, z_2 *and* z_3 *satisfy the condition* (2.136) *includes the curve* (2.135) *into the family of deformed curves on the helicoid* (2.134) *with a given precision.* □

Theorem 2.23 ([Najdanović *et al.* (2021a)]). *The field* $\mathbf{z}(t)$ *given by* (2.139) *where* $z_1(t)$ *is arbitrary real continuous differentiable function, and* $z_3(t)$ *is given in* (2.141), *is infinitesimal bending field of the helix* (2.246) *so that all bent curves are on the helicoid* (2.134) *with a given precision.* □

Example 2.9. Let be $z_1(t) = (1 + 2c^2) \sin t \cos^2 t$. Then, we obtain after necessary integration (for $c_1 = 0$) $z_3(t) = c \cos^2 t$, and $z_2(t) = (1 + 2c^2) \sin^2 t \cos t + \cos t$.

The corresponding infinitesimal bending field which given helix leaves at the helicoid with a given precision is

$$\mathbf{z}(t) = ((1 + 2c^2) \sin t \cos^2 t, (1 + 2c^2) \sin^2 t \cos t + \cos t, c \cos^2 t). \tag{2.142}$$

By a simple check, we conclude that this is an infinitesimal bending field and the condition (2.90) is valid.

2.5.5 *Torus Knots*

Let

$$C: \begin{aligned} x(u) &= (\cos(qu) + a) \cdot \cos(pu) \\ y(u) &= (\cos(qu) + a) \cdot \sin(pu) \\ z(u) &= -\sin(qu) \end{aligned} \tag{2.143}$$

be (p, q) torus knot, $0 \leq u \leq 2\pi$, which lies on the torus

$$S: (a - \sqrt{x^2 + y^2})^2 + z^2 = 1^2.$$

Let

$$C_\epsilon: \begin{aligned} x_\epsilon(u) &= (\cos(qu) + a) \cdot \cos(pu) + \epsilon z_1(u) \\ y_\epsilon(u) &= (\cos(qu) + a) \cdot \sin(pu) + \epsilon z_2(u) \\ z_\epsilon(u) &= -\sin(qu) + \epsilon z_3(u) \end{aligned} \tag{2.144}$$

be infinitesimal bending of C, where $\mathbf{z}(u) = (z_1(u), z_2(u), z_3(u))$ is infinitesimal bending field. If we put

$$F(x, y, z) = (a - \sqrt{x^2 + y^2})^2 + z^2 - 1^2,$$

then the torus knot satisfies the implicit equation of the torus S, $F(x(u), y(u), z(u)) = 0$. We request that the infinitesimal bending is approximately (with a given precision) on the torus S, i.e. we set the condition (2.90). This means it must be valid

$$\left(a - \sqrt{\begin{array}{c}((\cos(qu) + a)\cos(pu) + \epsilon z_1(u))^2 \\ +((\cos(qu) + a)\sin(pu) + \epsilon z_2(u))^2\end{array}}\right)^2 + (-\sin(qu) + \epsilon z_3(u))^2 - 1 = 0,$$

wherefrom, after some calculations and using Maclaurin series, one obtains

$$\cos(pu)\cos(qu)z_1(u) + \sin(pu)\cos(qu)z_2(u) - \sin(qu)z_3(u) = 0. \tag{2.145}$$

Thus, we proved the following theorem.

Theorem 2.24 ([Rančić *et al.* (2019)]). *The necessary and sufficient condition for the field* $\mathbf{z}(u) = (z_1(u), z_2(u), z_3(u))$ *to include* (p, q) *torus knot* (2.143) *into the family of deformed curves on the torus with a given precision is that Eq.* (2.145) *is satisfied.*

From Eq. (2.145), we obtain

$$z_3(u) = [\cos(pu)z_1(u) + \sin(pu)z_2(u)]\cot(qu). \tag{2.146}$$

Therefore, we are looking for the field

$$\mathbf{z}(u) = (z_1(u), z_2(u), [\cos(pu)z_1(u) + \sin(pu)z_2(u)]\cot(qu))$$

which satisfies the necessary and sufficient condition for infinitesimal bending, i.e.

$$\dot{\mathbf{r}}(u) \cdot \dot{\mathbf{z}}(u) = 0.$$

Since

$$\begin{aligned}\dot{\mathbf{r}}(u) &= (-q\sin(qu)\cos(pu) - p(\cos(qu)+a)\sin(pu),\\ &\quad -q\sin(qu)\sin(pu) + p(\cos(qu)+a)\cos(pu), -q\cos(qu))\\ \dot{\mathbf{z}}(u) &= \Bigg(\dot{z}_1(u), \dot{z}_2(u), (-p\sin(pu)z_1(u) + \cos(pu)\dot{z}_1(u)\\ &\quad + p\cos(pu)z_2(u) + \sin(pu)\dot{z}_2(u))\cot(qu) + (\cos(pu)z_1(u)\\ &\quad - q\sin(pu)z_2(u))\frac{1}{\sin^2(qu)}\Bigg)\end{aligned}$$

using one of the functions $z_1(u)$ and $z_2(u)$ arbitrarily, for instance, $z_1(u)$, we get the other by solving the following linear differential equation:

$$A(u)\dot{z}_2(u) + B(u)z_2(u) + C(u) = 0, \tag{2.147}$$

where

$$\begin{aligned}A(u) &= -q\sin(qu)\sin(pu) + p(\cos(qu)+a)\cos(pu)\\ &\quad - q\cos(qu)\cot(qu)\sin(pu), &(2.148)\\ B(u) &= -pq\cos(pu)\cos(qu)\cot(qu) + q^2\frac{\cos(qu)}{\sin^2(qu)}\sin pu, &(2.149)\\ C(u) &= [-q\sin(qu)\cos(pu) - p(\cos(qu)+a)\sin(pu)]\dot{z}_1(u)\\ &\quad - q\cos(qu)[-p\sin(pu)z_1(u) + \cos(pu)\dot{z}_1(u)]\cot(qu)\\ &\quad + q^2\frac{\cos(qu)}{\sin^2(qu)}\cos(pu)z_1(u). &(2.150)\end{aligned}$$

Therefore, the following theorem holds.

Theorem 2.25 ([Rančić *et al.* (2019)]). *Vector field*

$$\mathbf{z}(u) = (z_1(u), z_2(u), [\cos(pu)z_1(u) + \sin(pu)z_2(u)]\cot(qu)), \tag{2.151}$$

where $z_1(u)$ is arbitrary real continuous differentiable function and $z_2(u)$ is given by

$$z_2(u) = e^{-\int\frac{B(u)}{A(u)}du}\left[c - \int\frac{C(u)}{A(u)}e^{\int\frac{B(u)}{A(u)}du}du\right], \tag{2.152}$$

c is a constant, $A(u)$, $B(u)$ and $C(u)$ are given in Eqs. (2.148), (2.149) *and* (2.150), *respectively, is infinitesimal bending field of (p, q) torus knot under which all bent curves are on the initial torus with a given precision.*

Example 2.10. In particular, let $z_3(u) = 0$. Then, the condition (2.145) reduces to $\cos(pu)z_1(u) + \sin(pu)z_2(u) = 0$, wherefrom, we have $z_1(u) = -\tan(pu)z_2(u)$ and

$$\mathbf{z}(u) = (-\tan(pu)z_2(u), z_2(u), 0). \tag{2.153}$$

From the condition $\dot{\mathbf{r}}(u) \cdot \dot{\mathbf{z}}(u) = 0$, we obtain homogenous linear differential equation

$$\dot{z}_2(u) + \left(\frac{q\sin(qu)}{\cos(qu) + a} + \frac{p\sin(pu)}{\cos(pu)} \right) z_2(u) = 0$$

whose solution is

$$z_2(u) = c(\cos(qu) + a)\cos(pu), \tag{2.154}$$

c is a constant. Finally,

$$\mathbf{z}(u) = (-c\tan(pu)(\cos(qu) + a)\cos(pu), c(\cos(qu) + a)\cos(pu), 0). \tag{2.155}$$

It is easy to check that this vector field satisfies the following conditions: $\dot{\mathbf{r}} \cdot \dot{\mathbf{z}} = 0$ and $F(x_\epsilon(u), y_\epsilon(u), z_\epsilon(u)) = o(\epsilon)$.

2.6 Variations of Geometric Magnitudes

Under infinitesimal bending, geometric magnitudes of the curve are changed which is described with variations of these geometric magnitudes.

Definition 2.9 ([Vekua (1959)]). Let $\mathcal{A} = \mathcal{A}(u)$ be a magnitude which characterizes a geometric property on the curve C and $\tilde{\mathcal{A}}(u, \epsilon) = \mathcal{A}_\epsilon(u)$ the corresponding magnitude on the curve C_ϵ being infinitesimal bending of the curve C, and set

$$\Delta\mathcal{A} = \mathcal{A}_\epsilon - \mathcal{A} = \epsilon\,\delta\mathcal{A} + \epsilon^2\,\delta^2\mathcal{A} + \cdots + \epsilon^n\,\delta^n\mathcal{A} + \cdots \tag{2.156}$$

The coefficients $\delta\mathcal{A}, \delta^2\mathcal{A}, \ldots, \delta^n\mathcal{A}, \ldots$ are **the first, the second,..., the n-th variation** of the geometric magnitude $\mathcal{A}$, respectively under the infinitesimal bending C_ϵ of the curve C.

Let us mark some properties of variations according to [Najdanović and Velimirović (2017a)]:

I. For variations of the product of geometric magnitudes it is effective the equation

$$\delta^n \mathcal{AB} = \sum_{i=0}^{n} \delta^i \mathcal{A}\, \delta^{n-i}\mathcal{B}, \quad n \geq 0, \quad (\delta^0 \mathcal{A} \stackrel{def}{=} \mathcal{A}). \tag{2.157}$$

According to Definition (2.9), variations of geometric magnitudes $\mathcal{A}$ and $\mathcal{B}$, as well as of the product $\mathcal{AB}$ are:

$$\Delta\mathcal{A} = \mathcal{A}_\epsilon - \mathcal{A} = \epsilon\,\delta\mathcal{A} + \epsilon^2\,\delta^2\mathcal{A} + \cdots + \epsilon^n\,\delta^n\mathcal{A} + \cdots \tag{2.158}$$

$$\Delta\mathcal{B} = \mathcal{B}_\epsilon - \mathcal{B} = \epsilon\,\delta\mathcal{B} + \epsilon^2\,\delta^2\mathcal{B} + \cdots + \epsilon^n\,\delta^n\mathcal{B} + \cdots \tag{2.159}$$

$$\Delta\mathcal{AB} = \mathcal{A}_\epsilon\mathcal{B}_\epsilon - \mathcal{AB} = \epsilon\,\delta(\mathcal{AB}) + \epsilon^2\,\delta^2(\mathcal{AB}) + \cdots + \epsilon^n\,\delta^n(\mathcal{AB}) + \cdots \tag{2.160}$$

respectively. On the other hand, the following equality is satisfied:

$$\Delta\mathcal{AB} = \mathcal{A}_\epsilon\mathcal{B}_\epsilon - \mathcal{AB} = \mathcal{A}_\epsilon\mathcal{B}_\epsilon - \mathcal{A}_\epsilon B + \mathcal{A}_\epsilon B - \mathcal{AB} = \mathcal{A}_\epsilon(\mathcal{B}_\epsilon - B) + (\mathcal{A}_\epsilon - A)B \tag{2.161}$$

Substituting (2.158) and (2.159) into Eq. (2.161), we obtain

$$\Delta\mathcal{AB} = \epsilon(A\delta B + B\delta A) + \epsilon^2(A\delta^2 B + \delta A\,\delta B + B\delta^2 A) + \cdots + \epsilon^n \sum_{i=0}^{n} \delta^i \mathcal{A}\,\delta^{n-i}\mathcal{B} + \cdots \tag{2.162}$$

As the left sides of the last equation and of Eq. (2.160) are equal, Eq. (2.157) is valid.

II. An arbitrary order variation of a derivative is the derivative of the variation, i.e.

$$\delta^n \left(\frac{d\mathcal{A}}{du}\right) = \frac{d(\delta^n \mathcal{A})}{du}, \quad n \geq 0. \tag{2.163}$$

For

$$\frac{d\mathcal{A}}{du} = \mathcal{B}, \tag{2.164}$$

using (2.156), we obtain the equation

$$\begin{aligned}\triangle\frac{d\mathcal{A}}{du} &= \triangle\mathcal{B} = \epsilon\,\delta\mathcal{B} + \epsilon^2\,\delta^2\mathcal{B} + \cdots + \epsilon^n\,\delta^n\mathcal{B} + \cdots = \epsilon\,\delta\frac{d\mathcal{A}}{du}\\ &\quad + \epsilon^2\,\delta^2\frac{d\mathcal{A}}{du} + \cdots + \epsilon^n\,\delta^n\frac{d\mathcal{A}}{du} + \cdots\end{aligned} \tag{2.165}$$

We also have

$$\begin{aligned}\triangle\frac{d\mathcal{A}}{du} &= \triangle\mathcal{B} = \mathcal{B}_\epsilon(u) - \mathcal{B}(u) = \frac{d\mathcal{A}_\epsilon(u)}{du} - \frac{d\mathcal{A}(u)}{du}\\ &= \frac{d}{du}[\mathcal{A}_\epsilon(u) - \mathcal{A}(u)] = \frac{d\triangle\mathcal{A}}{du}\\ &= \frac{d(\epsilon\,\delta\mathcal{A} + \epsilon^2\,\delta^2\mathcal{A} + \cdots + \epsilon^n\,\delta^n\mathcal{A} + \cdots)}{du}\\ &= \epsilon\frac{d(\delta\mathcal{A})}{du} + \epsilon^2\frac{d(\delta^2\mathcal{A})}{du} + \cdots + \epsilon^n\frac{d(\delta^n\mathcal{A})}{du} + \cdots\end{aligned} \tag{2.166}$$

By comparing Eqs. (2.165) and (2.166), we confirm validity of Eq. (2.163). The same case is for the differential, i.e.

III. $\delta^n(d\mathcal{A}) = d(\delta^n\mathcal{A}), \quad n \geq 0.$

2.6.1 *The First Variation*

In the case when we consider infinitesimal bending of the first order, it is useful to study only the first variations of geometric magnitudes. So, we can represent the magnitude $\mathcal{A}_\epsilon$ as

$$\mathcal{A}_\epsilon = \mathcal{A} + \epsilon\,\delta\mathcal{A},$$

by neglecting the ϵ^n-terms, $n \geq 2$.

The first variation, or shortly the variation, is of course given by

$$\delta\mathcal{A}=\frac{d}{d\epsilon}\mathcal{A}_\epsilon(u)\big|_{\epsilon=0}, \tag{2.167}$$

i.e.

$$\delta\mathcal{A}=\lim_{\epsilon\to 0}\frac{\Delta\mathcal{A}}{\epsilon}=\lim_{\epsilon\to 0}\frac{\mathcal{A}_\epsilon(u)-\mathcal{A}(u)}{\epsilon}. \tag{2.168}$$

Also, we have:

$$\begin{gathered}\mathbf{Ia.}\,\delta(\mathcal{A}\mathcal{B})=\mathcal{A}\delta\mathcal{B}+\mathcal{B}\delta\mathcal{A},\quad \mathbf{IIa.}\,\delta\left(\frac{\partial\mathcal{A}}{\partial u}\right)=\frac{\partial(\delta\mathcal{A})}{\partial u},\\ \mathbf{IIIa.}\,\delta(d\mathcal{A})=d(\delta\mathcal{A}).\end{gathered} \tag{2.169}$$

If $\delta\mathcal{A}=0$, geometric property $\mathcal{A}=\mathcal{A}(u)$ is stationary under infinitesimal bending. According to Theorem 2.2, we have

$$\delta(ds)=0. \tag{2.170}$$

Therefore, under infinitesimal bending of a curve, the variation of the line element ds is equal to zero, i.e. the line element is stationary under infinitesimal bending.

Below, we will describe the behavior of some geometric magnitudes under infinitesimal bending (2.22) of a curve $C:\mathbf{r}=\mathbf{r}(s)$ parameterized by the arc length, according to [Najdanović (2015)]. Infinitesimal bending field is given in Eq. (2.23), i.e.

$$\mathbf{z}=z\mathbf{t}+z_1\mathbf{n}_1+z_2\mathbf{n}_2.$$

Theorem 2.26. *Under infinitesimal bending of the curve C, a unit vector of the orthonormal basis and its variation are orthogonal.*

Proof. Let $\mathbf{t}_\epsilon$ be the unit tangent vector of deformed curve C_ϵ. Then, it holds $\mathbf{t}_\epsilon\cdot\mathbf{t}_\epsilon=1$, i.e.

$$(\mathbf{t}+\epsilon\,\delta\mathbf{t})\cdot(\mathbf{t}+\epsilon\,\delta\mathbf{t})=1$$

wherefrom, we obtain $\mathbf{t}\cdot\delta\mathbf{t}=0$, after neglecting the terms of order higher than 1 with respect to ϵ. Similarly, we show the statement for $\mathbf{n}_1$ and $\mathbf{n}_2$. □

Note that the above statement is true for an arbitrary unit vector. The rigidity of the line element ds can also be proved in this way.

Theorem 2.27. *Under infinitesimal bending of the curve C, variation of the line element ds is equal to zero, i.e.*

$$\delta(ds) = 0. \tag{2.171}$$

Proof. Since

$$d\mathbf{r} = \mathbf{r}'(s)\, ds = \mathbf{t}\, ds,$$

after scalar product of previous equation with $\mathbf{t}$, we obtain

$$\mathbf{t} \cdot d\mathbf{r} = ds.$$

Applying **Ia**, **IIIa** and Theorem 2.26, we have

$$\begin{aligned}\delta(ds) &= \delta\mathbf{t} \cdot d\mathbf{r} + \mathbf{t} \cdot \delta(d\mathbf{r}) = \delta\mathbf{t} \cdot \mathbf{r}'\, ds + \mathbf{t} \cdot d(\delta\mathbf{r}) = \delta\mathbf{t} \cdot \mathbf{t}\, ds + \mathbf{t} \cdot d\mathbf{z} \\ &= \mathbf{t} \cdot (dz\, \mathbf{t} + z\, d\mathbf{t} + dz_1\, \mathbf{n}_1 + z_1\, d\mathbf{n}_1 + dz_2\, \mathbf{n}_2 + z_2\, d\mathbf{n}_2)\end{aligned}$$

Using Frenet–Serret equations, we obtain

$$\delta(ds) = (z' - kz_1)\, ds, \tag{2.172}$$

which, due to (2.25), gives (2.171). □

Theorem 2.28. *Under infinitesimal bending of the curve C, variation of the unit tangent vector is*

$$\delta\mathbf{t} = (z_1' - \tau z_2 + kz)\, \mathbf{n}_1 + (z_2' + \tau z_1)\, \mathbf{n}_2. \tag{2.173}$$

Proof. Since $\delta\mathbf{t} = \delta\mathbf{r}' = (\delta\mathbf{r})' = \mathbf{z}'$, using (2.23), (2.25) and Frenet–Serret equations, we obtain (2.173). □

Theorem 2.29. *Under infinitesimal bending of the curve C, variations of the unit principal normal and binormal vectors are respectively*

$$\delta\mathbf{n}_1 = -(kz + z_1' - \tau z_2)\mathbf{t} + \frac{1}{k}(k\tau z + z_2'' - \tau^2 z_2 + 2\tau z_1' + \tau' z_1)\mathbf{n}_2, \tag{2.174}$$

$$\delta\mathbf{n}_2 = -(z_2' + \tau z_1)\mathbf{t} - \frac{1}{k}(k\tau z + z_2'' - \tau^2 z_2 + 2\tau z_1' + \tau' z_1)\mathbf{n}_1. \tag{2.175}$$

Proof. The unit normal vector of deformed curve C_ϵ satisfies the condition $\mathbf{n}_{1\epsilon} \cdot \mathbf{n}_{1\epsilon} = 1$, i.e. $(\mathbf{n}_1 + \epsilon\delta\mathbf{n}_1) \cdot (\mathbf{n}_1 + \epsilon\delta\mathbf{n}_1) = 1$ which gives

$$\mathbf{n}_1 \cdot \delta\mathbf{n}_1 = 0. \tag{2.176}$$

Also, it is valid $\mathbf{n}_{1\epsilon} \cdot \mathbf{t}_\epsilon = 0$, i.e. $(\mathbf{n}_1 + \epsilon\delta\mathbf{n}_1) \cdot (\mathbf{t} + \epsilon\delta\mathbf{t}) = 0$, where from we have

$$\mathbf{t} \cdot \delta\mathbf{n}_1 = -\mathbf{n}_1 \cdot \delta\mathbf{t} = -(z_1' - \tau z_2 + kz). \tag{2.177}$$

Further, we take a variation of the first equation in (2.5),

$$\delta\mathbf{t}' = \mathbf{n}_1\delta k + k\delta\mathbf{n}_1. \tag{2.178}$$

Dotting with $\mathbf{n}_2$, we obtain $\mathbf{n}_2 \cdot \delta\mathbf{n}_1 = \frac{1}{k}\mathbf{n}_2 \cdot \delta\mathbf{t}'$. To evaluate $\delta\mathbf{t}'$, we apply commutativity of the variation and the derivative and obtain $\delta\mathbf{t}' = (\delta\mathbf{t})'$. Based on (2.173), Frenet–Serret equations and $z' = kz_1$ (due to (2.25)), one obtains

$$\begin{aligned}\delta\mathbf{t}' = &-k(kz + z_1' - \tau z_2)\,\mathbf{t} + (k'z + z_1'' + (k^2 - \tau^2)z_1 \\ &- 2\tau z_2' - \tau' z_2)\mathbf{n}_1 + (k\tau z + 2\tau z_1' + \tau' z_1 + z_2'' - \tau^2 z_2)\,\mathbf{n}_2.\end{aligned} \tag{2.179}$$

Now we have

$$\mathbf{n}_2 \cdot \delta\mathbf{n}_1 = \frac{1}{k}(k\tau z + 2\tau z_1' + \tau' z_1 + z_2'' - \tau^2 z_2) \tag{2.180}$$

From (2.176), (2.177), and (2.180), we obtain (2.174). Similarly, from the conditions $\mathbf{n}_{2\epsilon} \cdot \mathbf{n}_{2\epsilon} = 1$, $\mathbf{n}_{2\epsilon} \cdot \mathbf{t}_\epsilon = 0$ and $\mathbf{n}_{2\epsilon} \cdot \mathbf{n}_{1\epsilon} = 0$ we obtain (2.175). □

Theorem 2.30. *Under infinitesimal bending of the curve C, variation of the curvature is*

$$\delta k = k'z + z_1'' + (k^2 - \tau^2)z_1 - 2\tau z_2' - \tau' z_2. \tag{2.181}$$

Proof. Dotting Eq. (2.178) with $\mathbf{n}_1$ and using Theorem 2.26, we obtain $\delta k = \mathbf{n}_1 \cdot \delta \mathbf{t}'$. This leads to (2.181) after using (2.179). $\square$

Corollary 2.3. *Under infinitesimal bending of a plane curve, variation of the curvature is*

$$\delta k = k' z + z_1'' + k^2 z_1. \tag{2.182}$$

Theorem 2.31. *Under infinitesimal bending of the curve* C, *variation of the torsion is*

$$\delta \tau = z\tau' + k(z_2' + 2\tau z_1) + \left\{ \frac{1}{k} \left[2\tau z_1' + \tau' z_1 + z_2'' - \tau^2 z_2 \right] \right\}'. \tag{2.183}$$

Proof. Let us take a variation of the Frenet–Serret equation for $\mathbf{n}_1'$ and dot with $\mathbf{n}_2$. We have

$$\delta \tau = k \mathbf{n}_2 \cdot \delta \mathbf{t} + \mathbf{n}_2 \cdot \delta \mathbf{n}_1'. \tag{2.184}$$

We now rewrite the second term on the right-hand side as

$$\mathbf{n}_2 \cdot \delta \mathbf{n}_1' = (\mathbf{n}_2 \cdot \delta \mathbf{n}_1)' - \mathbf{n}_2' \cdot \delta \mathbf{n}_1 = (\mathbf{n}_2 \cdot \delta \mathbf{n}_1)', \tag{2.185}$$

after using the third Frenet–Serret equation. Since $\mathbf{r}'' = k\mathbf{n}_1$, we have $\mathbf{t}' = k\mathbf{n}_1$, i.e. $\mathbf{n}_1' = \frac{1}{k}\,\mathbf{t}'$. Further, $\delta \mathbf{n}_1 = \delta(\frac{1}{k})\,\mathbf{t}' + \frac{1}{k}\,\delta \mathbf{t}'$,

$$\mathbf{n}_2 \cdot \delta \mathbf{n}_1 = \mathbf{n}_2 \cdot \left[\delta \left(\frac{1}{k} \right) k\mathbf{n}_1 + \frac{1}{k}\,\delta \mathbf{t}' \right] = \frac{1}{k} \mathbf{n}_2 \cdot \delta \mathbf{t}'. \tag{2.186}$$

From (2.184), (2.185), and (2.186), we obtain

$$\delta \tau = k \mathbf{n}_2 \cdot \delta \mathbf{t} + \left(\frac{1}{k} \mathbf{n}_2 \cdot \delta \mathbf{t}' \right)' \tag{2.187}$$

Substituting (2.173) and (2.179) into (2.187) and using (2.25), we obtain (2.183). $\square$

Corollary 2.4. *Under infinitesimal bending of a plane curve, variation of the torsion is*

$$\delta \tau = k z_2' + \left(\frac{1}{k} z_2'' \right)'. \tag{2.188}$$

Based on the theorems that precede, corresponding geometric magnitudes of deformed curves under infinitesimal bending (of the first order) are:

$$
\begin{aligned}
\mathbf{t}_\epsilon &= \mathbf{t} + \epsilon\big[(z_1' - \tau z_2 + kz)\,\mathbf{n}_1 + (z_2' + \tau z_1)\,\mathbf{n}_2\big], \\
\mathbf{n}_{1\epsilon} &= \mathbf{n}_1 + \epsilon\left[-(kz + z_1' - \tau z_2)\mathbf{t} + \frac{1}{k}(k\tau z + z_2'' - \tau^2 z_2 + 2\tau z_1' \right. \\
&\qquad \left. + \tau' z_1)\mathbf{n}_2\right], \\
\mathbf{n}_{2\epsilon} &= \mathbf{n}_2 + \epsilon\left[-(z_2' + \tau z_1)\mathbf{t} - \frac{1}{k}(k\tau z + z_2'' - \tau^2 z_2 + 2\tau z_1' + \tau' z_1)\mathbf{n}_1\right], \\
k_\epsilon &= k + \epsilon[k'z + z_1'' + (k^2 - \tau^2)z_1 - 2\tau z_2' - \tau' z_2], \\
\tau_\epsilon &= \tau + \epsilon\left\{z\tau' + k(z_2' + 2\tau z_1) + \left[\frac{1}{k}\left(2\tau z_1' + \tau' z_1 + z_2'' - \tau^2 z_2\right)\right]'\right\}.
\end{aligned}
$$

after neglecting the terms of order higher than 1 with respect to ϵ.

It is useful to observe the variation as a function of the rotational field $\mathbf{y}$. The corresponding formulas can be obtained similarly to the previous ones given as a function of the field $\mathbf{z}$, but they have a much simpler form, as follows:

Theorem 2.32. *Under infinitesimal bending of the curve C, the following variations are given as functions of the rotational field $\mathbf{y}$:*

$$
\begin{aligned}
\delta\mathbf{t} &= y_2\mathbf{n}_1 - y_1\mathbf{n}_2, \\
\delta\mathbf{n}_1 &= -y_2\mathbf{t} + \frac{1}{k}(y_2\tau - y_1')\mathbf{n}_2, \\
\delta\mathbf{n}_2 &= 7y_1\mathbf{t} - \frac{1}{k}(y_2\tau - y_1')\mathbf{n}_1, \\
\delta k &= y_2' + y_1\tau, \\
\delta\tau &= \left[\frac{1}{k}(y_2\tau - y_1')\right]' - ky_1.
\end{aligned}
\tag{2.189}
$$

2.6.1.1 *Area of the Region Bounded by a Plane Curve*

We will examine the change of a region bounded by a plane curve (2.47), $K : \rho = \rho(\theta)$, $\theta \in [0, 2\pi]$, under infinitesimal bending of this curve (staying plane).

Theorem 2.33 ([Velimirović (2001)]). *Area of the region determined by a plane curve being infinitesimally bent staying plane is stationary.*

Proof. Area determined by a curve K, (2.47), is

$$P = \frac{1}{2}\int_0^{2\pi} \rho^2(\theta)\, d\theta. \tag{2.190}$$

Area bounded by deformed curve K_ϵ is

$$P_\epsilon = \frac{1}{2}\int_0^{2\pi} \rho_\epsilon^2(\theta)\, d\theta. \tag{2.191}$$

As from

$$\mathbf{r}_\epsilon = x_\epsilon \mathbf{i} + y_\epsilon \mathbf{j} = (x + \epsilon z_1)\mathbf{i} + (y + \epsilon z_2)\mathbf{j}$$
$$\rho_\epsilon^2 = (x + \epsilon z_1)^2 + (y + \epsilon z_2)^2 = \rho^2 + 2\epsilon \mathbf{r} \cdot \mathbf{z} + \epsilon^2 \|\mathbf{z}\|^2$$

area determined by deformed curve is

$$P_\epsilon = \frac{1}{2}\int_0^{2\pi} \rho^2(\theta)\, d\theta + \epsilon \int_0^{2\pi} \mathbf{r} \cdot \mathbf{z}\, d\theta + \frac{\epsilon^2}{2}\int_0^{2\pi} \|\mathbf{z}\|^2\, d\theta,$$

and variation of the area is

$$\delta P = \int_0^{2\pi} \mathbf{r} \cdot \mathbf{z}\, d\theta. \tag{2.192}$$

Having in mind, Theorem 2.9

$$\mathbf{r} \cdot \mathbf{z} = (\rho(\theta)\cos\theta\, \mathbf{i} + \rho(\theta)\sin\theta\, \mathbf{j}) \cdot (\rho(\theta)\sin\theta \mathbf{i} - \rho(\theta)\cos\theta \mathbf{j}) = 0,$$

wherefrom, we have

$$\delta P = 0,$$

i.e. area of the region in the plane, bounded by a curve that stays plane under infinitesimal bending, is stationary. □

2.6.1.2 *Curvature of a Curve on the Hyperbolic Paraboloid*

Let us examine the variation of the curvature k of the curve

$$C : \mathbf{r}(t) = \mathbf{r}(u(t), v(t)) \tag{2.193}$$

on the hyperbolic paraboloid

$$S : \mathbf{r}(u, v) = (u, v, uv), \tag{2.194}$$

under infinitesimal bending determined by the field (2.80). The curvatures of the curves

$$C_\epsilon : \mathbf{r}_\epsilon = \mathbf{r}(t) + \epsilon\mathbf{z}(t) \tag{2.195}$$

are given by the formula

$$k_\epsilon = \frac{||\dot{\mathbf{r}_\epsilon} \times \ddot{\mathbf{r}_\epsilon}||}{||\dot{\mathbf{r}_\epsilon}||^3}. \tag{2.196}$$

By direct calculation, we get

$$||\dot{\mathbf{r}_\epsilon} \times \ddot{\mathbf{r}_\epsilon}||^2 = ||\dot{\mathbf{r}} \times \ddot{\mathbf{r}}||^2 + 2\epsilon P_1(t) + \epsilon^2 P_2(t) + \epsilon^3 P_3(t) + \epsilon^4 P_4(t),$$

where

$$\begin{aligned}
P_1(t) &= (\dot{v}(uv)^{\cdot\cdot} - \ddot{v}(uv)^{\cdot})A(t) + (\dot{u}(uv)^{\cdot\cdot} - \ddot{u}(uv)^{\cdot})C(t) \\
&\quad + (\dot{u}\ddot{v} - \ddot{u}\dot{v})D(t); \\
P_2(t) &= 2(\dot{v}(uv)^{\cdot\cdot} - \ddot{v}(uv)^{\cdot})B(t) + A^2(t) + C^2(t) + D^2(t); \\
P_3(t) &= 2A(t)B(t); \quad P_4(t) = B^2(t); \\
A(t) &= (\ddot{u}\dot{v} - \dot{u}\ddot{v})z_2 + (4\dot{u}\dot{v} + \ddot{u}v)\dot{z_2} - \dot{u}v\ddot{z_2}; \\
B(t) &= \ddot{u}z_2\dot{z_2} + 2\dot{u}\dot{z_2}^2 - \dot{u}z_2\ddot{z_2}; \\
C(t) &= (2\dot{u}^2 - u\ddot{u})\dot{z_2} + u\dot{u}\ddot{z_2}; \quad D(t) = \dot{u}\ddot{z_2} - \ddot{u}\dot{z_2};
\end{aligned} \tag{2.197}$$

The function $z_2(t)$ is determined by Eq. (2.88). Also,

$$\|\dot{\mathbf{r}}_\epsilon\|^2 = \|\dot{\mathbf{r}}\|^2 + \epsilon^2[(\dot{u}z_2 + u\dot{z}_2)^2 + \dot{z}_2{}^2]. \tag{2.198}$$

It is valid

$$\lim_{\epsilon\to 0} \frac{k_\epsilon^2 - k^2}{\epsilon} = \frac{d}{d\epsilon}k_\epsilon^2\Big|_{\epsilon=0} = \frac{2P_1(t)}{\|\dot{\mathbf{r}}\|^6}, \tag{2.199}$$

and also,

$$\lim_{\epsilon\to 0} \frac{k_\epsilon^2 - k^2}{\epsilon} = \lim_{\epsilon\to 0} \frac{k_\epsilon - k}{\epsilon} \cdot (k_\epsilon + k) = \delta k \cdot 2k, \tag{2.200}$$

where δk is the variation of the curvature k of curve (2.193). By comparing (2.199) and (2.200), using $k \neq 0$ we obtain

$$\delta k = \frac{P_1(t)}{k\|\dot{\mathbf{r}}\|^6}. \tag{2.201}$$

Let us verify the case $k = 0$. From (2.199) and (2.200), we conclude $P_1(t) = 0$. Now it is

$$\delta k = \lim_{\epsilon\to 0} \frac{k_\epsilon}{\epsilon} = \frac{\sqrt{\epsilon^2 P_2(t) + \epsilon^3 P_3(t) + \epsilon^4 P_4(t)}}{\epsilon[\sqrt{\|\dot{\mathbf{r}}\|^2 + \epsilon^2((\dot{u}z_2 + u\dot{z}_2)^2 + \dot{z}_2{}^2)}]^3}$$

i.e.

$$\delta k = \frac{\sqrt{P_2(t)}}{\|\dot{\mathbf{r}}\|^3}. \tag{2.202}$$

Therefore, it holds

Theorem 2.34 ([Najdanović and Velimirović (2018)]). *The first variation of the curvature k of curve* (2.193) *on the hyperbolic paraboloid* (2.194) *under infinitesimal bending* (2.80) *does not have to be zero and it is given by Eq.* (2.201) *for $k \neq 0$,* i.e. (2.202), *for $k = 0$, where P_1 and P_2 are the functions given in* (2.197).

2.6.2 The Second Variation

In the case when we consider infinitesimal bending of the second order, we study the first and the second variations of geometric magnitudes. For this reason, we can represent the magnitude $\mathcal{A}_\epsilon$ as

$$\mathcal{A}_\epsilon = \mathcal{A} + \epsilon\, \delta\mathcal{A} + \epsilon^2 \delta^2 \mathcal{A},$$

by neglecting the terms with ϵ^n, $n \geq 3$. The previous considered properties of the variation are reduced to

Ib. $\delta\mathcal{AB} = \mathcal{A}\,\delta\mathcal{B} + \mathcal{B}\,\delta\mathcal{A}, \quad \delta^2\mathcal{AB} = \mathcal{A}\,\delta^2\mathcal{B} + \mathcal{B}\,\delta^2\mathcal{A} + \delta\mathcal{A}\,\delta\mathcal{B}$

IIb. $\delta\left(\frac{d\mathcal{A}}{du}\right) = \frac{d(\delta\mathcal{A})}{du}, \quad \delta^2\left(\frac{d\mathcal{A}}{du}\right) = \frac{d(\delta^2\mathcal{A})}{du}$

IIIb. $\delta(d\mathcal{A}) = d(\delta\mathcal{A}), \quad \delta^2(d\mathcal{A}) = d(\delta^2\mathcal{A})$

According to Theorem 2.6, we have

$$\delta(ds) = 0, \quad \delta^2(ds) = 0. \tag{2.203}$$

Therefore, under second-order infinitesimal bending of a curve, the first and the second variation of the line element ds are equal to zero.

Let us describe the behavior of some geometric magnitudes under second-order infinitesimal bending (2.43) of the curve $C : \mathbf{r} = \mathbf{r}(s)$ parameterized by arc length, according to [Najdanović and Velimirović (2017a)]. Infinitesimal bending fields are given in (2.44).

Theorem 2.35. *Under second-order infinitesimal bending of the curve C, the first and the second variation of the unit tangent vector respectively are:*

$$\delta\mathbf{t} = (k\overset{(1)}{z} + \overset{(1)}{z}{}_1' - \tau\overset{(1)}{z}{}_2)\,\mathbf{n}_1 + (\overset{(1)}{z}{}_2' + \tau\overset{(1)}{z}{}_1)\,\mathbf{n}_2, \tag{2.204}$$

$$\delta^2\mathbf{t} = (\overset{(2)}{z}{}' - k\overset{(2)}{z}{}_1)\,\mathbf{t} + (k\overset{(2)}{z} + \overset{(2)}{z}{}_1' - \tau\overset{(2)}{z}{}_2)\,\mathbf{n}_1 + (\overset{(2)}{z}{}_2' + \tau\overset{(2)}{z}{}_1)\,\mathbf{n}_2, \tag{2.205}$$

Proof. According to **IIb**, we have $\delta\mathbf{t} = \delta\mathbf{r}' = (\delta\mathbf{r})' = \overset{(1)}{\mathbf{z}}{}'$ and $\delta^2\mathbf{t} = \delta^2\mathbf{r}' = (\delta^2\mathbf{r})' = \overset{(2)}{\mathbf{z}}{}'$. Then, using Eqs. (2.44) and (2.46) and the Frenet–Serret equations, we obtain that Eqs. (2.204) and (2.205) are valid. □

Obviously, using Eqs. (2.46), (2.204) and (2.205), we obtain the connection between the first and the second variation of $\mathbf{t}$:

$$\delta^2\mathbf{t} = -\frac{1}{2}\|\delta\mathbf{t}\|^2\,\mathbf{t} + (k\overset{(2)}{z} + \overset{(2)}{z_1}{}' - \tau\overset{(2)}{z_2})\,\mathbf{n}_1 + (\overset{(2)}{z_2}{}' + \tau\overset{(2)}{z_1})\,\mathbf{n}_2. \quad (2.206)$$

Theorem 2.36. *Under second-order infinitesimal bending of the curve C, the first variation of the unit normal and binormal vectors respectively are:*

$$\begin{aligned}\delta\mathbf{n}_1 = &-\Big(k\overset{(1)}{z} + \overset{(1)}{z_1}{}' - \tau\overset{(1)}{z_2}\Big)\mathbf{t} + \frac{1}{k}\Big(k\tau\overset{(1)}{z} + 2\tau\overset{(1)}{z_1}{}' + \tau'\overset{(1)}{z_1}\\ &+\overset{(1)}{z_2}{}'' - \tau^2\overset{(1)}{z_2}\Big)\mathbf{n}_2,\end{aligned} \quad (2.207)$$

$$\delta\mathbf{n}_2 = -\Big(\overset{(1)}{z_2}{}' + \tau\overset{(1)}{z_1}\Big)\mathbf{t} - \frac{1}{k}\Big(k\tau\overset{(1)}{z} + 2\tau\overset{(1)}{z_1}{}' + \tau'\overset{(1)}{z_1} + \overset{(1)}{z_2}{}'' - \tau^2\overset{(1)}{z_2}\Big)\mathbf{n}_1, \quad (2.208)$$

Proof. The unit normal vector of deformed curve C_ϵ satisfies the condition $(\mathbf{n}_1 + \epsilon\delta\mathbf{n}_1 + \epsilon^2\delta^2\mathbf{n}_1)\cdot(\mathbf{n}_1 + \epsilon\delta\mathbf{n}_1 + \epsilon^2\delta^2\mathbf{n}_1) = 1$. This gives the following equations:

$$\mathbf{n}_1\cdot\delta\mathbf{n}_1 = 0, \quad (2.209)$$

$$\mathbf{n}_1\cdot\delta^2\mathbf{n}_1 = -\frac{1}{2}\delta\mathbf{n}_1\cdot\delta\mathbf{n}_1. \quad (2.210)$$

Also, the unit normal vector of C_ϵ is perpendicular to the unit tangent vector, i.e. $(\mathbf{n}_1 + \epsilon\delta\mathbf{n}_1 + \epsilon^2\delta^2\mathbf{n}_1)\cdot(\mathbf{t} + \epsilon\delta\mathbf{t} + \epsilon^2\delta^2\mathbf{t}) = 0$, wherefrom, we conclude

$$\mathbf{t}\cdot\delta\mathbf{n}_1 = -\mathbf{n}_1\cdot\delta\mathbf{t} = -\Big(k\overset{(1)}{z} + \overset{(1)}{z_1}{}' - \tau\overset{(1)}{z_2}\Big), \quad (2.211)$$

$$\mathbf{t}\cdot\delta^2\mathbf{n}_1 = -\mathbf{n}_1\cdot\delta^2\mathbf{t} - \delta\mathbf{n}_1\cdot\delta\mathbf{t}. \quad (2.212)$$

Further, we take a first variation of the first equation of (2.5),

$$\delta\mathbf{t}' = \mathbf{n}_1\delta k + k\delta\mathbf{n}_1. \quad (2.213)$$

Dotting this equation with $\mathbf{n}_2$, we obtain

$$\mathbf{n}_2\cdot\delta\mathbf{n}_1 = \frac{1}{k}\mathbf{n}_2\cdot\delta\mathbf{t}'.$$

To compute $\delta\mathbf{t}'$, we apply commutativity of the variation and the derivative. For this reason, we obtain $\delta\mathbf{t}' = (\delta\mathbf{t})'$. Based on Eq. (2.204), the Frenet–Serret equations and $\overset{(1)}{z}{}' = k\overset{(1)}{z}_1$, one obtains

$$\begin{aligned}\delta\mathbf{t}' = &-k\Big(k\overset{(1)}{z} + \overset{(1)}{z}_1{}' - \tau\overset{(1)}{z}_2\Big)\mathbf{t} + \Big(k'\overset{(1)}{z} + \overset{(1)}{z}_1{}'' + (k^2-\tau^2)\overset{(1)}{z}_1 - 2\tau\overset{(1)}{z}_2{}' \\ &- \tau'\overset{(1)}{z}_2\Big)\mathbf{n}_1 + \Big(k\tau\overset{(1)}{z} + 2\tau\overset{(1)}{z}_1{}' + \tau'\overset{(1)}{z}_1 + \overset{(1)}{z}_2{}'' - \tau^2\overset{(1)}{z}_2\Big)\mathbf{n}_2.\end{aligned} \tag{2.214}$$

Now we have

$$\mathbf{n}_2\cdot\delta\mathbf{n}_1 = \frac{1}{k}\Big(k\tau\overset{(1)}{z} + 2\tau\overset{(1)}{z}_1{}' + \tau'\overset{(1)}{z}_1 + \overset{(1)}{z}_2{}'' - \tau^2\overset{(1)}{z}_2\Big) \tag{2.215}$$

From Eqs. (2.209), (2.211), and (2.215) we obtain that Eq. (2.207) is valid. Similarly, from the conditions $\mathbf{n}_{2\epsilon}\cdot\mathbf{n}_{2\epsilon} = 1$, $\mathbf{n}_{2\epsilon}\cdot\mathbf{t}_\epsilon = 0$ and $\mathbf{n}_{2\epsilon}\cdot\mathbf{n}_{1\epsilon} = 0$ we confirm the validity of Eq. (2.208). □

Theorem 2.37. *Under second-order infinitesimal bending of the curve C, a unit vector of the orthonormal basis and its first variation are orthogonal, i.e.*

$$\mathbf{t}\cdot\delta\mathbf{t} = 0, \quad \mathbf{n}_1\cdot\delta\mathbf{n}_1, \quad \mathbf{n}_2\cdot\delta\mathbf{n}_2 = 0. \tag{2.216}$$

A unit vector of the orthonormal basis and its second variation are not orthogonal, i.e.

$$\mathbf{t}\cdot\delta^2\mathbf{t} = -\frac{1}{2}\|\delta\mathbf{t}\|^2 = \overset{(2)}{z}{}' - k\overset{(2)}{z}_1, \tag{2.217}$$

$$\begin{aligned}\mathbf{n}_1\cdot\delta^2\mathbf{n}_1 = -\frac{1}{2}\Bigg[&\Big(k\overset{(1)}{z} + \overset{(1)}{z}_1{}' - \tau\overset{(1)}{z}_2\Big)^2 + \frac{1}{k^2}\Big(k\tau\overset{(1)}{z} + 2\tau\overset{(1)}{z}_1{}' + \tau'\overset{(1)}{z}_1 + \overset{(1)}{z}_2{}'' \\ &- \tau^2\overset{(1)}{z}_2\Big)^2\Bigg],\end{aligned} \tag{2.218}$$

$$\begin{aligned}\mathbf{n}_2\cdot\delta^2\mathbf{n}_2 = -\frac{1}{2}\Bigg[&\Big(\overset{(1)}{z}_2{}' + \tau\overset{(1)}{z}_1\Big)^2 + \frac{1}{k^2}\Big(k\tau\overset{(1)}{z} + 2\tau\overset{(1)}{z}_1{}' + \tau'\overset{(1)}{z}_1 + \overset{(1)}{z}_2{}'' \\ &- \tau^2\overset{(1)}{z}_2\Big)^2\Bigg]\end{aligned} \tag{2.219}$$

Proof. The previous results come directly from Eqs. (2.204), (2.207), (2.208), (2.210) and the same for $\mathbf{n}_2$ as (2.210). □

Theorem 2.38. *Under second-order infinitesimal bending of the curve C, the second variation of the unit normal vector is*

$$\delta^2 \mathbf{n}_1 = g_t \mathbf{t} + g_{n_1} \mathbf{n}_1 + g_{n_2} \mathbf{n}_2, \tag{2.220}$$

where

$$\begin{aligned} g_t = &-\left(k \overset{(2)}{z} + \overset{(2)}{z}_1{}' - \tau \overset{(2)}{z}_2\right) - \frac{1}{k}\left(\overset{(1)}{z}_2{}' + \tau \overset{(1)}{z}_1\right)\left(k\tau \overset{(1)}{z} + 2\tau \overset{(1)}{z}_1{}'\right. \\ &\left. + \tau' \overset{(1)}{z}_1 + \overset{(1)}{z}_2{}'' - \tau^2 \overset{(1)}{z}_2\right), \end{aligned} \tag{2.221}$$

$$\begin{aligned} g_{n_1} = &-\frac{1}{2}\left[\left(k \overset{(1)}{z} + \overset{(1)}{z}_1{}' - \tau \overset{(1)}{z}_2\right)^2 + \frac{1}{k^2}\left(k\tau \overset{(1)}{z} + 2\tau \overset{(1)}{z}_1{}' + \tau' \overset{(1)}{z}_1\right.\right. \\ &\left.\left. + \overset{(1)}{z}_2{}'' - \tau^2 \overset{(1)}{z}_2\right)^2\right], \end{aligned} \tag{2.222}$$

$$\begin{aligned} g_{n_2} = &\frac{1}{k}\left[\tau\left(k \overset{(2)}{z} + \overset{(2)}{z}_1{}' - \tau \overset{(2)}{z}_2\right) + \left(\overset{(2)}{z}_2{}' + \tau \overset{(2)}{z}_1\right)' - \frac{1}{k}\left(k' \overset{(1)}{z} + \overset{(1)}{z}_1{}''\right.\right. \\ &\left. + (k^2 - \tau^2)\overset{(1)}{z}_1 - 2\tau \overset{(1)}{z}_2{}' - \tau' \overset{(1)}{z}_2\right)\left(k\tau \overset{(1)}{z} + 2\tau \overset{(1)}{z}_1{}'\right. \\ &\left.\left. + \tau' \overset{(1)}{z}_1 + \overset{(1)}{z}_2{}'' - \tau^2 \overset{(1)}{z}_2\right)\right]. \end{aligned} \tag{2.223}$$

Proof. From the equation $\mathbf{n}_1 \cdot \mathbf{t} = 0$, using the second variation, we obtain

$$\mathbf{t} \cdot \delta^2 \mathbf{n}_1 = -\mathbf{n}_1 \cdot \delta^2 \mathbf{t} - \delta \mathbf{n}_1 \cdot \delta \mathbf{t}.$$

If we make use of Eqs. (2.204), (2.205) and (2.207), we get

$$\begin{aligned} \mathbf{t} \cdot \delta^2 \mathbf{n}_1 = &-\left(k \overset{(2)}{z} + \overset{(2)}{z}_1{}' - \tau \overset{(2)}{z}_2\right) - \frac{1}{k}\left(\overset{(1)}{z}_2{}' + \tau \overset{(1)}{z}_1\right)\left(k\tau \overset{(1)}{z} + 2\tau \overset{(1)}{z}_1{}'\right. \\ &\left. + \tau' \overset{(1)}{z}_1 + \overset{(1)}{z}_2{}'' - \tau^2 \overset{(1)}{z}_2\right). \end{aligned} \tag{2.224}$$

Let us take the second variation of the equation $\mathbf{t}' = k\mathbf{n}_1$ and dot with $\mathbf{n}_2$. We obtain

$$\mathbf{n}_2 \cdot \delta^2 \mathbf{n}_1 = \frac{1}{k}(\mathbf{n}_2 \cdot \delta^2 \mathbf{t}' - \delta k \mathbf{n}_2 \cdot \delta \mathbf{n}_1).$$

To evaluate $\delta^2\mathbf{t}'$, we use $\delta^2\mathbf{t}' = (\delta^2\mathbf{t})'$, the expression for $\delta^2\mathbf{t}$, (2.205), and Frenet–Serret equations. Dotting with $\mathbf{n}_2$, we obtain

$$\mathbf{n}_2 \cdot \delta^2\mathbf{t}' = \tau\left(k\overset{(2)}{z} + \overset{(2)}{z}_1{}' - \tau\overset{(2)}{z}_2\right) + \left(\overset{(2)}{z}_2{}' + \tau\overset{(2)}{z}_1\right)'.$$

Finally, we have

$$\begin{aligned}\mathbf{n}_2 \cdot \delta^2\mathbf{n}_1 = \frac{1}{k}\Bigg[&\tau\left(k\overset{(2)}{z} + \overset{(2)}{z}_1{}' - \tau\overset{(2)}{z}_2\right) + \left(\overset{(2)}{z}_2{}' + \tau\overset{(2)}{z}_1\right)' - \frac{1}{k}\Big(k'\overset{(1)}{z} + \overset{(1)}{z}_1{}''\\ &+ (k^2-\tau^2)\overset{(1)}{z}_1 - 2\tau\overset{(1)}{z}_2{}' - \tau'\overset{(1)}{z}_2\Big)\Big(k\tau\overset{(1)}{z} + 2\tau\overset{(1)}{z}_1{}'\\ &+ \tau'\overset{(1)}{z}_1 + \overset{(1)}{z}_2{}'' - \tau^2\overset{(1)}{z}_2\Big)\Bigg].\end{aligned} \tag{2.225}$$

Equations (2.218), (2.224), and (2.225) give the normal, the tangent, and the binormal component of $\delta^2\mathbf{n}_1$, respectively. □

Theorem 2.39. *Under second-order infinitesimal bending of the curve C, the first and the second variation of the curvature are respectively*

$$\delta k = k'\overset{(1)}{z} + \overset{(1)}{z}_1{}'' + (k^2-\tau^2)\overset{(1)}{z}_1 - 2\tau\overset{(1)}{z}_2{}' - \tau'\overset{(1)}{z}_2, \tag{2.226}$$

$$\begin{aligned}\delta^2 k = &-\frac{k}{2}\left(\tau\overset{(1)}{z}_1 + \overset{(1)}{z}_2{}'\right)^2 + \frac{1}{2k}\left(k\tau\overset{(1)}{z} + 2\tau\overset{(1)}{z}_1{}' + \tau'\overset{(1)}{z}_1 + \overset{(1)}{z}_2{}'' - \tau^2\overset{(1)}{z}_2\right)^2\\ &+ \left(k\overset{(2)}{z} + \overset{(2)}{z}_1{}' - \tau\overset{(2)}{z}_2\right)' - \tau\left(\overset{(2)}{z}_2{}' + \tau\overset{(2)}{z}_1\right).\end{aligned} \tag{2.227}$$

Proof. Equation (2.226) is proved analogously to the proof of Theorem 2.30. Let us take the second variation of the first Frenet–Serret equation. We obtain it holds $\delta^2\mathbf{t}' = \mathbf{n}_1\delta^2 k + k\delta^2\mathbf{n}_1 + \delta k\delta\mathbf{n}_1$. After scalar product with $\mathbf{n}_1$, we obtain

$$\delta^2 k = \mathbf{n}_1 \cdot \delta^2\mathbf{t}' - k\mathbf{n}_1 \cdot \delta^2\mathbf{n}_1 - \delta k\mathbf{n}_1 \cdot \delta\mathbf{n}_1 \tag{2.228}$$

Since $\delta^2\mathbf{t}' = (\delta^2\mathbf{t})'$, applying Eq. (2.205), Frenet–Serret equations and dotting with $\mathbf{n}_1$, we get

$$\mathbf{n}_1 \cdot \delta^2\mathbf{t}' = k\left(\overset{(2)}{z}{}' - k\overset{(2)}{z}_1\right) + \left(k\overset{(2)}{z} + \overset{(2)}{z}_1{}' - \tau\overset{(2)}{z}_2\right)' - \tau\left(\overset{(2)}{z}_2{}' + \tau\overset{(2)}{z}_1\right). \tag{2.229}$$

Using the previous equation, Theorem 2.37, and the second equation in Eq. (2.46), we complete the proof of this theorem. □

Directly from Theorem 2.39, in the case of a plane curve ($\tau = 0$), we obtain the next corollary.

Corollary 2.5. *Under second-order infinitesimal bending of the plane curve, the first and the second variation of the curvature are respectively*

$$\delta k = k' \overset{(1)}{z} + \overset{(1)}{z}_1{}'' + k^2 \overset{(1)}{z}_1, \tag{2.230}$$

$$\delta^2 k = -\frac{k}{2}\left(\overset{(1)}{z}_2{}'\right)^2 + \frac{1}{2k}\left(\overset{(1)}{z}_2{}''\right)^2 + \left(k \overset{(2)}{z} + \overset{(2)}{z}_1{}'\right)'. \tag{2.231}$$

Theorem 2.40. *Under second-order infinitesimal bending of the curve C, the first and the second variation of the torsion are respectively*

$$\delta \tau = \tau' \overset{(1)}{z} + k\left(\overset{(1)}{z}_2{}' + 2\tau \overset{(1)}{z}_1\right) + \left\{\frac{1}{k}\left[2\tau \overset{(1)}{z}_1{}' + \tau' \overset{(1)}{z}_1 + \overset{(1)}{z}_2{}'' - \tau^2 \overset{(1)}{z}_2\right]\right\}', \tag{2.232}$$

$$\begin{aligned}
\delta^2 \tau = {} & \left\{\frac{1}{k}\left[\tau\left(k \overset{(2)}{z} + \overset{(2)}{z}_1{}' - \tau \overset{(2)}{z}_2\right) + \left(\overset{(2)}{z}_2{}' + \tau \overset{(2)}{z}_1\right)' - \frac{1}{k}\left(k' \overset{(1)}{z} + \overset{(1)}{z}_1{}''\right.\right.\right. \\
& \left. + (k^2 - \tau^2)\overset{(1)}{z}_1 - 2\tau \overset{(1)}{z}_2{}' - \tau' \overset{(1)}{z}_2\right)\left(k\tau \overset{(1)}{z} + 2\tau \overset{(1)}{z}_1{}' + \tau' \overset{(1)}{z}_1\right. \\
& \left.\left.\left. + \overset{(1)}{z}_2{}'' - \tau^2 \overset{(1)}{z}_2\right)\right]\right\}' - \frac{\tau}{2}\left[\left(k \overset{(1)}{z} + \overset{(1)}{z}_1{}' - \tau \overset{(1)}{z}_2\right)^2 - \left(\overset{(1)}{z}_2{}' + \tau \overset{(1)}{z}_1\right)^2\right] \\
& + k\left(\overset{(2)}{z}_2{}' + \tau \overset{(2)}{z}_1\right) + \left(k' \overset{(1)}{z} + \overset{(1)}{z}_1{}'' + (k^2 - \tau^2)\overset{(1)}{z}_1 - 2\tau \overset{(1)}{z}_2{}'\right. \\
& \left. - \tau' \overset{(1)}{z}_2\right)\left(\overset{(1)}{z}_2{}' + \tau \overset{(1)}{z}_1\right).
\end{aligned} \tag{2.233}$$

Proof. Analogous to Theorem 2.31, using the expressions for δt and δn_1, (2.204) and (2.207), respectively, as well as the first equation in (2.46), we obtain the first variation of the torsion.

For the second variation of the torsion, let us take the second variation of the second Frenet–Serret equation. We have

$$\delta^2 \mathbf{n}_1' = -\delta^2 k \mathbf{t} - k\delta^2 \mathbf{t} - \delta k \delta t + \delta^2 \tau \mathbf{n}_2 + \tau \delta^2 \mathbf{n}_2 + \delta \tau \delta \mathbf{n}_2. \tag{2.234}$$

Dotting with $\mathbf{n}_2$, we obtain

$$\delta^2\tau = (\mathbf{n}_2\cdot\delta^2\mathbf{n}_1)' + \tau\mathbf{n}_1\cdot\delta^2\mathbf{n}_1 + k\mathbf{n}_2\cdot\delta^2\mathbf{t} + \delta k\,\mathbf{n}_2\cdot\delta\mathbf{t} - \tau\mathbf{n}_2\cdot\delta^2\mathbf{n}_2. \quad (2.235)$$

We used $\mathbf{n}_2\cdot\delta\mathbf{n}_2 = 0$ and also

$$\mathbf{n}_2\cdot\delta^2\mathbf{n}_1' = \mathbf{n}_2\cdot(\delta^2\mathbf{n}_1)' = (\mathbf{n}_2\cdot\delta^2\mathbf{n}_1)' - \mathbf{n}_2'\cdot\delta^2\mathbf{n}_1 = (\mathbf{n}_2\cdot\delta^2\mathbf{n}_1)' + \tau\mathbf{n}_1\cdot\delta^2\mathbf{n}_1,$$

wherefrom, we have the second variation. □

Corollary 2.6. *Under second-order infinitesimal bending of the plane curve C, the first and the second variation of the torsion are respectively*

$$\delta\tau = k\overset{(1)}{z}_2{}' + \left(\frac{1}{k}\overset{(1)}{z}_2{}''\right)', \quad (2.236)$$

$$\begin{aligned}\delta^2\tau = &\left\{\frac{1}{k}\left[\overset{(2)}{z}_2{}'' - \frac{1}{k}\left(k'\overset{(1)}{z} + \overset{(1)}{z}_1{}'' + k^2\overset{(1)}{z}_1\right)\overset{(1)}{z}_2{}''\right]\right\}' \\ &+ k\overset{(2)}{z}_2{}' + \left(k'\overset{(1)}{z} + \overset{(1)}{z}_1{}'' + k^2\overset{(1)}{z}_1\right)\overset{(1)}{z}_2{}'. \end{aligned} \quad (2.237)$$

Theorem 2.41. *Under second-order infinitesimal bending of the curve C, the second variation of the binormal vector is*

$$\delta^2\mathbf{n}_2 = f_t\mathbf{t} + f_{n_1}\mathbf{n}_1 + f_{n_2}\mathbf{n}_2, \quad (2.238)$$

where

$$\begin{aligned} f_t = &-\left(\overset{(2)}{z}_2{}' + \tau\overset{(2)}{z}_1\right) + \frac{1}{k}\left(k\overset{(1)}{z} + \overset{(1)}{z}_1{}' - \tau\overset{(1)}{z}_2\right)\Big(k\tau\overset{(1)}{z} + 2\tau\overset{(1)}{z}_1{}' \\ &+ \tau'\overset{(1)}{z}_1 + \overset{(1)}{z}_2{}'' - \tau^2\overset{(1)}{z}_2\Big), \end{aligned} \quad (2.239)$$

$$\begin{aligned} f_{n_1} = &\frac{1}{\tau}\Big\{kg_t + g_{n_1}' - \tau g_{n_2} + k\left(k\overset{(2)}{z} + \overset{(2)}{z}_1{}' - \tau\overset{(2)}{z}_2\right) + \Big(k'\overset{(1)}{z} + \overset{(1)}{z}_1{}'' \\ &+ (k^2 - \tau^2)\overset{(1)}{z}_1 - 2\tau\overset{(1)}{z}_2{}' - \tau'\overset{(1)}{z}_2\Big)\left(k\overset{(1)}{z} + \overset{(1)}{z}_1{}' - \tau\overset{(1)}{z}_2\right) + \frac{1}{k} \end{aligned}$$

$$\cdot \left[\tau' \overset{(1)}{z} + 2k\tau \overset{(1)}{z}_1 + k \overset{(1)}{z}_2{}' + \left(\frac{1}{k} \left(2\tau \overset{(1)}{z}_1{}' + \tau' \overset{(1)}{z}_1 + \overset{(1)}{z}_2{}'' - \tau^2 \overset{(1)}{z}_2 \right) \right)' \right]$$

$$\cdot \left(k\tau \overset{(1)}{z} + 2\tau \overset{(1)}{z}_1{}' + \tau' \overset{(1)}{z}_1 + \overset{(1)}{z}_2{}'' - \tau^2 \overset{(1)}{z}_2 \right) \Bigg\}, \tag{2.240}$$

$$f_{n_2} = -\frac{1}{2} \left[\left(\overset{(1)}{z}_2{}' + \tau \overset{(1)}{z}_1 \right)^2 + \frac{1}{k^2} \left(k\tau \overset{(1)}{z} + 2\tau \overset{(1)}{z}_1{}' + \tau' \overset{(1)}{z}_1 + \overset{(1)}{z}_2{}'' - \tau^2 \overset{(1)}{z}_2 \right)^2 \right] \tag{2.241}$$

Proof. Using the second variation of the equation $\mathbf{n}_2 \cdot \mathbf{t} = 0$ and Eqs. (2.204), (2.205) and (2.208), we obtain (2.239). If we take the second variation of the second Frenet–Serret equation and dot with $\mathbf{n}_1$, we obtain

$$\mathbf{n}_1 \cdot \delta^2 \mathbf{n}_2 = \frac{1}{\tau} (\mathbf{n}_1 \cdot \delta^2 \mathbf{n}_1' + k \mathbf{n}_1 \cdot \delta^2 \mathbf{t} + \delta k \mathbf{n}_1 \cdot \delta \mathbf{t} - \delta \tau \mathbf{n}_1 \cdot \delta \mathbf{n}_2).$$

Since $\mathbf{n}_1 \cdot \delta^2 \mathbf{n}_1' = \mathbf{n}_1 \cdot (\delta^2 \mathbf{n}_1)'$, after using the Frenet–Serret equations, we obtain

$$\mathbf{n}_1 \cdot \delta^2 \mathbf{n}_1' = k g_t + g_{n_1}' - \tau g_{n_2},$$

where g_t, g_{n_1}, and g_{n_2} are the tangent, the normal and the binormal component of $\delta^2 \mathbf{n}_1$, respectively, according to Theorem 2.38. Now, we obtain $\mathbf{n}_1 \cdot \delta^2 \mathbf{n}_2 = f_{n_1}$. For $\mathbf{n}_2 \cdot \delta^2 \mathbf{n}_2 = f_{n_2}$, we use Eq. (2.219). □

Example 2.11. Let us examine second-order infinitesimal bending of the circle

$$\mathbf{r} = (\cos u, \sin u, 0), \qquad (\text{or } x^2 + y^2 = 1). \tag{2.242}$$

Here, $r = 1$ and $u = s$, i.e. the curve is parameterized by the arc length s and we have

$$\mathbf{t} = \mathbf{r}' = (-\sin s, \cos s, 0), \quad \mathbf{n}_1 = \frac{\mathbf{r}''}{\|\mathbf{r}''\|} = (-\cos s, -\sin s, 0),$$

$$\mathbf{n}_2 = \mathbf{t} \times \mathbf{n}_1 = (0, 0, 1).$$

According to (2.40), we have

$$\overset{(1)}{\mathbf{z}} = \int [p(s)(-\cos s\, \mathbf{i} - \sin s\, \mathbf{j}) + q(s)\, \mathbf{k}]\, ds + C_1,$$

where $p(s)$ and $q(s)$ are arbitrary functions. For instance, for $p(s) = C_1 = 0$, $q(s) = 1$, an infinitesimal bending field of the first order is

$$\overset{(1)}{\mathbf{z}} = (0, 0, s) = s\,\mathbf{k} = s\,\mathbf{n}_2. \tag{2.243}$$

Now, according to Eq. (2.41) for $r(s) = C_2 = 0$ and $g(s) = 1$, we obtain:

$$\overset{(2)}{\mathbf{z}} = \int \left(\frac{1}{2}\sin s\,\mathbf{i} - \frac{1}{2}\cos s\,\mathbf{j} + \mathbf{k}\right) ds,$$

$$\overset{(2)}{\mathbf{z}} = \left(-\frac{1}{2}\cos s, -\frac{1}{2}\sin s, s\right) = -\frac{1}{2}\cos s\,\mathbf{i} - \frac{1}{2}\sin s\,\mathbf{j} + s\,\mathbf{k},$$

or after determination of this vector field via the vectors of the Frenet–Serret frame $\{\mathbf{t}, \mathbf{n}_1, \mathbf{n}_2\}$,

$$\overset{(2)}{\mathbf{z}} = \frac{1}{2}\mathbf{n}_1 + s\mathbf{n}_2. \tag{2.244}$$

Thus, the curves we got under infinitesimal bending of the second order of the circle with these infinitesimal bending fields present a family of helices

$$\mathbf{r}_\epsilon = \left(\cos s - \frac{1}{2}\cos s\,\epsilon^2, \sin s - \frac{1}{2}\sin s\,\epsilon^2, s\epsilon + s\epsilon^2\right) \tag{2.245}$$

that are not on the cylinder $x^2+y^2 = 1$. Let us examine if the relation (2.11) is valid for $m = 2$. According to $u = s$, we have $ds^2 = du^2$ and also

$$ds_\epsilon^2 = d\mathbf{r}_\epsilon \cdot d\mathbf{r}_\epsilon = du^2 + 2\epsilon^3 du^2 + \frac{5}{4}\epsilon^4 du^2,$$

wherefrom, we get

$$ds_\epsilon^2 - ds^2 = 2\epsilon^3 du^2 + \frac{5}{4}\epsilon^4 du^2 = o(\epsilon^2),$$

i.e. the fields (2.243) and (2.244) are corresponding fields of second-order infinitesimal bending of the circle (2.242).

Let us compute the variation of the curvature. Obviously, $k = \frac{1}{r} = \|\mathbf{r}''\| = 1$, $\tau = 0$ (plane curve).

Since

$$\overset{(1)}{z} = \overset{(1)}{z}_1 = \overset{(2)}{z} = 0, \quad \overset{(1)}{z}_2 = \overset{(2)}{z}_2 = s, \quad \overset{(2)}{z}_1 = \frac{1}{2},$$

according to (2.230) and (2.231), we obtain $\delta k = 0$ and $\delta^2 k = -\frac{1}{2}$.

Example 2.12. Let us observe a helix

$$\mathbf{r} = (a\cos u, a\sin u, bu), \quad u \in \mathcal{R}, \quad a \neq 0, \quad b \neq 0,$$

which lies on a cylinder $x^2 + y^2 = a^2$. If we measure s from the point $\mathbf{r}(u) = (a, 0, 0)$, we get $s = \int_0^u \sqrt{a^2 \sin^2 u + a^2 \cos^2 u + b^2}\, du = \sqrt{a^2 + b^2}\, u$, i.e. $u = u(s) = \frac{s}{\sqrt{a^2+b^2}}$. Substituting this result into the starting equation, we obtain the parameterization of the helix by the arc length

$$\mathbf{r} = \left(a\cos\frac{s}{\sqrt{a^2+b^2}}, a\sin\frac{s}{\sqrt{a^2+b^2}}, \frac{bs}{\sqrt{a^2+b^2}}\right).$$

Specially, for $a = b = \frac{\sqrt{2}}{2}$, we have a helix

$$\mathbf{r} = \left(\frac{\sqrt{2}}{2}\cos s, \frac{\sqrt{2}}{2}\sin s, \frac{\sqrt{2}}{2}s\right). \tag{2.246}$$

The vectors of the Frenet–Serret frame $\{\mathbf{t}, \mathbf{n}_1, \mathbf{n}_2\}$ are

$$\mathbf{t} = \left(-\frac{\sqrt{2}}{2}\sin s, \frac{\sqrt{2}}{2}\cos s, \frac{\sqrt{2}}{2}\right), \quad \mathbf{n}_1 = (-\cos s, -\sin s, 0),$$

$$\mathbf{n}_2 = \left(\frac{\sqrt{2}}{2}\sin s, -\frac{\sqrt{2}}{2}\cos s, \frac{\sqrt{2}}{2}\right).$$

For $p(s) = 1$, $q(s) = r(s) = 0$, $g(s) = \frac{1}{2}$, $C_1 = C_2 = 0$, we get infinitesimal bending fields of the first and the second order respectively

$$\overset{(1)}{\mathbf{z}} = (-\sin s, \cos s, 0) = -\sin s\,\mathbf{i} + \cos s\,\mathbf{j}, \tag{2.247}$$

$$\overset{(2)}{\mathbf{z}} = \left(-\frac{\sqrt{2}}{2}\cos s, -\frac{\sqrt{2}}{2}\sin s, 0\right) = -\frac{\sqrt{2}}{2}\cos s\,\mathbf{i} - \frac{\sqrt{2}}{2}\sin s\,\mathbf{j}, \tag{2.248}$$

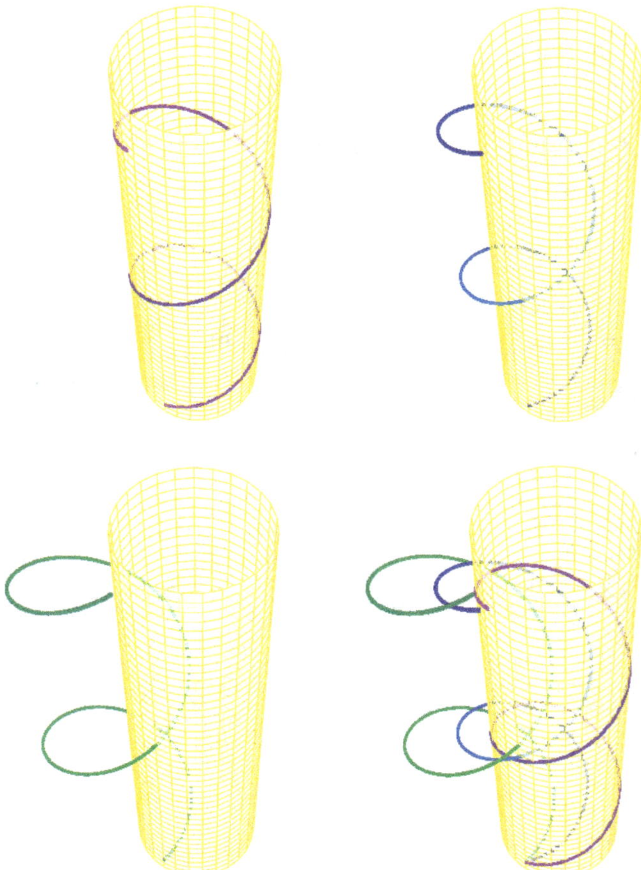

Fig. 2.15. Infinitesimal bending of the second order of a helix for $\epsilon \in \{0, 0.75, 1.5\}$.

i.e. in the basis $\{\mathbf{t}, \mathbf{n}_1, \mathbf{n}_2\}$:

$$\overset{(1)}{\mathbf{z}} = \frac{\sqrt{2}}{2}\mathbf{t} - \frac{\sqrt{2}}{2}\mathbf{n}_2, \quad \overset{(2)}{\mathbf{z}} = \frac{\sqrt{2}}{2}\mathbf{n}_1. \tag{2.249}$$

The curve obtained under this infinitesimal bending is

$$\mathbf{r}_\epsilon = \left(\frac{\sqrt{2}}{2}\cos s - \epsilon \sin s - \epsilon^2\frac{\sqrt{2}}{2}\cos s, \right.$$

$$\left. \frac{\sqrt{2}}{2}\sin s + \epsilon \cos s - \epsilon^2\frac{\sqrt{2}}{2}\sin s, \frac{\sqrt{2}}{2}s\right). \tag{2.250}$$

Graphical presentation of the family of curves C_ϵ is given in Fig. 2.15. For the sake of illustration, we have chosen large parameters for ϵ. More realistically, one has to choose a much smaller ϵ since we assume $o(\epsilon^2)$ in the definition of the infinitesimal bending.

It is easy to check the equations (2.45), so we conclude that the vector fields (2.247) and (2.248) are infinitesimal bending fields of the first and the second order respectively of the given helix. Moreover, it is valid $ds_\epsilon^2 - ds^2 = \frac{\epsilon^4}{2} du^2 = o(\epsilon^3)$, which means that this infinitesimal bending is the bending of the third order (for $\overset{(3)}{\mathbf{z}} = \mathbf{0}$).

For the helix (2.246), we have $k = \tau = \frac{\sqrt{2}}{2}$. According to Eq. (2.249), the relations

$$\overset{(1)}{z}_1 = \overset{(2)}{z} = \overset{(2)}{z}_2 = 0, \quad \overset{(1)}{z} = \overset{(2)}{z}_1 = -\overset{(1)}{z}_2 = \frac{\sqrt{2}}{2}$$

are valid. Using Theorem 2.39, we obtain $\delta k = \delta^2 k = 0$.

2.7 Willmore Energy

Let a curve be given by

$$C : \mathbf{r} = \mathbf{r}(s), \quad s \in \mathcal{I}, \quad (\mathbf{r} : \mathcal{I} \mapsto \mathcal{R}^3), \tag{2.251}$$

where s is the arc length. The Willmore energy of the curve C is given with the following equation

$$\mathcal{W} = \frac{1}{2} \int_{\mathcal{I}} k^2 \, ds, \tag{2.252}$$

k is the curvature of C.

We will examine the change of the Willmore energy according to [Najdanović (2015); Najdanović and Velimirović (2017b)].

2.7.1 *The First Variation of Willmore Energy*

The next theorem is related to determination of the Willmore energy of a curve under infinitesimal bending of the first order.

Theorem 2.42. *Under infinitesimal bending of the curve C, (2.251), variation of its Willmore energy is*

$$\delta\mathcal{W} = \int_{\mathcal{I}} \left[\left(k'' + \frac{1}{2}k^3 - k\tau^2 \right) z_1 + (2k'\tau + k\tau')z_2 \right] ds$$

$$+ \int_{\mathcal{I}} \left[\frac{1}{2}k^2 z - k'z_1 + kz_1' - 2k\tau z_2 \right]' ds, \tag{2.253}$$

where k and τ are the curvature and the torsion of C, respectively.

Proof. The Willmore energy of deformed curve is

$$\mathcal{W}_\epsilon = \frac{1}{2}\int_{\mathcal{I}} k_\epsilon^2\, ds_\epsilon = \frac{1}{2}\int_{\mathcal{I}} (k + \epsilon\,\delta k)^2 [ds + \epsilon\,\delta(ds)], \tag{2.254}$$

i.e.

$$\mathcal{W}_\epsilon = \mathcal{W} + \epsilon \left[\int_{\mathcal{I}} k\,\delta k\, ds + \frac{1}{2}\int_{\mathcal{I}} k^2\,\delta(ds) \right]. \tag{2.255}$$

According to Theorem 2.27, we obtain

$$\mathcal{W}_\epsilon = \mathcal{W} + \epsilon \int_{\mathcal{I}} k\,\delta k\, ds, \tag{2.256}$$

i.e.

$$\delta\mathcal{W} = \int_{\mathcal{I}} k\,\delta k\, ds, \tag{2.257}$$

Applying (2.181), we get

$$\delta\mathcal{W} = \int_{\mathcal{I}} k \left[k'z + z_1'' + (k^2 - \tau^2)z_1 - 2\tau z_2' - \tau' z_2 \right] ds. \tag{2.258}$$

Since $(kz_1')' = k'z_1' + kz_1''$, we have

$$kz_1'' = (kz_1')' - k'z_1'. \tag{2.259}$$

Also, it is $(k'z_1)' = k''z_1 + k'z_1'$, i.e.

$$-k'z_1' = k''z_1 - (k'z_1)'. \tag{2.260}$$

From (2.259) and (2.260), we have

$$kz_1'' = (kz_1')' + k''z_1 - (k'z_1)'. \tag{2.261}$$

Further, $(-2\tau z_2 k)' = -2\tau' z_2 k - 2\tau z_2' k - 2\tau z_2 k'$, or

$$-2\tau z_2' k = (-2\tau z_2 k)' + (2\tau k)' z_2. \tag{2.262}$$

Also, using (2.25), we obtain

$$kk'z = \left(\frac{1}{2}k^2 z\right)' - \frac{1}{2}k^2 z' = \left(\frac{1}{2}k^2 z\right)' - \frac{1}{2}k^3 z_1. \tag{2.263}$$

Substituting (2.261), (2.262), and (2.263) into (2.258), we have (2.253). □

Corollary 2.7. *The Willmore energy of deformed curve under infinitesimal bending is*

$$\mathcal{W}_\epsilon = \mathcal{W} + \epsilon\left\{\int_{\mathcal{I}} \left[\left(k'' + \frac{1}{2}k^3 - k\tau^2\right) z_1 + (2k'\tau + k\tau')z_2\right] ds \right.$$
$$\left. + \int_{\mathcal{I}} \left[\frac{1}{2}k^2 z - k'z_1 + kz_1' - 2k\tau z_2\right]' ds\right\}. \tag{2.264}$$

Corollary 2.8. *Under infinitesimal bending of a plane curve, variation of its Willmore energy is*

$$\delta\mathcal{W} = \int_{\mathcal{I}} \left(k'' + \frac{1}{2}k^3\right) z_1\, ds + \int_{\mathcal{I}} \left[\frac{1}{2}k^2 z - k'z_1 + kz_1'\right]' ds. \tag{2.265}$$

The expression (2.253) and corollaries hold for both open and closed curves C and require from their user only the computation of curvature and torsion of C, in addition to knowing the bending field.

Remark 2.2. We can see that the second integral in (2.253) is an integral of a total derivative. With the condition that the vector field

$\mathbf{z}$ has support in the interior of $\mathcal{I} = [0, L]$, we obtain the well-known Euler-Lagrange equations

$$k'' + \frac{1}{2}k^3 - k\tau^2 = 0, \tag{2.266}$$

$$2k'\tau + k\tau' = 0. \tag{2.267}$$

These are the governing equations of the curve that minimize the integral (2.253). Integrating the second one gives

$$\tau k^2 = const, \tag{2.268}$$

which determines τ as a function of k. If we substitute τ into (2.266) we get a second-order differential equation for k which is, obviously, integrable.

Note that, by applying the rotational field $\mathbf{y}$ and Eq. (2.189), we arrive at the following simpler notation for the first variation of the Willmore energy.

Theorem 2.43. *Under infinitesimal bending of the curve C, (2.251), the variation of its Willmore energy is*

$$\delta\mathcal{W} = \int_{\mathcal{I}} (k\tau y_1 - k' y_2)\, ds + \int_{\mathcal{I}} (k y_2)'\, ds, \tag{2.269}$$

where k and τ are the curvature and the torsion of C, respectively, y_1 and y_2 are the components of the rotational field given by Eq. (2.34)

2.7.2 The Second Variation of Willmore Energy

Let us consider the Willmore energy of the curve C (2.251):

$$\mathcal{W} = \frac{1}{2}\int_{\mathcal{I}} k^2\, ds, \tag{2.270}$$

under infinitesimal bending of the second-order. The Willmore energy of deformed curve is

$$\mathcal{W}_\epsilon = \frac{1}{2}\int_{\mathcal{I}} k_\epsilon^2\, ds_\epsilon = \frac{1}{2}\int_{\mathcal{I}} (k + \epsilon\, \delta k + \epsilon^2\, \delta^2 k)^2 (ds + \epsilon\, \delta\, ds + \epsilon^2\, \delta^2\, ds).$$

Since $\delta\, ds = \delta^2 ds = 0$, we have

$$\mathcal{W}_\epsilon = \mathcal{W} + \epsilon \int_{\mathcal{I}} k\, \delta k\, ds + \epsilon^2 \int_{\mathcal{I}} \left(\frac{1}{2} (\delta k)^2 + k \delta^2 k \right) ds,$$

which is obtained after neglecting the terms with ϵ^3 and ϵ^4. Therefore,

$$\delta \mathcal{W} = \int_{\mathcal{I}} k\, \delta k\, ds, \quad \delta^2 \mathcal{W} = \int_{\mathcal{I}} \left(\frac{1}{2} (\delta k)^2 + k \delta^2 k \right) ds. \tag{2.271}$$

Theorem 2.44. *Under second-order infinitesimal bending of the curve C, the first variation of the Willmore energy is*

$$\begin{aligned} \delta \mathcal{W} = & \int_{\mathcal{I}} \left[\left(k'' + \frac{1}{2} k^3 - k\tau^2 \right) \overset{(1)}{z}_1 + (2k'\tau + k\tau') \overset{(1)}{z}_2 \right] ds \\ & + \int_{\mathcal{I}} \left[\frac{1}{2} k^2 \overset{(1)}{z} - k' \overset{(1)}{z}_1 + k \overset{(1)}{z}_1{}' - 2k\tau \overset{(1)}{z}_2 \right]' ds. \end{aligned} \tag{2.272}$$

Proof. Applying the results (2.226) and (2.289), we get

$$\delta \mathcal{W} = \int_{\mathcal{I}} k \left[k' \overset{(1)}{z} + \overset{(1)}{z}_1{}'' + (k^2 - \tau^2) \overset{(1)}{z}_1 - 2\tau \overset{(1)}{z}_2{}' - \tau' \overset{(1)}{z}_2 \right] ds, \tag{2.273}$$

wherefrom, similarly to the case in Theorem 2.42, we obtain (2.272). □

Directly from the previous theorem, in the case of the plane curve ($\tau = 0$), we obtain the next corollary.

Corollary 2.9. *Under second-order infinitesimal bending of the plane curve, the first variation of the Willmore energy is*

$$\delta \mathcal{W} = \int_{\mathcal{I}} \left(k'' + \frac{1}{2} k^3 \right) \overset{(1)}{z}_1\, ds + \int_{\mathcal{I}} \left[\frac{1}{2} k^2 \overset{(1)}{z} - k' \overset{(1)}{z}_1 + k \overset{(1)}{z}_1{}' \right]' ds.$$

Theorem 2.45. *Under second-order infinitesimal bending of the curve C, the second variation of the Willmore energy is*

$$\delta^2 \mathcal{W} = I_1 + I_2, \tag{2.274}$$

where

$$
\begin{aligned}
I_1 = \int_{\mathcal{I}} \frac{1}{2}\Big[& k'\overset{(1)}{z} + \overset{(1)}{z}_1{}'' + (k^2-\tau^2)\overset{(1)}{z}_1 - 2\tau\overset{(1)}{z}_2{}' - \tau'\overset{(1)}{z}_2\Big]^2 \\
& - \frac{3}{4}\Big(k\tau\overset{(1)}{z}_1 + k\overset{(1)}{z}_2{}'\Big)^2 + \frac{1}{2}\Big(k\tau\overset{(1)}{z} + 2\tau\overset{(1)}{z}_1{}' + \tau'\overset{(1)}{z}_1 \\
& + \overset{(1)}{z}_2{}'' - \tau^2\overset{(1)}{z}_2\Big)^2 - \frac{1}{4}\Big(k^2\overset{(1)}{z} + k\overset{(1)}{z}_1{}' - k\tau\overset{(1)}{z}_2\Big)^2 ds, \qquad (2.275) \\
I_2 = \int_{\mathcal{I}} \Big[& \Big(k'' + \frac{1}{2}k^3 - k\tau^2\Big)\overset{(2)}{z}_1 + (2k'\tau + k\tau')\overset{(2)}{z}_2\Big] ds \\
& + \int_{\mathcal{I}} \Big[\frac{1}{2}k^2\overset{(2)}{z} - k'\overset{(2)}{z}_1 + k\overset{(2)}{z}_1{}' - 2k\tau\overset{(2)}{z}_2\Big]' ds
\end{aligned}
$$

Proof. Based on Eqs. (2.226), (2.227) and (2.289) we get

$$
\begin{aligned}
\delta^2\mathcal{W} = \int_{\mathcal{I}} \Big\{ & \frac{1}{2}\Big[k'\overset{(1)}{z} + \overset{(1)}{z}_1{}'' + (k^2-\tau^2)\overset{(1)}{z}_1 - 2\tau\overset{(1)}{z}_2{}' - \tau'\overset{(1)}{z}_2\Big]^2 \\
& + k\Big[-\frac{k}{2}\Big(\tau\overset{(1)}{z}_1 + \overset{(1)}{z}_2{}'\Big)^2 + \frac{1}{2k}\Big(k\tau\overset{(1)}{z} + 2\tau\overset{(1)}{z}_1{}' + \tau'\overset{(1)}{z}_1 + \overset{(1)}{z}_2{}'' \\
& - \tau^2\overset{(1)}{z}_2\Big)^2 + \Big(k\overset{(2)}{z} + \overset{(2)}{z}_1{}' - \tau\overset{(2)}{z}_2\Big)' - \tau\Big(\overset{(2)}{z}_2{}' + \tau\overset{(2)}{z}_1\Big)\Big]\Big\} ds.
\end{aligned}
$$

We can express the previous equation as $\delta^2\mathcal{W} = I_1 + I_2$ where

$$
\begin{aligned}
I_1 = \int_{\mathcal{I}} \frac{1}{2}\Big\{ & \Big[k'\overset{(1)}{z} + \overset{(1)}{z}_1{}'' + (k^2-\tau^2)\overset{(1)}{z}_1 - 2\tau\overset{(1)}{z}_2{}' - \tau'\overset{(1)}{z}_2\Big]^2 \\
& - \Big(k\tau\overset{(1)}{z}_1 + k\overset{(1)}{z}_2{}'\Big)^2 + \Big(k\tau\overset{(1)}{z} + 2\tau\overset{(1)}{z}_1{}' + \tau'\overset{(1)}{z}_1 + \overset{(1)}{z}_2{}'' - \tau^2\overset{(1)}{z}_2\Big)^2\Big\} ds, \\
I_2 = \int_{\mathcal{I}} \Big[& k\Big(k\overset{(2)}{z} + \overset{(2)}{z}_1{}' - \tau\overset{(2)}{z}_2\Big)' - k\tau\Big(\overset{(2)}{z}_2{}' + \tau\overset{(2)}{z}_1\Big)\Big] ds.
\end{aligned}
$$

Since

$$
\begin{aligned}
k\Big(k\overset{(2)}{z} + \overset{(2)}{z}_1{}' - \tau\overset{(2)}{z}_2\Big)' = \Big[k\Big(k\overset{(2)}{z} + \overset{(2)}{z}_1{}' - \tau\overset{(2)}{z}_2\Big)\Big]' - kk'\overset{(2)}{z} - \Big(k'\overset{(2)}{z}_1\Big)' \\
+ k''\overset{(2)}{z}_1 + k'\tau\overset{(2)}{z}_2,
\end{aligned}
$$

and also

$$k\tau \overset{(2)}{z}_2{}' = \left(k\tau \overset{(2)}{z}_2\right)' - (k\tau)'\overset{(2)}{z}_2,$$

we obtain

$$\begin{aligned} I_2 = &\int_I \left[-kk'\overset{(2)}{z} + (k''-k\tau^2)\overset{(2)}{z}_1 + (2k'\tau + k\tau')\overset{(2)}{z}_2 \right] \\ &+ \int_I \left[k\left(k\overset{(2)}{z} + \overset{(2)}{z}_1{}' - \tau\overset{(2)}{z}_2\right) - k'\overset{(2)}{z}_1 - k\tau\overset{(2)}{z}_2 \right]' ds. \end{aligned}$$

Furthermore, it holds

$$kk'\overset{(2)}{z} = \left(\frac{1}{2}k^2\overset{(2)}{z}\right)' - \frac{1}{2}k^2\overset{(2)}{z}{}',$$

which, after using the conditions (2.46) gives

$$kk'\overset{(2)}{z} = \left(\frac{1}{2}k^2\overset{(2)}{z}\right)' - \frac{1}{2}k^3\overset{(2)}{z}_1 + \frac{1}{4}k^2\left[\left(k\overset{(1)}{z} + \overset{(1)}{z}_1{}' - \tau\overset{(1)}{z}_2\right)^2 + \left(\tau\overset{(1)}{z}_1 + \overset{(1)}{z}_2{}'\right)^2\right].$$

If we put the last summand from the previous equation into the expression for I_1, and the rest into I_2, we confirm the validity of Eqs. (2.274), (2.275) after a bit of calculation. □

Corollary 2.10. *Under second-order infinitesimal bending of the plane curve, the second variation of the Willmore energy is*

$$\delta^2\mathcal{W} = I_1 + I_2,$$

where

$$\begin{aligned} I_1 = &\int_I \frac{1}{2}\left[k'\overset{(1)}{z} + \overset{(1)}{z}_1{}'' + k^2\overset{(1)}{z}_1\right]^2 - \frac{3}{4}\left(k\overset{(1)}{z}_2{}'\right)^2 + \frac{1}{2}\left(\overset{(1)}{z}_2{}''\right)^2 \\ &- \frac{1}{4}\left(k^2\overset{(1)}{z} + k\overset{(1)}{z}_1{}'\right)^2 ds, \\ I_2 = &\int_I \left(k'' + \frac{1}{2}k^3\right)\overset{(2)}{z}_1\, ds + \int_I \left[\frac{1}{2}k^2\overset{(2)}{z} - k'\overset{(2)}{z}_1 + k\overset{(2)}{z}_1{}'\right]' ds. \end{aligned}$$

The previous theorems hold for both open and closed curves C and require only the computation of curvature and torsion of C, in addition to knowing the bending fields.

Remark 2.3. If we compare the expressions for the first variation and the part I_2 of the second variation of the Willmore energy, we conclude the Euler–Lagrange equations are the same in both cases. That means that the same governing equations for the curve minimize the integrals $\delta\mathcal{W}$ and I_2.

2.7.3 *Willmore Energy of Knots*

In order to obtain a family of closed curves in the case of knots, we keep in mind the condition $\mathbf{z}(0) = \mathbf{z}(L)$, $(\mathcal{I} = [0, L])$, for infinitesimal bending field, and also, suppose that the curve, as well as infinitesimal bending field are sufficiently smooth. We have the next theorems.

Theorem 2.46. *Under first-order infinitesimal bending of the knot C, the (first) variation of the Willmore energy is:*

$$\delta\mathcal{W} = \int_{\mathcal{I}} \left[\left(k'' + \frac{1}{2}k^3 - k\tau^2 \right) z_1 + (2k'\tau + k\tau')z_2 \right] ds,$$

i.e.

$$\delta\mathcal{W} = \int_{\mathcal{I}} (k\tau y_1 - k' y_2)\, ds,$$

where k and τ are the curvature and the torsion of C, respectively, z_1 and z_2 are components of the bending field $\mathbf{z}$, and y_1 and y_2 are components of the rotational field given by Eq. (2.34).

Theorem 2.47. *Under second-order infinitesimal bending of the knot C, the first and the second variation of the Willmore energy are:*

$$\delta\mathcal{W} = \int_I \left[\left(k'' + \frac{1}{2}k^3 - k\tau^2 \right) \overset{(1)}{z_1} + (2k'\tau + k\tau')\overset{(1)}{z_2} \right], ds$$

$$\delta^2\mathcal{W} = \int_I \frac{1}{2} \left[k'\overset{(1)}{z} + \overset{(1)}{z_1}{}'' + (k^2 - \tau^2)\overset{(1)}{z_1} - 2\tau \overset{(1)}{z_2}{}' - \tau'\overset{(1)}{z_2} \right]^2$$

$$- \frac{3}{4}\left(k\tau \overset{(1)}{z_1} + k\overset{(1)}{z_2}{}' \right)^2 + \frac{1}{2}\left(k\tau \overset{(1)}{z} + 2\tau \overset{(1)}{z_1}{}' + \tau'\overset{(1)}{z_1} + \overset{(1)}{z_2}{}'' - \tau^2 \overset{(1)}{z_2} \right)^2$$

$$-\frac{1}{4}\left(k^2 \overset{(1)}{z} + k\overset{(1)}{z_1}{}' - k\tau \overset{(1)}{z_2}\right)^2 + \left(k'' + \frac{1}{2}k^3 - k\tau^2\right)\overset{(2)}{z_1}$$

$$+ (2k'\tau + k\tau')\overset{(2)}{z_2}\Big]\, ds.$$

Willmore energy calculation on bent knots will be illustrated on knot examples under second-order infinitesimal bending field.

The first example is a trefoil knot given by the parametric equations: $x = \sin(u) + 2\sin(2u)$, $y = \cos(u) - 2\cos(2u)$, $z = -\sin(3u)$. We will give trefoil knot bent by the second-order infinitesimal bending, given by Theorem 2.7, for $p(u) = 0$, $q(u) = 0$, $r(u) = \cos(3u)$ and $g(u) = \sin(6u)$, see Figs. 2.16 and 2.17.

The numerically calculated Willmore energy of the second-order infinitesimally bent trefoil knot is 3.923 and 3.986 for $\epsilon = 0$ and $\epsilon = 1$ respectively.

A second example is a knot, see Figs. 2.18 and 2.19, given by the parametric equations: $x = (\cos(3u) + 2) \cdot \cos(2u)$, $y = (\cos(3u) + 2) \cdot \sin(2u)$, $z = -\sin(3u)$, which is a kind of torus knot, obtained for $p = 2$, $q = 3$, topologically equivalent to the trefoil knot.

Bending fields are defined by: $p(u) = \cos(6u)$ and $q(u) = \sin(6u)$.

The numerically calculated Willmore energy of the second-order infinitesimally bent $p = 2$, $q = 3$ torus knot is 5.1028, 5.3538 and 8.4124 for $\epsilon = 0$, $\epsilon = 0.6$ and $\epsilon = 1.2$ respectively.

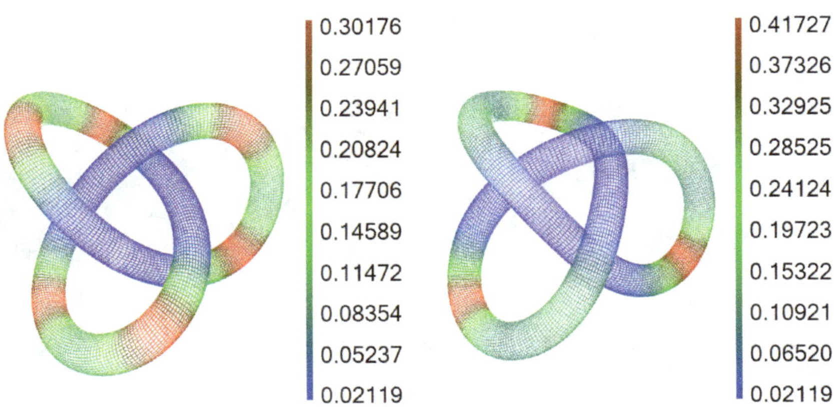

Fig. 2.16. Trefoil knot: basic and second-order inf. bent with $\epsilon = 1$.

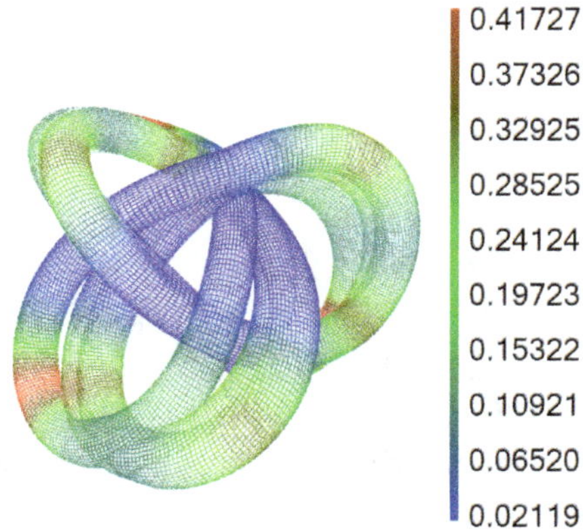

Fig. 2.17. Trefoil knot: both knots together.

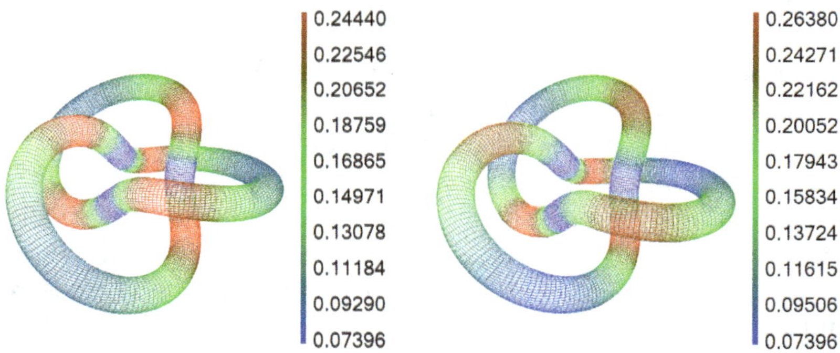

Fig. 2.18. $p = 2$, $q = 3$ torus knot: basic and second order inf. bent with $\epsilon = 0.6$

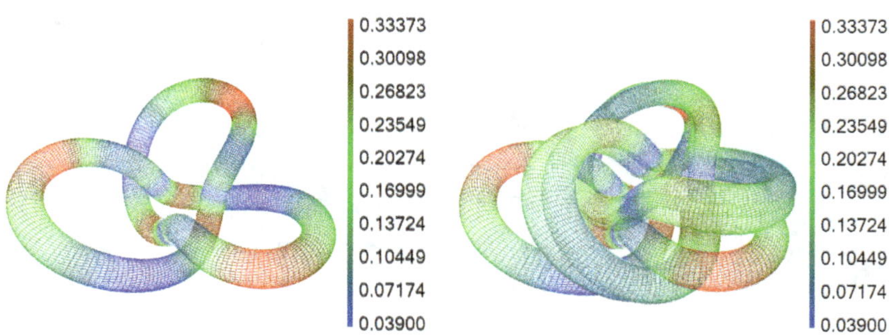

Fig. 2.19. $p = 2$, $q = 3$ torus knot: second order inf. bent with $\epsilon = 1.2$ and all knots together

2.8 Total Curvature

The total curvature of the curve $C : \mathbf{r} = \mathbf{r}(s)$, $s \in \mathcal{I}$, parameterized by arc length, is defined by the following equation:

$$\mathcal{K} = \int_{\mathcal{I}} k(s)\, ds. \tag{2.276}$$

We will examine the change of the total curvature according to [Najdanović *et al.* (2019)].

2.8.1 *The First Variation of the Total Curvature*

Let us determine the variation of the total curvature under infinitesimal bending of a curve. According to (2.168), it is valid

$$\begin{aligned}\delta\mathcal{K} &= \lim_{\epsilon\to 0} \frac{\mathcal{K}_\epsilon - \mathcal{K}}{\epsilon} = \lim_{\epsilon\to 0} \frac{\int_{\mathcal{I}} k_\epsilon\, ds_\epsilon - \int_{\mathcal{I}} k\, ds}{\epsilon} \\ &= \lim_{\epsilon\to 0} \frac{\int_{\mathcal{I}} (k + \epsilon\delta k)(ds + \epsilon\delta(ds)) - \int_{\mathcal{I}} k\, ds}{\epsilon}.\end{aligned}$$

Since $\delta(ds) = 0$, we have

$$\delta\mathcal{K} = \int_{\mathcal{I}} \delta k\, ds. \tag{2.277}$$

Theorem 2.48. *Under infinitesimal bending of the curve C, the variation of its total curvature is given by*

$$\delta\mathcal{K} = \int_{\mathcal{I}} (-\tau^2 z_1 + \tau' z_2) ds + \int_{\mathcal{I}} (kz + z_1' - 2\tau z_2)' ds, \tag{2.278}$$

where k and τ are the curvature and the torsion of C, respectively.

Proof. Using (2.181), we have that the variation of the total curvature is

$$\delta\mathcal{K} = \int_{\mathcal{I}} [k'z + z_1'' + (k^2 - \tau^2) z_1 - 2\tau z_2' - \tau' z_2]\, ds. \tag{2.279}$$

The condition (2.25) holds under infinitesimal bending, i.e. $z' = kz_1$. Using this fact, we get

$$k'z = (kz)' - kz' = (kz)' - k^2 z_1. \tag{2.280}$$

Also, it is valid

$$\tau z_2' = (\tau z_2)' - \tau' z_2. \tag{2.281}$$

If we substitute the previous two equations into (2.279), we obtain (2.278). □

In terms of the rotational field $\mathbf{y}$, we have:

Theorem 2.49. *Under infinitesimal bending of the curve C, the variation of its total curvature is given by*

$$\delta\mathcal{K} = \int_{\mathcal{I}} y_1 \tau \, ds + \int_{\mathcal{I}} y_2' \, ds, \tag{2.282}$$

τ is the torsion of C, and y_1 and y_2 are the components of the rotational field given by Eq. (2.34).

2.8.2 The Second Variation of the Total Curvature

The total curvature of deformed curve under second-order infinitesimal bending is

$$\mathcal{K}_\epsilon = \int_{\mathcal{I}} k_\epsilon \, ds_\epsilon = \int_{\mathcal{I}} (k + \epsilon\,\delta k + \epsilon^2\,\delta^2 k)(ds + \epsilon\,\delta\,ds + \epsilon^2\,\delta^2\,ds).$$

Since $\delta\,ds = \delta^2 ds = 0$, we have

$$\mathcal{K}_\epsilon = \mathcal{K} + \epsilon \int_{\mathcal{I}} k\,\delta k\,ds + \epsilon^2 \int_{\mathcal{I}} \delta^2 k\,ds,$$

which is obtained after neglecting the terms with ϵ^3 and ϵ^4. Therefore,

$$\delta\mathcal{K} = \int_{\mathcal{I}} \delta k\,ds, \delta^2\mathcal{K} = \int_{\mathcal{I}} \delta^2 k\,ds. \tag{2.283}$$

Theorem 2.50. *Under second-order infinitesimal bending of the curve C, the first variation of the total curvature is*

$$\delta\mathcal{K} = \int_{\mathcal{I}} \Big(-\tau^2 \overset{(1)}{z_1} + \tau' \overset{(1)}{z_2} \Big) ds + \int_{\mathcal{I}} \Big(k \overset{(1)}{z} + \overset{(1)}{z_1}{}' - 2\tau \overset{(1)}{z_2} \Big)' ds, \tag{2.284}$$

where k and τ are the curvature and the torsion of C, respectively.

Proof. Using (2.226) we obtain (2.284) after a bit of calculation. □

Remark 2.4. From Eq. (2.284) we can see that the second summand is an integral of a total derivative. Also, under the constraint that the vector field $\mathbf{z}$ has support in the interior of $\mathcal{I}$, we obtain the Euler-Lagrange equations $-\tau^2 = 0$, $\tau' = 0$, that minimize the integral (2.284) for the curves. Obviously, $\tau = 0$, so the first variation of the total curvature of a plane curve vanishes in that case.

Theorem 2.51. *Under second-order infinitesimal bending of the curve C, the second variation of the total curvature is*

$$\delta^2\mathcal{K} = \mathcal{I}_1 + \mathcal{I}_2, \tag{2.285}$$

where

$$\begin{aligned}\mathcal{I}_1 = \int_{\mathcal{I}} \Bigg[&-\frac{k}{2}\left(\tau \overset{(1)}{z}_1 + \overset{(1)}{z}_2{}'\right)^2 + \frac{1}{2k}\Bigg(k\tau \overset{(1)}{z} + 2\tau \overset{(1)}{z}_1{}' + \tau' \overset{(1)}{z}_1 \\ &+ \overset{(1)}{z}_2{}'' - \tau^2 \overset{(1)}{z}_2\Bigg)^2\Bigg]\, ds,\end{aligned} \tag{2.286}$$

$$\mathcal{I}_2 = \int_{\mathcal{I}} \left(-\tau^2 \overset{(2)}{z}_1 + \tau' \overset{(2)}{z}_2\right) ds + \int_{\mathcal{I}} \left(k \overset{(2)}{z} + \overset{(2)}{z}_1{}' - \tau \overset{(2)}{z}_2\right)' ds.$$

Proof. Based on Eqs. (2.227) and (2.283), we get

$$\begin{aligned}\delta^2\mathcal{K} = \int_{\mathcal{I}} \Bigg[&-\frac{k}{2}\left(\tau \overset{(1)}{z}_1 + \overset{(1)}{z}_2{}'\right)^2 + \frac{1}{2k}\Bigg(k\tau \overset{(1)}{z} + 2\tau \overset{(1)}{z}_1{}' + \tau' \overset{(1)}{z}_1 + \overset{(1)}{z}_2{}'' \\ &- \tau^2 \overset{(1)}{z}_2\Bigg)^2 + \left(k \overset{(2)}{z} + \overset{(2)}{z}_1{}' - \tau \overset{(2)}{z}_2\right)' - \tau\left(\overset{(2)}{z}_2{}' + \tau \overset{(2)}{z}_1\right)\Bigg]\, ds.\end{aligned}$$

We can express the previous equation as $\delta^2\mathcal{K} = \mathcal{I}_1 + \mathcal{I}_2$ where

$$\begin{aligned}\mathcal{I}_1 = \int_{\mathcal{I}} \Bigg[&-\frac{k}{2}\left(\tau \overset{(1)}{z}_1 + \overset{(1)}{z}_2{}'\right)^2 + \frac{1}{2k}\Bigg(k\tau \overset{(1)}{z} + 2\tau \overset{(1)}{z}_1{}' \\ &+ \tau' \overset{(1)}{z}_1 + \overset{(1)}{z}_2{}'' - \tau^2 \overset{(1)}{z}_2\Bigg)^2\Bigg]\, ds,\end{aligned}$$

$$\mathcal{I}_2 = \int_{\mathcal{I}} \left[\left(k \overset{(2)}{z} + \overset{(2)}{z_1}{}' - \tau \overset{(2)}{z_2} \right)' - \tau \left(\overset{(2)}{z_2}{}' + \tau \overset{(2)}{z_1} \right) \right] ds.$$

Since

$$\tau \overset{(2)}{z_2}{}' = \left(\tau \overset{(2)}{z_2} \right)' - \tau' \overset{(2)}{z_2}$$

we obtain

$$\mathcal{I}_2 = \int_{\mathcal{I}} \left(-\tau^2 \overset{(2)}{z_1} + \tau' \overset{(2)}{z_2} \right) ds + \int_{\mathcal{I}} \left(k \overset{(2)}{z} + \overset{(2)}{z_1}{}' - \tau \overset{(2)}{z_2} \right)' ds.$$

□

Remark 2.5. If we compare the expressions for the first variation and the part $\mathcal{I}_2$ of the second variation of the total curvature, we conclude the Euler–Lagrange equations $-\tau^2 = 0$, $\tau' = 0$ are the same in both cases. That means that the same governing equations, which determine a plane curve, minimize the integrals $\delta\mathcal{K}$ and $\mathcal{I}_2$.

In the case of a closed plane curve with conditions $\overset{(1)}{\mathbf{z}}(0) = \overset{(1)}{\mathbf{z}}(L)$, $\overset{(2)}{\mathbf{z}}(0) = \overset{(2)}{\mathbf{z}}(L)$, $(\mathcal{I} = [0, L])$ we obtain

$$\delta^2\mathcal{K} = \int_{\mathcal{I}} \frac{1}{2k} \left[\left(\overset{(1)}{z_2}{}'' \right)^2 - k^2 \left(\overset{(1)}{z_2}{}' \right)^2 \right] ds,$$

and we conclude that $\delta\mathcal{K} = \delta^2\mathcal{K} = 0$ for any infinitesimal bending field of the second order and any infinitesimal bending field of the first order whose component $\overset{(1)}{z_2}$ is a constant.

2.8.3 *Total Curvature of Knots*

Theorem 2.52. *Under first-order infinitesimal bending of the knot C, variation of its total curvature is*

$$\delta\mathcal{K} = \int_{\mathcal{I}} (-\tau^2 z_1 + \tau' z_2) ds,$$

i.e.

$$\delta\mathcal{K} = \int_{\mathcal{I}} \tau y_1 \, ds,$$

where τ is the torsion of C, z_1 and z_2 are components of the bending field, and y_1 is a component of the rotational field given in Eq. (2.34).

Theorem 2.53. *Under second-order infinitesimal bending of the knot C, the first and the second variation of its total curvature are respectively*

$$\delta\mathcal{K} = \int_{\mathcal{I}} \Big(-\tau^2 \overset{(1)}{z}_1 + \tau' \overset{(1)}{z}_2 \Big) ds,$$

$$\delta^2\mathcal{K} = \int_{\mathcal{I}} \Bigg[-\frac{k}{2}\Big(\tau \overset{(1)}{z}_1 + \overset{(1)}{z}_2{}'\Big)^2 + \frac{1}{2k}\Big(k\tau \overset{(1)}{z} + 2\tau \overset{(1)}{z}_1{}' + \tau' \overset{(1)}{z}_1 + \overset{(1)}{z}_2{}'' - \tau^2 \overset{(1)}{z}_2\Big)^2\Bigg] ds, + \int_{\mathcal{I}} \Big(-\tau^2 \overset{(2)}{z}_1 + \tau' \overset{(2)}{z}_2 \Big) ds,$$

where k is the curvature and τ is the torsion of C.

Trefoil knot

A trefoil knot, see Figs. 2.20–2.23, is given by the parametric equations: $x = \sin(u) + 2\sin(2u)$, $y = \cos(u) - 2\cos(2u)$, $z = -\sin(3u)$. We will give the trefoil knot infinitesimal bending examples of the first and second order. An example of the first-order infinitesimally bent trefoil knot colored by local value of curvature is given in Figs. 2.20 and 2.21. The bending fields are defined by: $p(u) = \cos(3u)$ and $q(u) = \sin(6u)$.

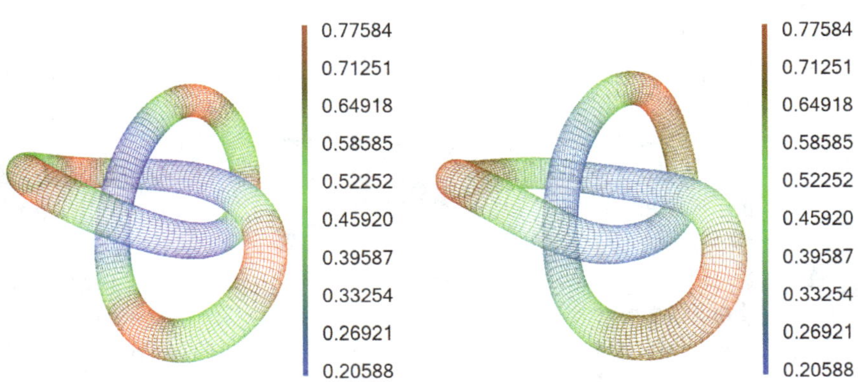

Fig. 2.20. Trefoil knot: basic and first order inf. bent with $\epsilon = 0.5$

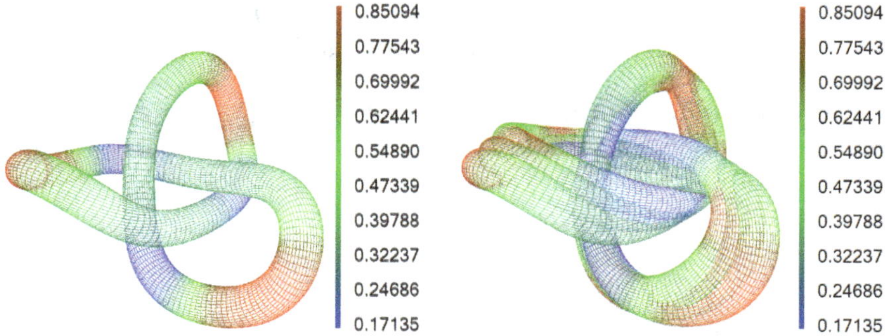

Fig. 2.21. Trefoil knot: first order inf. bent with $\epsilon = 1.$ and all knots together.

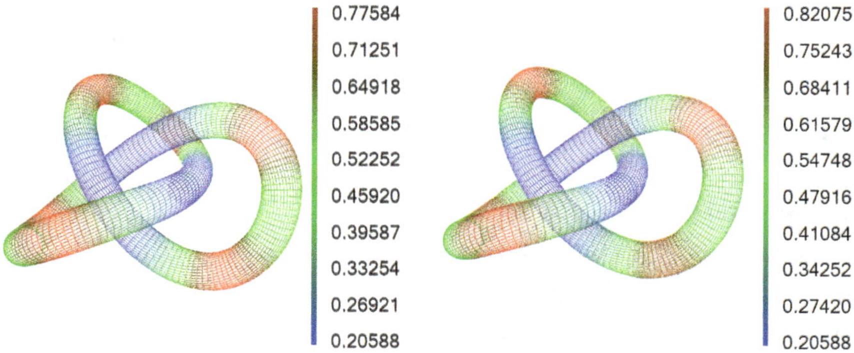

Fig. 2.22. Trefoil knot: basic and second order inf. bent with $\epsilon = 0.5$

The numerically calculated total curvature of the first-order infinitesimally bent trefoil knot curve is 13.950, 14.318, and 15.302 for $\epsilon = 0$, $\epsilon = 0.5$, and $\epsilon = 1$ respectively.

An example of the second-order infinitesimally bent trefoil knot is given in Figs. 2.22 and 2.23. The bending fields are defined by: $p(u) = 0$ and $q(u) = 0$, $r(u) = \cos(3u)$ and $g(u) = \sin(3u)$.

The numerically calculated total curvature of the second-order infinitesimally bent trefoil knot is 13.971 and 14.266 for $\epsilon = 0.5$ and $\epsilon = 1$ respectively.

An interesting example of the second-order bending field for trefoil knot is a field defined by $p(u) = 0$ and $q(u) = 0$, $r(u) = 0$ and $g(u) = -\frac{z'}{\mathbf{n}_2 z}$, where z' is the first derivative of the z trefoil component and $\mathbf{n}_2 z$ is z component of the unit binormal vector. For that bending field

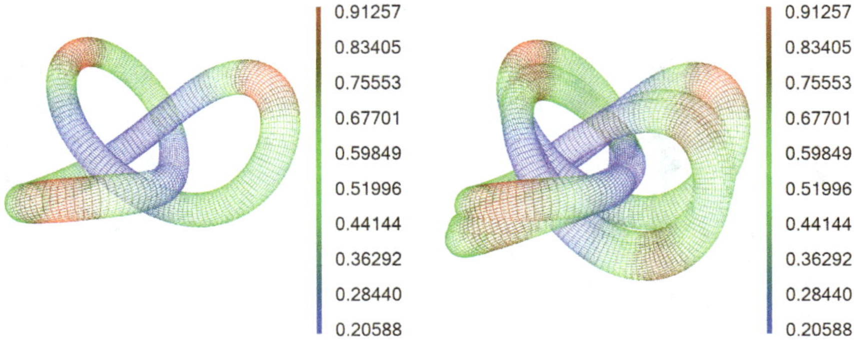

Fig. 2.23. Trefoil knot: second order inf. bent with $\epsilon = 1.$ and all knots together.

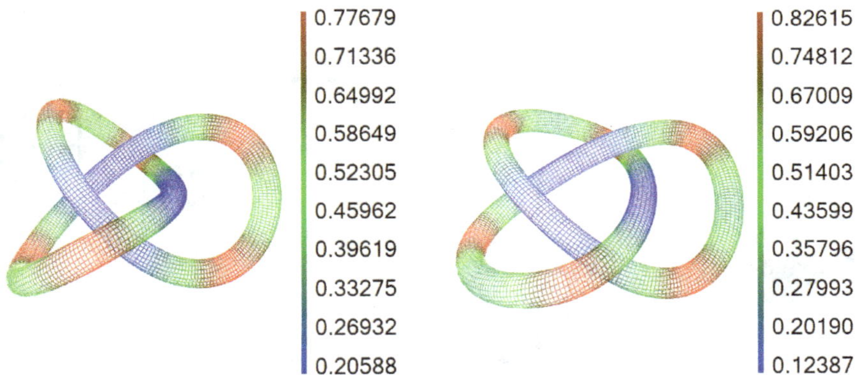

Fig. 2.24. Trefoil knot: basic and second order inf. bent with $\epsilon = 0.6$

bent curve is flattened, see Figs. 2.24 and 2.25 and value of the total curvature is decreasing. The numerically calculated total curvature of the second-order infinitesimally bent trefoil knot is 13.066, 12.791, and 12.631 for $\epsilon = 0.6$, $\epsilon = 0.8$ and $\epsilon = 0.99$ respectively.

In Figs. 2.26–2.27, we have $p3q2$ torus knot given by the parametric equations: $x = (\cos(2u) + 2) \cdot \cos(3u)$, $y = (\cos(2u) + 2) \cdot \sin(3u)$, $z = -\sin(2u)$.

Bending fields are defined by: $p(u) = 0$ and $q(u) = 0$, $r(u) = -\cos(6u)$ and $g(u) = \sin(6u)$.

Numerically calculated total curvature of the $p = 3$, $q = 2$ torus knot is 18.6791, 19.02214 and 22.4726 for $\epsilon = 0$, $\epsilon = 0.8$, and $\epsilon = 1.6$, respectively.

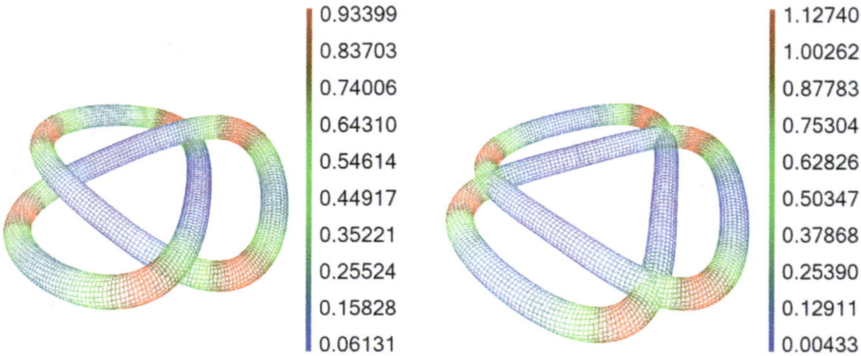

Fig. 2.25. Trefoil knot: second order inf. bent with $\epsilon = 0.8$. and $\epsilon = 0.99$.

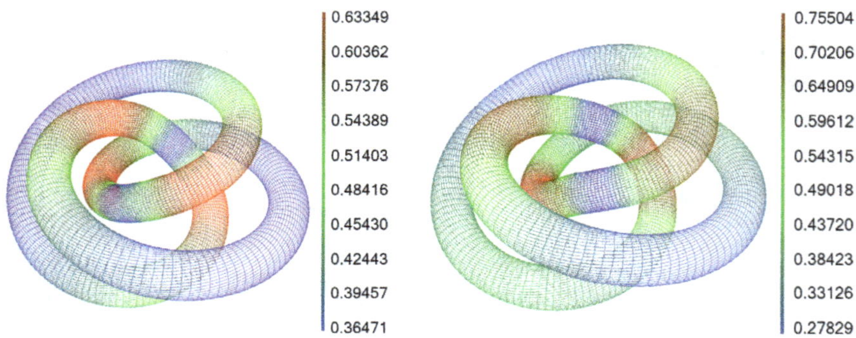

Fig. 2.26. $p = 3$, $q = 2$ torus knot: basic and infinitesimally bent with $\epsilon = 0.8$.

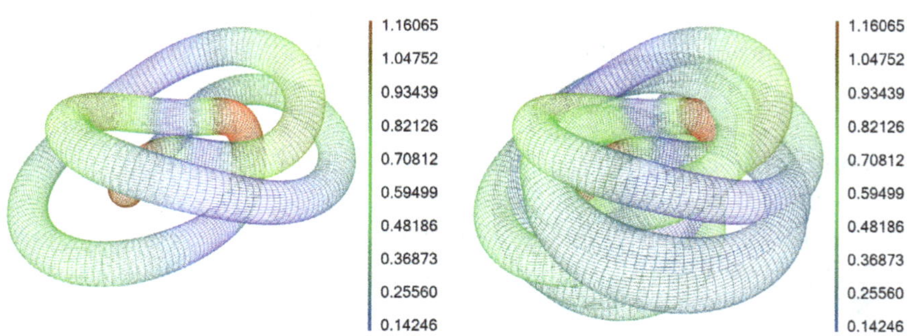

Fig. 2.27. $p = 3$, $q = 2$ torus knot: infinitesimally bent with $\epsilon = 1.6$ and all knots together.

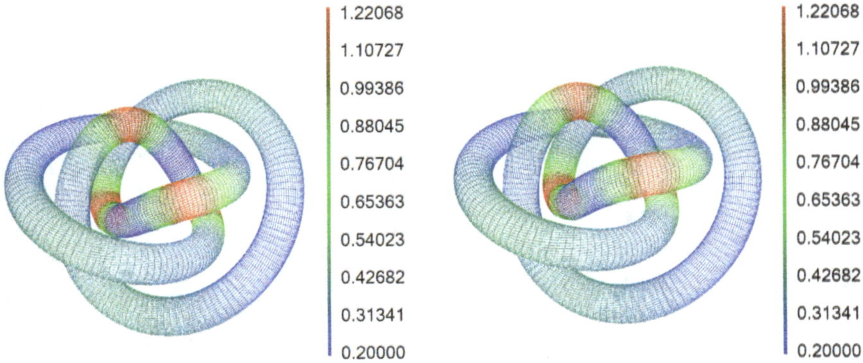

Fig. 2.28. Figure eight knot: basic and infinitesimally bent with $\epsilon = 0.75$.

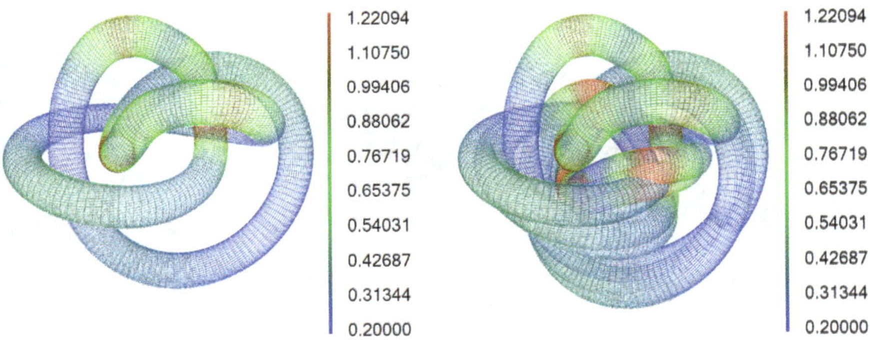

Fig. 2.29. Figure eight knot: infinitesimally bent with $\epsilon = 1.5$ and all knots together.

Figure eight knot

A figure eight knot is given by the parametric equations: $x = (2 + \cos(2u)) \cdot \cos(3u)$, $y = (2 + \cos(2u)) \cdot \sin(3u)$, $z = \sin(4u)$, see Figs. 2.28 and 2.29.

The bending fields are defined by: $p(u) = 0$ and $q(u) = 0$, $r(u) = -\cos(2u)$ and $g(u) = \sin(2u)$.

The numerically calculated total curvature of the figure eight knot is 20.7670, 20.8725, and 21.2715 for $\epsilon = 0$, $\epsilon = 0.75$ and $\epsilon = 1.5$, respectively.

2.9 Torsional Energy

The torsional energy of the curve $C : \mathbf{r} = \mathbf{r}(s)$, $s \in \mathcal{I}$ is

$$\mathcal{E}_\tau = \frac{1}{2} \int_{\mathcal{I}} \tau^2(s)\, ds. \tag{2.287}$$

We will examine the change of the torsional energy according to [Maksimivić *et al.* (2023); Rančić *et al.* (2024)].

2.9.1 *The First Variation of the Torsional Energy*

Theorem 2.54. *Under infinitesimal bending of the curve C, the variation of its torsional energy is:*

$$\begin{aligned}\delta\mathcal{E}_\tau = &\int_{\mathcal{I}} \left\{ \left[\frac{3}{2} k\tau^2 + \frac{\tau'^2}{k} + 2\tau \left(\frac{\tau'}{k}\right)' \right] z_1 \right.\\ &+ \left. \left[\frac{\tau'\tau^2}{k} - (k\tau)' - \left(\frac{\tau'}{k}\right)'' \right] z_2 \right\} ds\\ &+ \int_{\mathcal{I}} \left\{ \frac{\tau}{k} \left(2\tau z_1{}' + \tau' z_1 + z_2{}'' + (k^2 - \tau^2) z_2 \right) \right.\\ &- \left. \frac{2\tau\tau' z_1}{k} - \frac{\tau'}{k} z_2{}' + \left(\frac{\tau'}{k}\right)' z_2 + \frac{1}{2}\tau^2 z \right\}' ds,\end{aligned} \tag{2.288}$$

where k is the curvature and τ is the torsion of C.

Proof. During bending, the torsional energy changes and has the form

$$\mathcal{E}_{\tau_\epsilon} = \mathcal{E}_\tau + \epsilon \int_{\mathcal{I}} \tau\, \delta\tau\, ds,$$

after neglecting the terms with ϵ^k, $k \geq 2$. Therefore,

$$\delta\mathcal{E}_\tau = \int_{\mathcal{I}} \tau \delta\tau\, ds. \tag{2.289}$$

Using Eqs. (2.25) and (2.183) and applying the following transformations:

$$\frac{2\tau\tau'}{k}z_1{}' = \left(\frac{2\tau\tau'}{k}z_1\right)' - \left(\frac{2\tau\tau'}{k}\right)' z_1,$$

$$\frac{\tau'}{k}z_2{}'' = \left[\frac{\tau'}{k}z_2{}' - \left(\frac{\tau'}{k}\right)' z_2\right]' + \left(\frac{\tau'}{k}\right)'' z_2,$$

$$\tau\tau' z = \left(\frac{1}{2}\tau^2\right)' z = \left(\frac{1}{2}\tau^2 z\right)' - \frac{1}{2}\tau^2 z' = \left(\frac{1}{2}\tau^2 z\right)' - \frac{1}{2}\tau^2 k z_1,$$

we get Eq. (2.288). □

In terms of the rotational field $\mathbf{y}$ we have:

Theorem 2.55. *Under infinitesimal bending of the curve C, the variation of its torsional energy is given by*

$$\delta\mathcal{E}_\tau = -\int_{\mathcal{I}} \left\{ \left[\left(\frac{\tau'}{k}\right)' + k\tau \right] y_1 + \frac{\tau'\tau}{k} y_2 \right\} ds + \int_{\mathcal{I}} \left\{ \tau \left[\frac{1}{k}(y_2\tau - y_1') \right] + \frac{\tau'}{k} y_1 \right\} ds, \tag{2.290}$$

k and τ are the curvature and the torsion of C, respectively, y_1 and y_2 are the components of the rotational field given by Eq. (2.34).

2.9.2 *The Second Variation of the Torsional Energy*

During bending, the torsional energy changes and has the form

$$\mathcal{E}_{\tau\epsilon} = \frac{1}{2}\int_{\mathcal{I}} \tau_\epsilon^2 \, ds_\epsilon = \frac{1}{2}\int_{\mathcal{I}} (\tau + \epsilon\,\delta\tau + \epsilon^2\,\delta^2\tau)^2 (ds + \epsilon\,\delta\, ds + \epsilon^2\,\delta^2\, ds).$$

Since the variation of the line element is $\delta\, ds = \delta^2 ds = 0$, we have

$$\mathcal{E}_{\tau_\epsilon} = \mathcal{E}_\tau + \epsilon \int_{\mathcal{I}} \tau\,\delta\tau\, ds + \frac{1}{2}\epsilon^2 \int_{\mathcal{I}} [(\delta\tau)^2 + 2\tau\delta^2\tau]\, ds,$$

which is obtained after neglecting the terms with ϵ^3 and ϵ^4. Therefore,

$$\delta\mathcal{E}_\tau = \int_{\mathcal{I}} \tau\delta\tau\, ds, \quad \delta^2\mathcal{E}_\tau = \frac{1}{2}\int_{\mathcal{I}} [(\delta\tau)^2 + 2\tau\delta^2\tau]\, ds. \tag{2.291}$$

Theorem 2.56. *Under second-order infinitesimal bending of the curve C, the first variation of the torsional energy is:*

$$\begin{aligned}
\delta\mathcal{E}_\tau = &\int_{\mathcal{I}} \left\{ \left[\frac{3}{2}k\tau^2 + \frac{\tau'^2}{k} + 2\tau\left(\frac{\tau'}{k}\right)'\right] \overset{(1)}{z}_1 \right.\\
&\left. + \left[\frac{\tau'\tau^2}{k} - (k\tau)' - \left(\frac{\tau'}{k}\right)''\right] \overset{(1)}{z}_2 \right\} ds\\
&+ \int_{\mathcal{I}} \left\{ \frac{\tau}{k}\left(2\tau\overset{(1)}{z}_1{}' + \tau'\overset{(1)}{z}_1 + \overset{(1)}{z}_2{}'' + (k^2-\tau^2)\overset{(1)}{z}_2\right) - \frac{2\tau\tau'\overset{(1)}{z}_1}{k} \right.\\
&\left. - \frac{\tau'}{k}\overset{(1)}{z}_2{}' + \left(\frac{\tau'}{k}\right)'\overset{(1)}{z}_2 + \frac{1}{2}\tau^2\overset{(1)}{z} \right\}' ds,
\end{aligned} \tag{2.292}$$

where k is the curvature and τ is the torsion of C.

Theorem 2.57. *Under second-order infinitesimal bending of the curve C, the second variation of the torsional energy is:*

$$\delta^2\mathcal{E}_\tau = I_1 + I_2 + \int_{\mathcal{I}} \left\{ \tau c - \frac{\tau'}{k}\tau\overset{(2)}{z}_1 - \frac{\tau'\tau}{k}\overset{(2)}{z}_1 - \frac{\tau'}{k}\overset{(2)}{z}_2 + \left(\frac{\tau'}{k}\right)'\overset{(2)}{z}_2 \right\}' ds, \tag{2.293}$$

$$\begin{aligned}
I_1 = &\int_{\mathcal{I}} \left\{ \frac{1}{2}\left\{ \tau'\overset{(1)}{z} + 2k\tau\overset{(1)}{z}_1 - k'\overset{(1)}{z}_2 \right.\right.\\
&\left. + \left[\frac{1}{k}\left(2\tau\overset{(1)}{z}_1{}' + \tau'\overset{(1)}{z}_1 + \overset{(1)}{z}_2{}'' + (k^2-\tau^2)\overset{(1)}{z}_2\right)\right]' \right\}^2\\
&+ \frac{\tau'}{k^2}(k'\overset{(1)}{z} + \overset{(1)}{z}_1{}'' + (k^2-\tau^2)\overset{(1)}{z}_1 - 2\tau\overset{(1)}{z}_2{}' - \tau'\overset{(1)}{z}_2)(k\tau\overset{(1)}{z}\\
&\left. + 2\tau\overset{(1)}{z}_1{}' + \tau'\overset{(1)}{z}_1 + \overset{(1)}{z}_2{}'' - \tau^2\overset{(1)}{z}_2) + \tau f(\overset{(1)}{z}) \right\} ds,
\end{aligned} \tag{2.294}$$

$$I_2 = \int_{\mathcal{I}} \left\{ \left[\frac{3}{2} k\tau^2 + \frac{\tau'^2}{k} + 2\left(\frac{\tau'}{k}\right)' \tau \right] \overset{(2)}{z_1} + \left[\frac{\tau'\tau^2}{k} - \left(\frac{\tau'}{k}\right)'' - (k\tau)' \right] \overset{(2)}{z_2} \right\} ds, \tag{2.295}$$

$$f(\overset{(1)}{\mathbf{z}}) = -\tau\Big(k\overset{(1)}{z} + \overset{(1)}{z_1}{}' - \tau\overset{(1)}{z_2}\Big)^2 + \Big(k'\overset{(1)}{z} + \overset{(1)}{z_1}{}'' + (k^2-\tau^2)\overset{(1)}{z_1} - 2\tau\overset{(1)}{z_2}{}' - \tau'\overset{(1)}{z_2}\Big)\Big(\overset{(1)}{z_2}{}' + \tau\overset{(1)}{z_1}\Big), \tag{2.296}$$

$$c = \frac{1}{k}\left[\tau\Big(k\overset{(2)}{z} + \overset{(2)}{z_1}{}' - \tau\overset{(2)}{z_2}\Big) + \Big(\overset{(2)}{z_2}{}' + \tau\overset{(2)}{z_1}\Big)' - \frac{1}{k}\Big(k'\overset{(1)}{z} + \overset{(1)}{z_1}{}'' + (k^2-\tau^2)\overset{(1)}{z_1} - 2\tau\overset{(1)}{z_2}{}' - \tau'\overset{(1)}{z_2}\Big)\Big(k\tau\overset{(1)}{z} + 2\tau\overset{(1)}{z_1}{}' + \tau'\overset{(1)}{z_1} + \overset{(1)}{z_2}{}'' - \tau^2\overset{(1)}{z_2}\Big)\right] - \tau\overset{(2)}{z} + k\overset{(2)}{z_2}, \tag{2.297}$$

k is the curvature and τ is the torsion of C.

Proof. Based on Eq. (2.233), the second variation of the torsion is

$$\delta^2\tau = \left\{\frac{1}{k}\left[\tau\Big(k\overset{(2)}{z} + \overset{(2)}{z_1}{}' - \tau\overset{(2)}{z_2}\Big) + \Big(\overset{(2)}{z_2}{}' + \tau\overset{(2)}{z_1}\Big)' - \frac{1}{k}\Big(k'\overset{(1)}{z} + \overset{(1)}{z_1}{}'' + (k^2-\tau^2)\overset{(1)}{z_1} - 2\tau\overset{(1)}{z_2}{}' - \tau'\overset{(1)}{z_2}\Big)\Big(k\tau\overset{(1)}{z} + 2\tau\overset{(1)}{z_1}{}' + \tau'\overset{(1)}{z_1} + \overset{(1)}{z_2}{}'' - \tau^2\overset{(1)}{z_2}\Big)\right] - \tau\overset{(2)}{z} + k\overset{(2)}{z_2}\right\}' + \tau'\overset{(2)}{z} + 2k\tau\overset{(2)}{z_1} - k'\overset{(2)}{z_2} + f\Big(\overset{(1)}{\mathbf{z}}\Big), \tag{2.298}$$

where $f(\overset{(1)}{\mathbf{z}})$ is given by Eq. (2.296).

Applying Eqs. (2.232) and (2.298) into Eq. (2.291), we obtain

$$\delta^2 E_\tau = \int_{\mathcal{I}} \frac{1}{2}\left\{\tau'\overset{(1)}{z} + 2k\tau\overset{(1)}{z_1} - k'\overset{(1)}{z_2} + \left[\frac{1}{k}\Big(2\tau\overset{(1)}{z_1}{}' + \tau'\overset{(1)}{z_1} + \overset{(1)}{z_2}{}'' + (k^2-\tau^2)\overset{(1)}{z_2}\Big)\right]'\right\}^2 ds$$
$$+ \int_{\mathcal{I}} \left\{(\tau c)' - \frac{\tau'}{k}\left[\tau\Big(k\overset{(2)}{z} + \overset{(2)}{z_1}{}' - \tau\overset{(2)}{z_2}\Big) + \Big(\overset{(2)}{z_2}{}' + \tau\overset{(2)}{z_1}\Big)'\right.\right.$$

$$
\begin{aligned}
&-\frac{1}{k}\Big(k'\overset{(1)}{z} + \overset{(1)}{z}_1{}'' + (k^2-\tau^2)\overset{(1)}{z}_1 - 2\tau\overset{(1)}{z}_2{}' - \tau'\overset{(1)}{z}_2\Big)\\
&\times\Big(k\tau\overset{(1)}{z} + 2\tau\overset{(1)}{z}_1{}' + \tau'\overset{(1)}{z}_1 + \overset{(1)}{z}_2{}'' - \tau^2\overset{(1)}{z}_2\Big)\Big] + \tau\tau'\overset{(2)}{z} - k\tau'\overset{(2)}{z}_2\\
&+\tau\tau'\overset{(2)}{z} + 2k\tau^2\overset{(2)}{z}_1 - k'\tau\overset{(2)}{z}_2 + \tau f\Big(\overset{(1)}{\mathbf{z}}\Big)\Big\}\, ds,
\end{aligned}
$$

where we denote by c the following expression:

$$
\begin{aligned}
c = \frac{1}{k}\Big[&\tau\Big(k\overset{(2)}{z} + \overset{(2)}{z}_1{}' - \tau\overset{(2)}{z}_2\Big) + \Big(\overset{(2)}{z}_2{}' + \tau\overset{(2)}{z}_1\Big)' - \frac{1}{k}\Big(k'\overset{(1)}{z} + \overset{(1)}{z}_1{}''\\
&+ (k^2-\tau^2)\overset{(1)}{z}_1 - 2\tau\overset{(1)}{z}_2{}' - \tau'\overset{(1)}{z}_2\Big)\Big(k\tau\overset{(1)}{z} + 2\tau\overset{(1)}{z}_1{}' + \tau'\overset{(1)}{z}_1 + \overset{(1)}{z}_2{}''\\
&- \tau^2\overset{(1)}{z}_2\Big)\Big] - \tau\overset{(2)}{z} + k\overset{(2)}{z}_2.
\end{aligned}
$$

After several transformations, using condition (2.46) and extracting the terms containing the first-order bending field separately from the terms containing the second-order bending field, as well as the expression which is the total differential, we obtain the expression (2.293) for the second order variation of the torsional energy. □

Finally, for plane curves we have the following corollary.

Corollary 2.11. *The first and second variation of the torsional energy of a plane curve (not necessarily closed) are respectively:*

$$\delta\mathcal{E}_\tau = 0,$$

$$\delta^2\mathcal{E}_\tau = \int_{\mathcal{I}} \frac{1}{2}\left\{-k'\overset{(1)}{z}_2 + \left[\frac{1}{k}\left(\overset{(1)}{z}_2{}'' + k^2\overset{(1)}{z}_2\right)\right]'\right\}^2 ds.$$

We see that, unlike the first variation, the second variation of the torsional energy for a plane curve is non-zero in the general case. In particular, for $\overset{(1)}{z}_2 = 0$, the second variation is zero for any plane curve. In that case, we determine the other components of the fields $\overset{(1)}{\mathbf{z}}$ and $\overset{(2)}{\mathbf{z}}$ from the equations: $\overset{(1)}{z}{}' - k\overset{(1)}{z}_1 = 0$, $\overset{(2)}{z}{}' - k\overset{(2)}{z}_1 = -\frac{1}{2}(k\overset{(1)}{z} + \overset{(1)}{z}_1{}')^2$.

2.9.3 *Torsional Energy of Knots*

Theorem 2.58. *Under first-order infinitesimal bending of the knot C, the variation of the torsional energy is:*

$$\delta\mathcal{E}_\tau = \int_{\mathcal{I}} \left\{ \left[\frac{3}{2}k\tau^2 + \frac{\tau'^2}{k} + 2\tau\left(\frac{\tau'}{k}\right)' \right] z_1 + \left[\frac{\tau'\tau^2}{k} - (k\tau)' - \left(\frac{\tau'}{k}\right)'' \right] z_2 \right\} ds,$$

i.e.

$$\delta\mathcal{E}_\tau = -\int_{\mathcal{I}} \left\{ \left[\left(\frac{\tau'}{k}\right)' + k\tau \right] y_1 + \frac{\tau'\tau}{k} y_2 \right\} ds, \tag{2.299}$$

where k is the curvature, τ is the torsion, z_1 and z_2 are the components of the bending field, and y_1 and y_2 are the components of the rotational field given by Eq. (2.34).

Theorem 2.59. *Under second-order infinitesimal bending of the knot C, the first and the second variations of the torsional energy are respectively:*

$$\delta\mathcal{E}_\tau = \int_{\mathcal{I}} \left\{ \left[\frac{3}{2}k\tau^2 + \frac{\tau'^2}{k} + 2\tau\left(\frac{\tau'}{k}\right)' \right] \overset{(1)}{z_1} + \left[\frac{\tau'\tau^2}{k} - (k\tau)' - \left(\frac{\tau'}{k}\right)'' \right] \overset{(1)}{z_2} \right\} ds,$$

$$\delta^2\mathcal{E}_\tau = I_1 + I_2,$$

where k is the curvature, τ is the torsion of C, I_1 and I_2 are given by Eqs. (2.294) and (2.295).

Torus Knot $p = 2$, $q = 3$

Let us consider a torus knot for values $p = 2$, $q = 3$, and $a = 2$, which is defined by the following parametric equations: $x(u) = (\cos(3u) + 2) \cdot \cos(2u)$, $y(u) = (\cos(3u) + 2) \cdot \sin(2u)$, $z(u) = -\sin(3u)$.

The fields of the first- and second-order infinitesimal bending are given by $\overset{(1)}{\mathbf{z}}(u)$ ($\overset{(1)}{z_1}(u) = (\cos(3u) + 2) \cdot \sin(2u)$, $\overset{(1)}{z_2}(u) = -(\cos(3u) +$

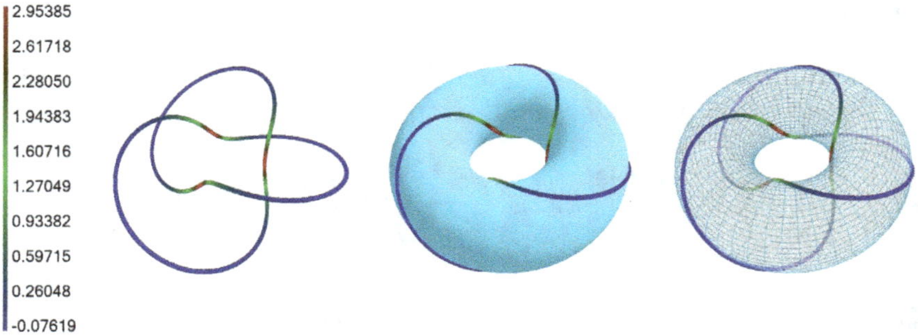

Fig. 2.30. Torus knot $p = 2$, $q = 3$, basic, no bending.

Fig. 2.31. Torus knot $p = 2$, $q = 3$, infinitesimally bent with $\epsilon = 1.0$.

$2) \cdot \cos(2u)$, $\overset{(1)}{z}_3(u) = 0)$ and $\overset{(2)}{\mathbf{z}}(u)$ $(\overset{(2)}{z}_1(u) = -0.5 \cdot (\cos(3u) + 2) \cdot \cos(2u)$, $\overset{(2)}{z}_2(u) = -0.5 \cdot (\cos(3u) + 2) \cdot \sin(2u)$, $\overset{(2)}{z}_3(u) = 0)$.

It is easy to check that infinitesimal bending fields satisfy conditions given by (2.37). The basic $p = 2, q = 3$ knot and bent curve are shown in Figs. 2.30–2.32. Knots are colored according to the values of torsion mapped to a color on a color scale.

We calculated the torsional energy of the $p = 2$, $q = 3$ knot, and it is 13.3482 for $\epsilon = 0$ and 12.9334 for $\epsilon = 1.0$.

Torus Knot $p = 2$, $q = 5$

Let us consider a torus knot for values $p = 2$, $q = 5$, and $a = 2$, which is defined by the following parametric equations: $x(u) = (\cos(5u) + 2) \cdot \cos(2u)$, $y(u) = (\cos(5u) + 2) \cdot \sin(2u)$, $z(u) = -\sin(5u)$.

Fig. 2.32. Torus knot $p = 2$, $q = 3$, basic and infinitesimally bent with $\epsilon = 1.0$.

Fig. 2.33. Torus knot $p = 2$, $q = 5$, basic, no bending.

The fields of the first- and second-order infinitesimal bending are given by $\overset{(1)}{\mathbf{z}}(u)$ ($\overset{(1)}{z}_1(u) = (\cos(5u) + 2) \cdot \sin(2u)$, $\overset{(1)}{z}_2(u) = -(\cos(5u) + 2) \cdot \cos(2u)$, $\overset{(1)}{z}_3(u) = 0$) and $\overset{(2)}{\mathbf{z}}(u)$ ($\overset{(2)}{z}_1(u) = -0.5 \cdot (\cos(5u) + 2) \cdot \cos(2u)$, $\overset{(2)}{z}_2(u) = -0.5 \cdot (\cos(5u) + 2) \cdot \sin(2u)$, $\overset{(2)}{z}_3(u) = 0$). The infinitesimal bending fields satisfy conditions given by (2.37). The basic $p = 2$, $q = 5$ knot and bent curve are shown in Figs. 2.33–2.35.

The knots are colored according to the values of torsion mapped to a color on a color scale. We calculated the torsional energy of the $p = 2$, $q = 5$ knot, and it is 10.521 for $\epsilon = 0$ and 10.522 for $\epsilon = 0.4$.

Torus Knot $p = 3$, $q = 4$

Let us consider a torus knot for values $p = 3$, $q = 4$, and $a = 2$, which is defined by the following parametric equations: $x(u) = (\cos(4u) + 2) \cdot \cos(3u)$, $y(u) = (\cos(4u) + 2) \cdot \sin(3u)$, $z(u) = -\sin(4u)$.

Fig. 2.34. Torus knot $p = 2$, $q = 5$, infinitesimally bent with $\epsilon = 0.4$.

Fig. 2.35. Torus Knot $p = 2$, $q = 5$, basic and infinitesimally bent with $\epsilon = 0.4$.

The fields of the first- and second-order infinitesimal bending are given by $\overset{(1)}{\mathbf{z}}(u)$ ($\overset{(1)}{z}_1(u) = -\sin(4u)$, $\overset{(1)}{z}_2(u) = 0$, $\overset{(1)}{z}_3(u) = -(\cos(4u) + 2) \cdot \cos(3u)$) and $\overset{(2)}{\mathbf{z}}(u)$ ($\overset{(2)}{z}_1(u) = -0.5 \cdot (\cos(4u) + 2) \cdot \cos(3u)$, $\overset{(2)}{z}_2(u) = -0.5 \cdot (\cos(4u) + 2) \cdot \sin(3u)$, $\overset{(2)}{z}_3(u) = 0$). It is easy to check that the infinitesimal bending fields satisfy the conditions given by (2.37). The basic $p = 3$, $q = 4$ knot and bent curve are shown in Figs. 2.36–2.38.

The knots are colored according to the values of torsion mapped to a color on a color scale. We calculated the torsional energy of the $p = 3, q = 4$ knot, and it is 21.8222 for $\epsilon = 0$ and 21.3129 for $\epsilon = 0.7$.

Torus Knot $p = 2$, $q = 7$

Let us consider a torus knot for values $p = 2$, $q = 7$, and $a = 2$, which is defined by the following parametric equations: $x(u) = (\cos(7u) + 2) \cdot \cos(2u)$, $y(u) = (\cos(7u) + 2) \cdot \sin(2u)$, $z(u) = -\sin(7u)$.

Fig. 2.36. Torus knot $p = 3$, $q = 4$, basic, no bending.

Fig. 2.37. Torus knot $p = 3$, $q = 4$, infinitesimally bent with $\epsilon = 0.7$.

Fig. 2.38. Torus knot $p = 3$, $q = 4$, basic and infinitesimally bent with $\epsilon = 0.7$.

The fields of the first- and second-order infinitesimal bending are given by $\overset{(1)}{\mathbf{z}}(u)$ ($\overset{(1)}{z}_1(u) = (\cos(7u) + 2) \cdot \sin(2u)$, $\overset{(1)}{z}_2(u) = -(\cos(7u) + 2) \cdot \cos(2u)$, $\overset{(1)}{z}_3(u) = 0$) and $\overset{(2)}{\mathbf{z}}(u)$ ($\overset{(2)}{z}_1(u) = -0.5 \cdot (\cos(7u) + 2) \cdot$

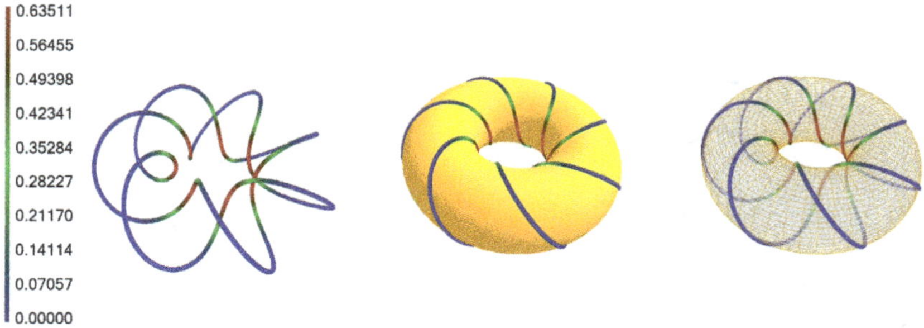

Fig. 2.39. Torus knot $p = 2$, $q = 7$, basic, no bending.

Fig. 2.40. Torus Knot $p = 2$, $q = 7$, infinitesimally bent with $\epsilon = 0.95$.

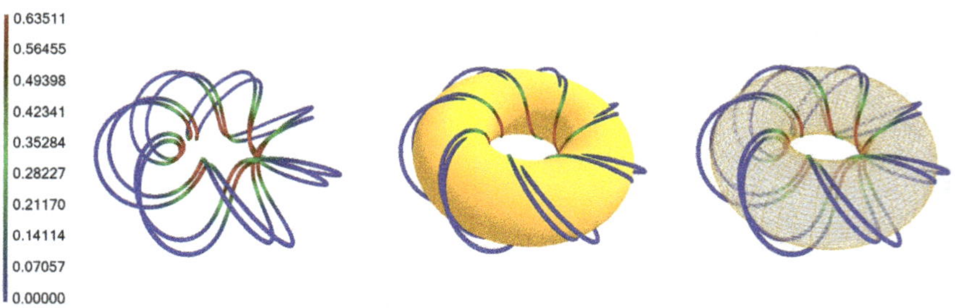

Fig. 2.41. Torus knot $p = 2$, $q = 7$, basic and infinitesimally bent with $\epsilon = 0.95$.

$\cos(2u)$, $\overset{(2)}{z}_2(u) = -0.5\cdot(\cos(7u)+2)\cdot\sin(2u)$, $\overset{(2)}{z}_3(u) = 0$). The infinitesimal bending fields satisfy the conditions given by (2.37). The basic $p = 2$, $q = 7$ knot and bent curve are shown in Figs. 2.39–2.41.

The knots are colored according to the values of torsion mapped to a color on a color scale. We calculated the torsional energy of the $p = 2$, $q = 7$ knot, and it is 8.657 for $\epsilon = 0$ and 8.591 for $\epsilon = 0.95$.

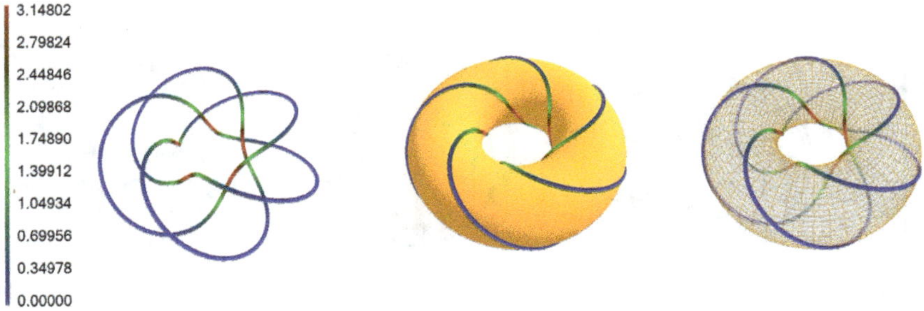

Fig. 2.42. Torus knot $p = 3$, $q = 5$, basic, no bending.

Fig. 2.43. Torus knot $p = 3$, $q = 5$, infinitesimally bent with $\epsilon = 0.85$.

Torus Knot $p = 3$, $q = 5$

Let us consider a torus knot for values $p = 3$, $q = 5$, and $a = 2$, which is defined by the following parametric equations: $x(u) = (\cos(5u) + 2) \cdot \cos(3u)$, $y(u) = (\cos(5u) + 2) \cdot \sin(3u)$, $z(u) = -\sin(5u)$.

The fields of the first- and second-order infinitesimal bending are given by $\overset{(1)}{\mathbf{z}}(u)$ ($\overset{(1)}{z_1}(u) = (\cos(5u) + 2) \cdot \sin(3u)$, $\overset{(1)}{z_2}(u) = -(\cos(5u) + 2) \cdot \cos(3u)$, $\overset{(1)}{z_3}(u) = 0$) and $\overset{(2)}{\mathbf{z}}(u)$ ($\overset{(2)}{z_1}(u) = -0.5 \cdot (\cos(5u) + 2) \cdot \cos(3u)$, $\overset{(2)}{z_2}(u) = -0.5 \cdot (\cos(5u) + 2) \cdot \sin(3u)$, $\overset{(2)}{z_3}(u) = 0$). The infinitesimal bending fields satisfy the conditions given by (2.37). The basic $p = 3$, $q = 5$ knot and bent curve are shown in Figs. 2.42–2.44.

The knots are colored according to the values of torsion mapped to a color on a color scale. We calculated the torsional energy of the $p = 3$, $q = 5$ knot, and it is 18.975 for $\epsilon = 0$ and 18.746 for $\epsilon = 0.85$.

Fig. 2.44. Torus knot $p = 3$, $q = 5$, basic and infinitesimally bent with $\epsilon = 0.85$.

2.10 Total Torsion

The total torsion of the curve $C : \mathbf{r} = \mathbf{r}(s)$, $s \in \mathcal{I}$ is

$$\mathcal{T} = \int_{\mathcal{I}} \tau(s) ds. \tag{2.300}$$

We will examine the change of the total torsion according to [Najdanović *et al.* (2021b)], [Najdanović *et al.* (2024)].

2.10.1 *The First Variation of the Total Torsion*

Theorem 2.60. *Under infinitesimal bending of the curve C, the variation of its total torsion is*

$$\delta\mathcal{T} = \int_{\mathcal{I}} (k\tau z_1 - k' z_2) ds + \int_{\mathcal{I}} \Big[z\tau + \frac{1}{k}(2\tau z_1' + \tau' z_1 + z_2'' + (k^2 - \tau^2) z_2) \Big]' ds, \tag{2.301}$$

where k and τ are the curvature and the torsion of C, respectively.

Proof. The variation of the total torsion is

$$\delta\mathcal{T} = \int_{\mathcal{I}} \delta\tau \, ds = \int_{\mathcal{I}} \Big\{ z\tau' + k(z_2' + 2\tau z_1) + \Big[\frac{1}{k} \left(2\tau z_1' + \tau' z_1 + z_2'' - \tau^2 z_2 \right) \Big]' \Big\} ds. \tag{2.302}$$

If we use the equation $kz_2' = (kz_2)' - k'z_2$, and also $z\tau' = (z\tau)' - z'\tau = (z\tau)' - k\tau z_1$, due to (2.25), we obtain

$$\delta\mathcal{T} = \int_{\mathcal{I}} \Big\{ (z\tau)' - k\tau z_1 + (kz_2)' - k'z_2 + 2k\tau z_1 + \left[\frac{1}{k}\left(2\tau z_1' + \tau' z_1 + z_2'' - \tau^2 z_2\right)\right]' \Big\}\, ds,$$

i.e. (2.301). □

In addition, the following applies:

Theorem 2.61. *Under infinitesimal bending of the curve C, the variation of its total torsion is*

$$\delta\mathcal{T} = -\int_{\mathcal{I}} k y_1\, ds + \int_{\mathcal{I}} \left[\frac{1}{k}(y_2\tau - y_1')\right]' ds, \tag{2.303}$$

where k and τ are the curvature and the torsion of C, respectively, y_1 and y_2 are the components of the rotational field given by Eq. (2.34).

About the total torsion of spherical curves we have the next theorem.

Theorem 2.62 ([Najdanović *et al.* (2024)]). *The variation of the total torsion of a closed curve C on the sphere S under infinitesimal bending which includes C into the family of closed curves that are on S with a given precision is equal to zero.*

Proof. In order to get a family of closed curves under infinitesimal bending of a closed spherical curve, we suppose that $\mathbf{z}$ is closed function.

According to Theorem 2.16 and Eq. (2.113) for infinitesimal bending field, we have

$$z - \text{ arbitrary closed } C^1 \text{ function}, \quad z_1 = \frac{1}{k}z', \quad z_2 = \frac{\tau}{k'}z', \tag{2.304}$$

for $k' \neq 0$. In this case, using (2.301) for $\mathcal{I} = [0, L]$, we get

$$\delta T = \int_0^L (k\tau z_1 - k'z_2)\, ds = 0. \tag{2.305}$$

However, this may not be the true. We are going to discuss general situation.

Let us assume that there are points $0 = s_0 < s_1 < \cdots < s_n = L$ where k' is zero and such that on (s_{i-1}, s_i) k' is either strictly nonzero or identically zero. Then, we have

$$\delta T = \int_0^L \delta\tau\, ds = \sum_{i=1}^{i=n} \int_{s_{i-1}}^{s_i} \delta\tau\, ds. \tag{2.306}$$

- If $k' \equiv 0$ on (s_{i-1}, s_i) then this segment of C is a part of a circle and hence is a plane curve. Thus $\tau \equiv 0$ on $[s_{i-1}, s_i]$ and

$$\int_{s_{i-1}}^{s_i} (k\tau z_1 - k' z_2) ds = 0. \tag{2.307}$$

- If $k' \neq 0$ on (s_{i-1}, s_i), then $\int_{s_{i-1}}^{s_i} \delta\tau\, ds = 0$.

Finally, in the sum (2.306), all terms will be equal to zero. Therefore, $\delta T = \int_0^L \delta\tau\, ds = 0$. □

Example 2.13. Let $C : \mathbf{r}(t) = (\sin t, 0, \cos t)$ be the circle on the unit sphere $||\mathbf{r}||^2 = 1$ with center at the origin. Obviously, $||\dot{\mathbf{r}}(t)|| = 1$. Suppose that $\mathbf{z}(t) = (\cos t, a(t)\sin(t), -\sin t)$ is an infinitesimal bending field, according to Example 2.6, where $a(t)$ is arbitrary real continuous differentiable function. In order to obtain closed curves under bending, let us suppose that $a(t)$ is a closed function. Our aim is to find the variation of the total torsion of the circle C under infinitesimal bending determined by the field $\mathbf{z}$.

Let us find

$$\tau_\epsilon = \frac{[\dot{\mathbf{r}}_\epsilon, \ddot{\mathbf{r}}_\epsilon, \dddot{\mathbf{r}}_\epsilon]}{||\dot{\mathbf{r}}_\epsilon \times \ddot{\mathbf{r}}_\epsilon||^2},$$

where

$$\mathbf{r}_\epsilon = (\sin t + \epsilon \cos t, \epsilon a(t) \sin t, \cos t - \epsilon \sin t).$$

We have

$$[\dot{\mathbf{r}}_\epsilon, \ddot{\mathbf{r}}_\epsilon, \dddot{\mathbf{r}}_\epsilon] = [\dot{\mathbf{r}}, \ddot{\mathbf{r}}, \dddot{\mathbf{r}}] + \epsilon(\dddot{a}(t)\sin t + 3\ddot{a}(t)\cos t - 2\dot{a}(t)\sin t) + \epsilon^2 \ldots,$$

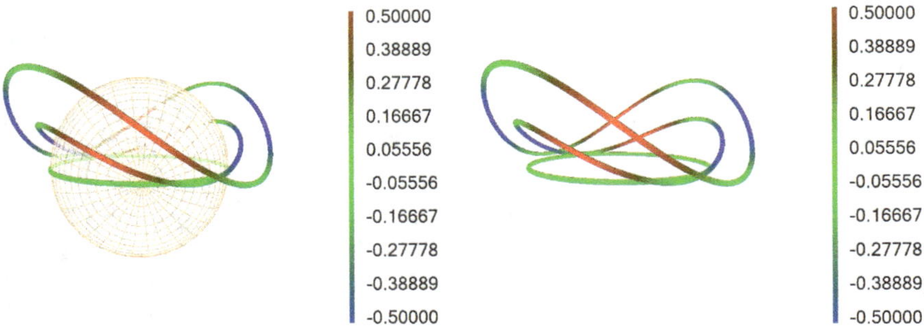

Fig. 2.45. $a(t) = \sin(t)$, $c = 2$, $r = 1$.

where $[\dot{\mathbf{r}}, \ddot{\mathbf{r}}, \dddot{\mathbf{r}}] = 0$ is valid for a circle. Also

$$\begin{aligned}\dot{\mathbf{r}}_\epsilon \times \ddot{\mathbf{r}}_\epsilon &= (\epsilon(\dot{a}(t)\sin t\cos t + \ddot{a}(t)\sin^2 t - a(t)) + \epsilon^2 ..., 1 + \epsilon^2,\\ &\quad \epsilon(a''(t)\sin t\cos t + \dot{a}(t)\cos^2 t + \dot{a}(t)) + \epsilon^2 ...)\\ \|\dot{\mathbf{r}}_\epsilon \times \ddot{\mathbf{r}}_\epsilon\|^2 &= \|\dot{\mathbf{r}} \times \ddot{\mathbf{r}}\|^2 + \epsilon^2[(\dot{a}(t)\sin t\cos t + \ddot{a}(t)\sin^2 t - a(t))^2 + 2\\ &\quad + (\ddot{a}(t)\sin t\cos t + \dot{a}(t)\cos^2 t + \dot{a}(t))^2] + \epsilon^3 \dots\end{aligned}$$

where $\|\dot{\mathbf{r}} \times \ddot{\mathbf{r}}\|^2 = 1$. Now, we have

$$\delta\tau = \frac{d}{d\epsilon}\tau_\epsilon|_{\epsilon=0} = \dddot{a}(t)\sin t + 3\ddot{a}(t)\cos t - 2\dot{a}(t)\sin t,$$

after some calculations and using Maclaurin series. Finally

$$\begin{aligned}\delta T &= \int_0^{2\pi} \delta\tau dt = \int_0^{2\pi} [\dddot{a}(t)\sin t + 3\ddot{a}(t)\cos t - 2\dot{a}(t)\sin t]\, dt\\ &= \ddot{a}(t)\sin t + 2\dot{a}(t)\cos t|_0^{2\pi} = 2\dot{a}(2\pi) - 2\dot{a}(0) = 0,\end{aligned}$$

because $\dot{a}(0) = \dot{a}(2\pi)$.

We are going to illustrate infinitesimal bending of spherical curves. In Figs. 2.45–2.49, colors are used to indicate the values of torsion at different points of the curves, together with color-values scale. We applied different bending fields according to Eq. (2.100). The maximum value used for ϵ is 1.5. The numerical value of the total torsion for basic and deformed curves in examples 1–4 is zero. In the fifth example, the numerical value of the total torsion of basic curve is zero, and of deformed curve with $\epsilon = 1.5$ is equal to -1.2783.

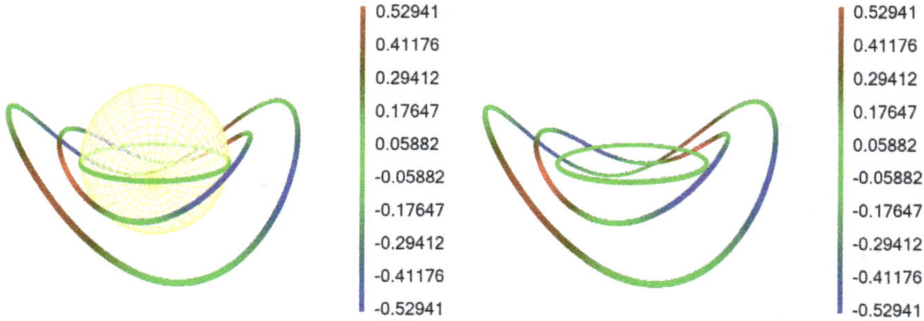

Fig. 2.46. $a(t) = \cos(t)$, $c = 2$, $r = 1$.

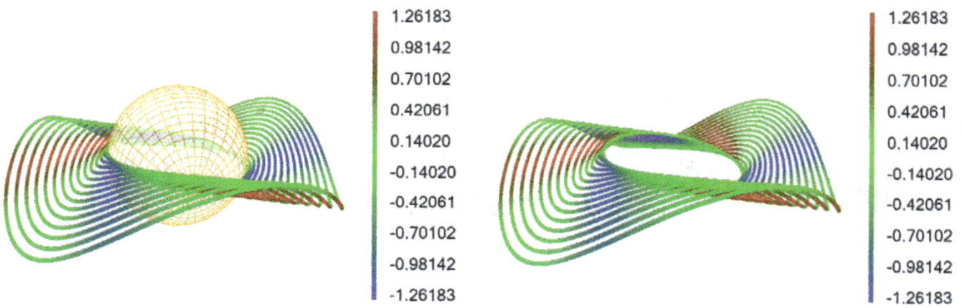

Fig. 2.47. $a(t) = \sin(t)\cos(t)$, $c = 2$, $r = 1$.

In this example, the corresponding bending field is obtained using the function $a(t)$ which is not closed, i.e. $a(0) = a(2\pi)$, but also $\dot{a}(0) \neq \dot{a}(2\pi)$.

2.10.2 *The Second Variation of the Total Torsion*

Let us consider the total torsion of the curve C under second-order infinitesimal bending.

The total torsion of deformed curve is

$$\mathcal{T}_\epsilon = \int_{\mathcal{I}} \tau_\epsilon \, ds_\epsilon = \int_{\mathcal{I}} (\tau + \epsilon\, \delta\tau + \epsilon^2\, \delta^2\tau)(ds + \epsilon\, \delta\, ds + \epsilon^2\, \delta^2\, ds).$$

Since $\delta\, ds = \delta^2 ds = 0$, we have

$$\mathcal{T}_\epsilon = \mathcal{T} + \epsilon \int_{\mathcal{I}} \delta\tau \, ds + \epsilon^2 \int_{\mathcal{I}} \delta^2\tau \, ds$$

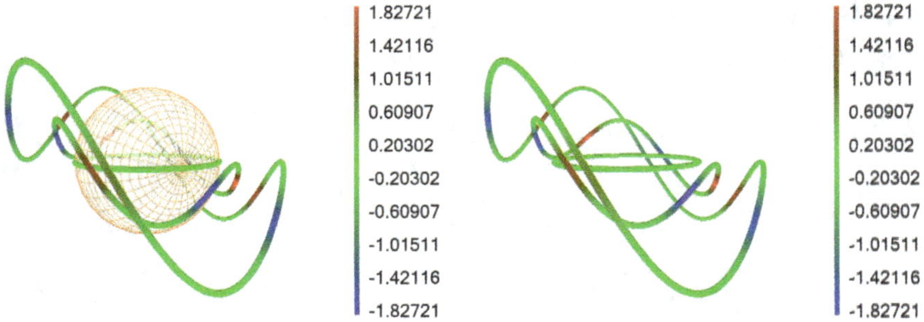

Fig. 2.48. $a(t) = \sin(2t)$, $c = 2$, $r = 1$.

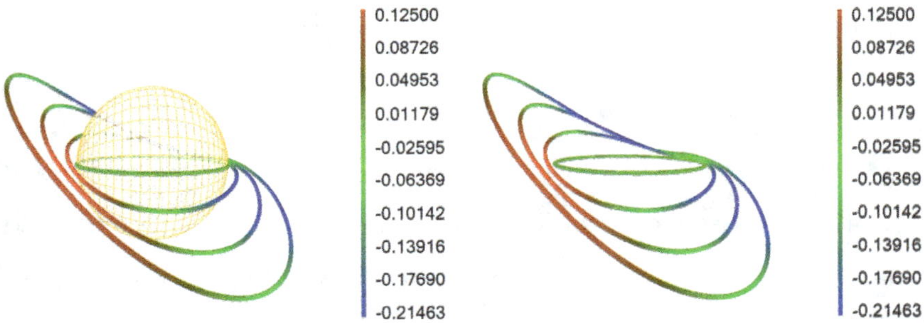

Fig. 2.49. $a(t) = \sin(t/2)$, $c = 2$, $r = 1$.

which is obtained after neglecting the terms with ϵ^3 and ϵ^4. Therefore,

$$\delta\mathcal{T} = \int_{\mathcal{I}} \delta\tau\, ds, \quad \delta^2\mathcal{T} = \int_{\mathcal{I}} \delta^2\tau\, ds. \tag{2.308}$$

Theorem 2.63. *Under second-order infinitesimal bending of the curve C, the first variation of the total torsion is*

$$\begin{aligned}\delta\mathcal{T} &= \int_{\mathcal{I}} \left(k\tau \overset{(1)}{z}_1 - k' \overset{(1)}{z}_2\right) ds \\ &\quad + \int_{\mathcal{I}} \left\{ \frac{1}{k}\left[2\tau \overset{(1)}{z}_1{}' + \tau' \overset{(1)}{z}_1 + \overset{(1)}{z}_2{}'' + (k^2 - \tau^2)\overset{(1)}{z}_2\right] + \tau \overset{(1)}{z} \right\}' ds,\end{aligned} \tag{2.309}$$

where k and τ are the curvature and the torsion of the curve C, respectively.

Theorem 2.64. *Under second-order infinitesimal bending of the curve C, the second variation of the total torsion is*

$$\begin{aligned}\delta^2\mathcal{T} = \int_{\mathcal{I}} \Bigg\{ \frac{1}{k}\Bigg[&\tau\Big(k\overset{(2)}{z} + \overset{(2)}{z}_1{}' - \tau\overset{(2)}{z}_2\Big) + \Big(\overset{(2)}{z}_2{}' + \tau\overset{(2)}{z}_1\Big)' - \frac{1}{k}\Big(k'\overset{(1)}{z} + \overset{(1)}{z}_1{}'' \\ &+ (k^2-\tau^2)\overset{(1)}{z}_1 - 2\tau\overset{(1)}{z}_2{}' - \tau'\overset{(1)}{z}_2\Big)\Big(k\tau\overset{(1)}{z} + 2\tau\overset{(1)}{z}_1{}' + \tau'\overset{(1)}{z}_1 + \overset{(1)}{z}_2{}'' \\ &- \tau^2\overset{(1)}{z}_2\Big)\Bigg] + k\overset{(2)}{z}_2\Bigg\}' ds + \int_{\mathcal{I}}\Big(k\tau\overset{(2)}{z}_1 - k'\overset{(2)}{z}_2\Big)\,ds + \int_{\mathcal{I}} g\Big(\overset{(1)}{\mathbf{z}}\Big)\,ds,\end{aligned} \tag{2.310}$$

where

$$\begin{aligned}g(\overset{(1)}{\mathbf{z}}) = &-\frac{\tau}{2}\Big[\Big(k\overset{(1)}{z} + \overset{(1)}{z}_1{}' - \tau\overset{(1)}{z}_2\Big)^2 - \Big(\overset{(1)}{z}_2{}' + \tau\overset{(1)}{z}_1\Big)^2\Big] \\ &+ \Big(k'\overset{(1)}{z} + \overset{(1)}{z}_1{}'' + (k^2-\tau^2)\overset{(1)}{z}_1 - 2\tau\overset{(1)}{z}_2{}' - \tau'\overset{(1)}{z}_2\Big)\Big(\overset{(1)}{z}_2{}' + \tau\overset{(1)}{z}_1\Big).\end{aligned} \tag{2.311}$$

Proof. The second variation of the total torsion is easy obtained after using the expression for $\delta^2\tau$, as well as $k\overset{(2)}{z}_2{}' = (k\overset{(2)}{z}_2)' - k'\overset{(2)}{z}_2$. We have

$$\begin{aligned}\delta^2\tau = \Bigg\{\frac{1}{k}\Bigg[&\tau\Big(k\overset{(2)}{z} + \overset{(2)}{z}_1{}' - \tau\overset{(2)}{z}_2\Big) + \Big(\overset{(2)}{z}_2{}' + \tau\overset{(2)}{z}_1\Big)' - \frac{1}{k}\Big(k'\overset{(1)}{z} + \overset{(1)}{z}_1{}'' \\ &+ (k^2-\tau^2)\overset{(1)}{z}_1 - 2\tau\overset{(1)}{z}_2{}' - \tau'\overset{(1)}{z}_2\Big)\Big(k\tau\overset{(1)}{z} + 2\tau\overset{(1)}{z}_1{}' \\ &+ \tau'\overset{(1)}{z}_1 + \overset{(1)}{z}_2{}'' - \tau^2\overset{(1)}{z}_2\Big)\Bigg] + k\overset{(2)}{z}_2\Bigg\}' + k\tau\overset{(2)}{z}_1 - k'\overset{(2)}{z}_2 + g\Big(\overset{(1)}{\mathbf{z}}\Big).\end{aligned} \tag{2.312}$$

wherefrom, we obtain Eq. (2.310). □

2.10.3 *Total Torsion of Knots*

Theorem 2.65. *Under first-order infinitesimal bending of the knot C, the variation of the total torsion is*

$$\delta\mathcal{T} = \int_{\mathcal{I}}(k\tau z_1 - k'z_2)ds,$$

i.e.

$$\delta\mathcal{T} = -\int_{\mathcal{I}} ky_1\,ds,$$

where k and τ are the curvature and the torsion of C, z_1 and z_2 are the components of the bending field, and y_1 is a component of the rotational field given in Eq. (2.34), respectively.

Theorem 2.66. *Under second-order infinitesimal bending of the knot C, the first and the second variation of its total torsion are respectively*

$$\delta\mathcal{T} = \int_{\mathcal{I}} \left(k\tau \overset{(1)}{z}_1 - k' \overset{(1)}{z}_2 \right) ds,$$

$$\delta^2\mathcal{T} = \int_{\mathcal{I}} \left(k\tau \overset{(2)}{z}_1 - k' \overset{(2)}{z}_2 \right) ds + \int_{\mathcal{I}} g\left(\overset{(1)}{\mathbf{z}} \right) ds,$$

where k is the curvature, τ is the torsion of C, and $g(\overset{(1)}{\mathbf{z}})$ is given in (2.311).

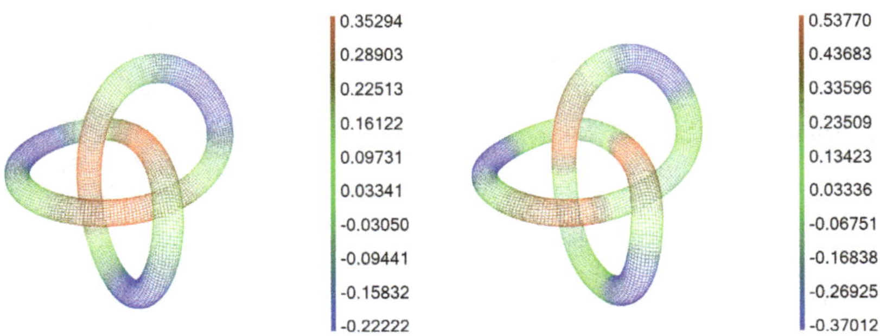

Fig. 2.50. Trefoil knot: basic and infinitesimally bent with $\epsilon = 0.55$.

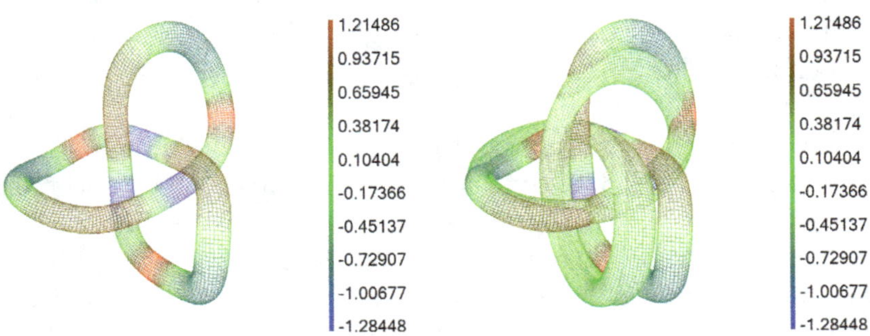

Fig. 2.51. Trefoil knot: infinitesimally bent with $\epsilon = 1.1$ and all knots together.

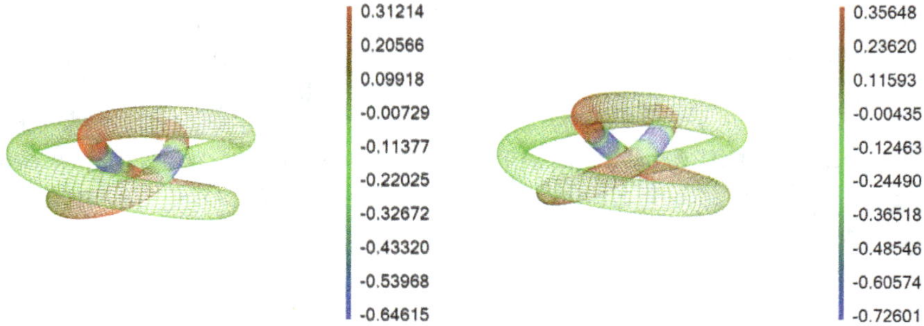

Fig. 2.52. $p = 3$, $q = 2$ torus knot: basic and infinitesimally bent with $\epsilon = 0.8$.

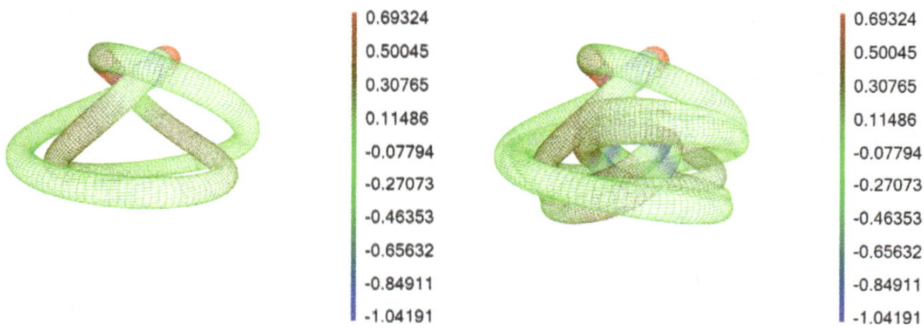

Fig. 2.53. $p = 3$, $q = 2$ torus knot: infinitesimally bent with $\epsilon = 1.6$ and all knots together.

Trefoil knot

A trefoil knot is given by the parametric equations: $x = \sin(u) + 2\sin(2u)$, $y = \cos(u) - 2\cos(2u)$, $z = -\sin(3u)$. The basic and infinitesimally bent trefoil knots are given in Figs. 2.50 and 2.51. The bending fields are defined by: $p(u) = 0$, $q(u) = 0$, $r(u) = \cos(3u)$ and $g(u) = \sin(6u)$.

The numerically calculated total torsion of the trefoil knot is 2.2250, 2.2354, and 2.3708 for $\epsilon = 0$, $\epsilon = 0.55$ and $\epsilon = 1.1$ respectively.

A $p3q2$ torus knot is given by the parametric equations: $x = (\cos(2u) + 2) \cdot \cos(3u)$, $y = (\cos(2u) + 2) \cdot \sin(3u)$, $z = -\sin(2u)$. The basic and infinitesimally bent knots are given in Figs. 2.52 and 2.53. The bending fields are defined by: $p(u) = 0$, $q(u) = 0$, $r(u) = 0$ and $g(u) = \sin(2u)$.

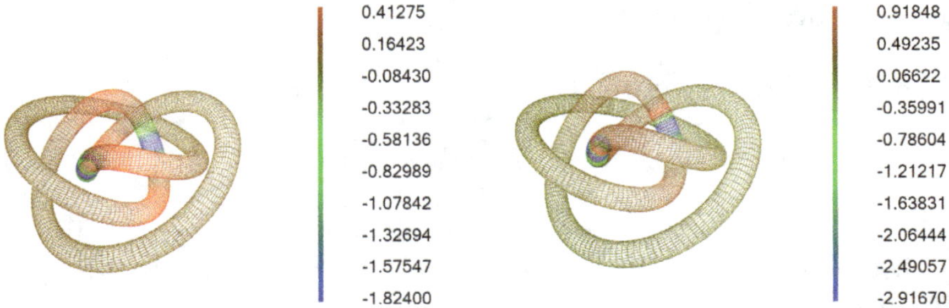

Fig. 2.54. Figure eight knot: basic and infinitesimally bent with $\epsilon = 0.76$.

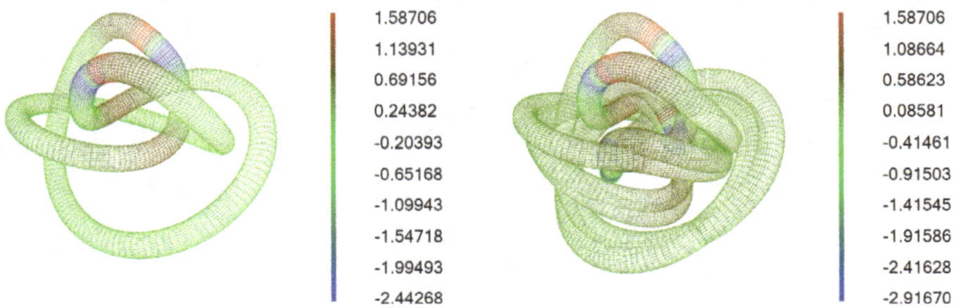

Fig. 2.55. Figure eight knot: infinitesimally bent with $\epsilon = 1.52$ and all knots together.

The numerically calculated total torsion of the $p = 3$, $q = 2$ torus knot is 0.6234, 0.6128 and 0.4725 for $\epsilon = 0$, $\epsilon = 0.8$ and $\epsilon = 1.6$ respectively.

Figure eight knot

A figure eight knot is given by the parametric equations: $x = (2 + \cos(2u)) \cdot \cos(3u)$, $y = (2 + \cos(2u)) \cdot \sin(3u)$, $z = \sin(4u)$. The basic and infinitesimally bent figure eight knot are given in Figs. 2.54 and 2.55.

The bending fields are defined by: $p(u) = 0$, $q(u) = 0$, $r(u) = -cos(2u)$ and $g(u) = \sin(2u)$.

The numerically calculated total torsion of the figure eight knot is -0.5423, -0.4843 and -0.3033 for $\epsilon = 0$, $\epsilon = 0.76$ and $\epsilon = 1.52$ respectively.

2.11 Total Normalcy

The total normalcy of the curve $C : \mathbf{r} = \mathbf{r}(s)$, $s \in \mathcal{I}$, is

$$\omega = \int_{\mathcal{I}} \sqrt{k^2(s) + \tau^2(s)}\, ds, \tag{2.313}$$

where k and τ are the curvature and the torsion of C, respectively.

We will examine the change of the total normalcy according to [Rančić *et al.* (2019)].

2.11.1 *The First Variation of the Total Normalcy*

According to (2.167), the variation of the total normalcy satisfies

$$\delta\omega = \frac{d}{d\epsilon} w_\epsilon \Big|_{\epsilon=0} = \frac{d}{d\epsilon} \int_{\mathcal{I}} \sqrt{(k+\epsilon\delta k)^2 + (\tau + \epsilon\delta\tau)^2}\, (ds + \epsilon\delta(ds)) \Big|_{\epsilon=0},$$

i.e.

$$\delta\omega = \int_{\mathcal{I}} \frac{k\delta k + \tau\delta\tau}{\sqrt{k^2+\tau^2}} ds. \tag{2.314}$$

We will consider separately the following two integrals:

$$I_1 = \int_{\mathcal{I}} \frac{k\delta k}{\sqrt{k^2+\tau^2}} ds, \quad I_2 = \int_{\mathcal{I}} \frac{\tau\delta\tau}{\sqrt{k^2+\tau^2}} ds.$$

Using (2.181), one obtains

$$\begin{aligned} I_1 &= \int_{\mathcal{I}} \frac{k}{\sqrt{k^2+\tau^2}} (k'z + z_1'' + (k^2-\tau^2)z_1 - 2\tau z_2' - \tau' z_2) ds \\ &= \int_{\mathcal{I}} \Bigg(\frac{kk'}{\sqrt{k^2+\tau^2}} z + \frac{k}{\sqrt{k^2+\tau^2}} z_1'' + \frac{k(k^2-\tau^2)}{\sqrt{k^2+\tau^2}} z_1 - \frac{2k\tau}{\sqrt{k^2+\tau^2}} z_2' \\ &\quad - \frac{k\tau'}{\sqrt{k^2+\tau^2}} z_2 \Bigg) ds. \end{aligned}$$

After a few transformations of the second and the fourth term in the previous integral, we obtain

$$I_1 = \int_{\mathcal{I}} \Bigg\{ \frac{kk'}{\sqrt{k^2+\tau^2}} z + \left[\left(\frac{k}{\sqrt{k^2+\tau^2}} \right)'' + \frac{k(k^2-\tau^2)}{\sqrt{k^2+\tau^2}} \right] z_1$$

$$+ \left[\left(\frac{2k\tau}{\sqrt{k^2+\tau^2}} \right)' - \frac{k\tau'}{\sqrt{k^2+\tau^2}} \right] z_2 \Bigg\} ds$$

$$+ \int_{\mathcal{I}} \left[\frac{k}{\sqrt{k^2+\tau^2}} z_1' - \left(\frac{k}{\sqrt{k^2+\tau^2}} \right)' z_1 - \frac{2k\tau}{\sqrt{k^2+\tau^2}} z_2 \right]' ds. \tag{2.315}$$

The second integral is according to (2.183)

$$I_2 = \int_{\mathcal{I}} \frac{\tau}{k^2+\tau^2} \Bigg\{ z\tau' + k(z_2' + 2\tau z_1) + \left[\frac{1}{k} \left(2\tau z_1' + \tau' z_1 + z_2'' - \tau^2 z_2 \right) \right]' \Bigg\}, ds$$

wherefrom, after a few transformations, we obtain

$$I_2 = \int_{\mathcal{I}} \Bigg\{ \frac{\tau\tau'}{\sqrt{k^2+\tau^2}} z + \left[\frac{2k\tau^2}{\sqrt{k^2+\tau^2}} + \left(\left(\frac{\tau}{\sqrt{k^2+\tau^2}} \right)' \frac{2\tau}{k} \right)' - \left(\frac{\tau}{\sqrt{k^2+\tau^2}} \right)' \frac{\tau'}{k} \right] z_1 + \left[- \left(\frac{k\tau}{\sqrt{k^2+\tau^2}} \right)' - \left(\left(\frac{\tau}{\sqrt{k^2+\tau^2}} \right)' \frac{1}{k} \right)'' + \left(\frac{\tau}{\sqrt{k^2+\tau^2}} \right)' \frac{\tau^2}{k} \right] z_2 \Bigg\} ds$$

$$+ \int_{\mathcal{I}} \Bigg\{ \frac{\tau}{k\sqrt{k^2+\tau^2}} (2\tau z_1' + \tau' z_1 + z_2'' - \tau^2 z_2) - \left(\frac{\tau}{\sqrt{k^2+\tau^2}} \right)' \frac{2\tau}{k} z_1 - \left(\frac{\tau}{\sqrt{k^2+\tau^2}} \right)' \frac{1}{k} z_2' + \frac{k\tau}{\sqrt{k^2+\tau^2}} z_2 + \left(\left(\frac{\tau}{\sqrt{k^2+\tau^2}} \right)' \frac{1}{k} \right)' z_2 \Bigg\}' ds \tag{2.316}$$

Since

$$\frac{kk' + \tau\tau'}{\sqrt{k^2+\tau^2}} z = (\sqrt{k^2+\tau^2})' z = (\sqrt{k^2+\tau^2} z)' - \frac{k(k^2+\tau^2)}{\sqrt{k^2+\tau^2}} z_1$$

after using the necessary and sufficient condition for the infinitesimal bending, $z' = kz_1$, we proved the next theorem.

Theorem 2.67. *Under first-order infinitesimal bending of the curve C, the variation of its total normalcy is*

$$\begin{aligned}\delta\omega = &\int_{\mathcal{I}}\Bigg\{\Bigg[\left(\frac{k}{\sqrt{k^2+\tau^2}}\right)'' + \left(\left(\frac{\tau}{\sqrt{k^2+\tau^2}}\right)'\frac{2\tau}{k}\right)' \\ &- \left(\frac{\tau}{\sqrt{k^2+\tau^2}}\right)'\frac{\tau'}{k}\Bigg]z_1 + \Bigg[\left(\frac{k}{\sqrt{k^2+\tau^2}}\right)'\tau \\ &- \left(\left(\frac{\tau}{\sqrt{k^2+\tau^2}}\right)'\frac{1}{k}\right)'' + \left(\frac{\tau}{\sqrt{k^2+\tau^2}}\right)'\frac{\tau^2}{k}\Bigg]z_2\Bigg\}\,ds \\ &+ \int_{\mathcal{I}}\Bigg\{\sqrt{k^2+\tau^2}\,z + \frac{k}{\sqrt{k^2+\tau^2}}z_1' - \left(\frac{k}{\sqrt{k^2+\tau^2}}\right)'z_1 \\ &- \frac{k\tau}{\sqrt{k^2+\tau^2}}z_2 + \frac{\tau}{k\sqrt{k^2+\tau^2}}(2\tau z_1' + \tau' z_1 + z_2'' - \tau^2 z_2) \\ &- \left(\frac{\tau}{\sqrt{k^2+\tau^2}}\right)'\frac{2\tau}{k}z_1 - \left(\frac{\tau}{\sqrt{k^2+\tau^2}}\right)'\frac{1}{k}z_2' \\ &+ \left(\left(\frac{\tau}{\sqrt{k^2+\tau^2}}\right)'\frac{1}{k}\right)'z_2\Bigg\}'\,ds, \end{aligned} \tag{2.317}$$

where k and τ are the curvature and the torsion of C, respectively.

In terms of the rotational field $\mathbf{y}$ we have:

Theorem 2.68. *Under first-order infinitesimal bending of the curve C, the variation of its total normalcy is given by*

$$\begin{aligned}\delta\omega = &\int_{\mathcal{I}}\Bigg\{-\left[\frac{1}{k}\left(\frac{\tau}{\sqrt{k^2+\tau^2}}\right)'\right]'y_1 - \Bigg[\frac{\tau}{k}\left(\frac{\tau}{\sqrt{k^2+\tau^2}}\right)' \\ &+ \left(\frac{k}{\sqrt{k^2+\tau^2}}\right)'\Bigg]y_2\Bigg\}\,ds + \int_{\mathcal{I}}\Bigg[\frac{ky_2}{\sqrt{k^2+\tau^2}} + \frac{\tau}{k\sqrt{k^2+\tau^2}} \\ &\cdot(y_2\tau - y_1') + \frac{1}{k}\left(\frac{\tau}{\sqrt{k^2+\tau^2}}\right)'y_1\Bigg]'\,ds, \end{aligned} \tag{2.318}$$

k and τ are the curvature and the torsion of C, respectively, y_1 and y_2 are the components of the rotational field given by Eq. (2.34).

2.11.2 Total Normalcy of Knots

Theorem 2.69. *Under first-order infinitesimal bending of the knot C, the variation of its total normalcy is*

$$\begin{aligned}\delta\omega = \int_I \Bigg\{ \Bigg[\left(\frac{k}{\sqrt{k^2+\tau^2}}\right)'' + \left(\left(\frac{\tau}{\sqrt{k^2+\tau^2}}\right)' \frac{2\tau}{k}\right)' \\ - \left(\frac{\tau}{\sqrt{k^2+\tau^2}}\right)' \frac{\tau'}{k} \Bigg] z_1 + \Bigg[\left(\frac{k}{\sqrt{k^2+\tau^2}}\right)' \tau \\ - \left(\left(\frac{\tau}{\sqrt{k^2+\tau^2}}\right)' \frac{1}{k}\right)'' + \left(\frac{\tau}{\sqrt{k^2+\tau^2}}\right)' \frac{\tau^2}{k} \Bigg] z_2 \Bigg\}\, ds, \quad (2.319)\end{aligned}$$

i.e.

$$\begin{aligned}\delta\omega = \int_I \Bigg\{ - \left[\frac{1}{k}\left(\frac{\tau}{\sqrt{k^2+\tau^2}}\right)'\right]' y_1 \\ - \left[\frac{\tau}{k}\left(\frac{\tau}{\sqrt{k^2+\tau^2}}\right)' + \left(\frac{k}{\sqrt{k^2+\tau^2}}\right)'\right] y_2 \Bigg\}\, ds, \quad (2.320)\end{aligned}$$

where k and τ are the curvature and the torsion of C, respectively.

We will calculate total normalcy for some infinitesimally bent knots. For the trefoil knot given by the parametric equations: $x = \sin(u) + 2\sin(2u)$, $y = \cos(u) - 2\cos(2u)$, $z = -\sin(3u)$, the basic and infinitesimally bent knots are given in Figs. 2.56 and 2.57.

The bending fields are defined by: $p(u) = \cos(3u)$ and $q(u) = \sin(3u)$.

The numerically calculated total normalcy of the trefoil knot curve is 15.626719, 15.875856 and 16.6498 for $\epsilon = 0$, $\epsilon = 0.5$ and $\epsilon = 1$ respectively.

A $p3q2$ torus knot is given by the parametric equations: $x = (\cos(2u)+2)\cdot\cos(3u)$, $y = (\cos(2u)+2)\cdot\sin(3u)$, $z = -\sin(2u)$. The basic and infinitesimally bent knots are given in Fig. 2.58.

The bending fields are defined by: $p(u) = \cos(6u)$ and $q(u) = \sin(6u)$.

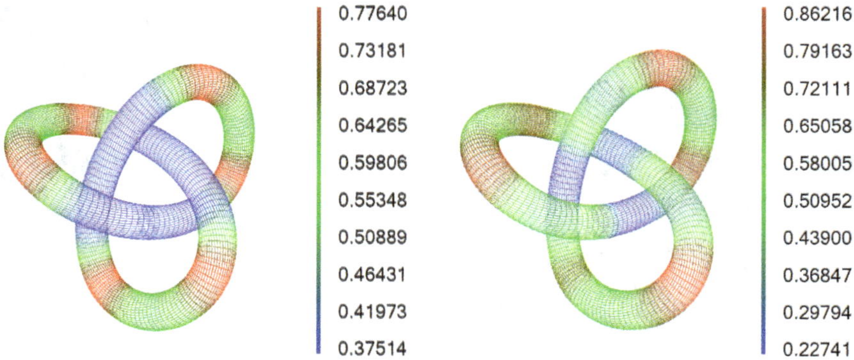

Fig. 2.56. Trefoil knot: basic and infinitesimally bent with $\epsilon = 0.5$.

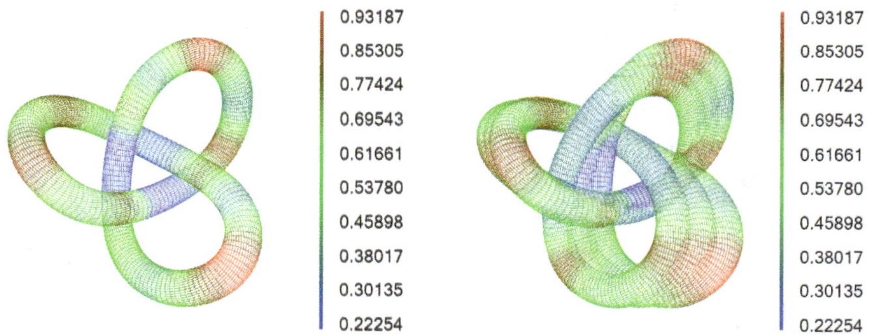

Fig. 2.57. Trefoil knot: infinitesimally bent with $\epsilon = 1$ and all knots together.

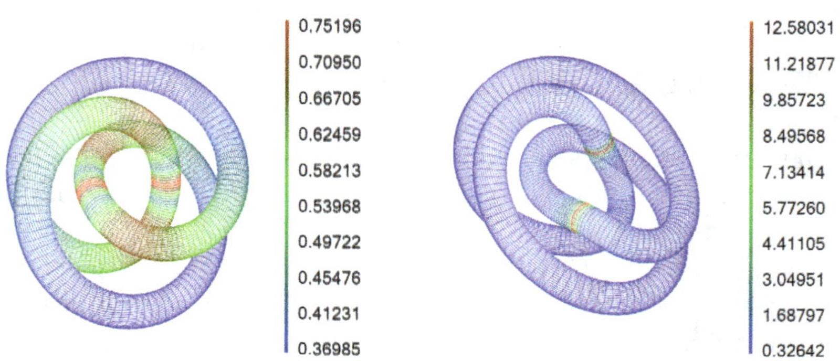

Fig. 2.58. $p = 3$, $q = 2$ torus knot: basic and infinitesimally bent with $\epsilon = 1.8$.

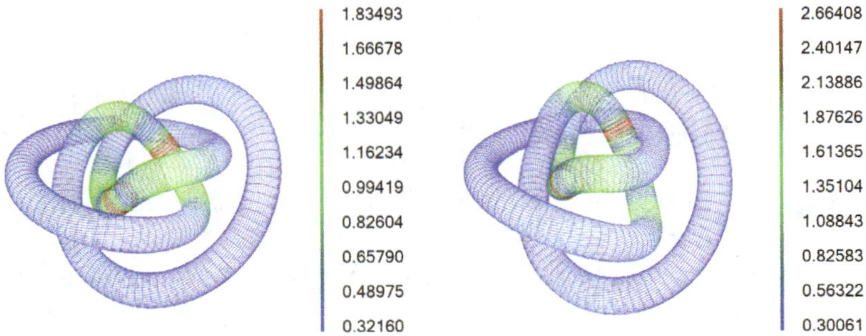

Fig. 2.59. Figure eight knot: basic and infinitesimally bent with $\epsilon = 1.3$.

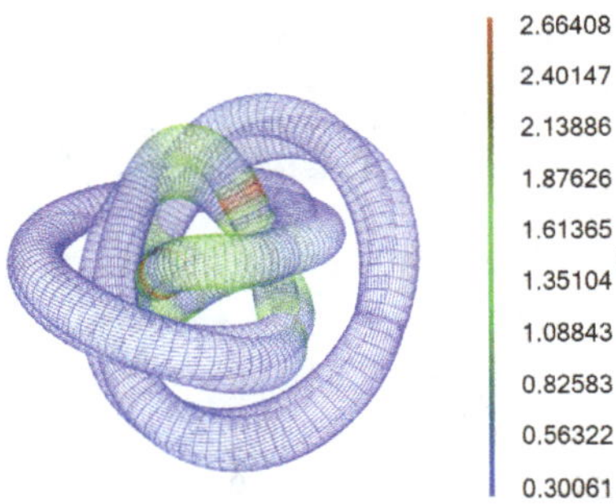

Fig. 2.60. Figure eight knot: basic and infinitesimally bent with $\epsilon = 1.3$ together.

The numerically calculated total normalcy of the $p = 3$, $q = 2$ torus knot is 19.558 and 32.0013 for $\epsilon = 0$ and $\epsilon = 1.8$ respectively.

A figure eight knot is given by the parametric equations: $x = (2 + \cos(2u)) \cdot \cos(3u)$, $y = (2 + \cos(2u)) \cdot \sin(3u)$, $z = \sin(4u)$. The basic and infinitesimally bent figure eight knot are given in Figs. 2.59 and 2.60.

The bending fields are defined by: $p(u) = \cos(6u)$ and $q(u) = \sin(6u)$.

The numerically calculated total normalcy of the figure eight knot is 23.724 and 28.583 for $\epsilon = 0$ and $\epsilon = 1.3$ respectively.

2.12 Möbius Energy of Knots

Let C be a knot in 3-space $\mathcal{R}^3$. We can present C as the image of the standard unit circle S in the plane under a smooth map $\mathbf{r} : S \to \mathcal{R}^3$.

Then, the Möbius energy of the knot is

$$E(C) = \int_S \int_S \left(\frac{1}{\|\mathbf{r}(s) - \mathbf{r}(t)\|^2} - \frac{1}{l(s,t)^2} \right) \|\dot{\mathbf{r}}(s)\| \, \|\dot{\mathbf{r}}(t)\| \, dsdt,$$

where $l(s,t)$ denotes the minimum distance along the knot C between points $\mathbf{r}(s)$ and $\mathbf{r}(t)$.

Let us assume the parameterization $\mathbf{r}$ be by arc length, then

$$\|\dot{\mathbf{r}}(s)\| \equiv \|\mathbf{r}'(s)\| = 1. \tag{2.321}$$

So, we have

$$E(C) = \int_S \int_S \left(\frac{1}{\|\mathbf{r}(s) - \mathbf{r}(t)\|^2} - \frac{1}{l(s,t)^2} \right) dsdt. \tag{2.322}$$

Let us observe infinitesimal bending of C:

$$C_\epsilon : \mathbf{r}_\epsilon = \mathbf{r}(s) + \epsilon \mathbf{z}(s).$$

The energy of the deformed knot C_ϵ will be

$$E(C_\epsilon) = \int_S \int_S \left(\frac{1}{\|\mathbf{r}_\epsilon(s) - \mathbf{r}_\epsilon(t)\|^2} - \frac{1}{l_\epsilon(s,t)^2} \right) \|\mathbf{r}_\epsilon'(s)\| \, \|\mathbf{r}_\epsilon'(t)\| \, dsdt,$$

where $l_\epsilon(s,t)$ is the distance long the deformed knot C_ϵ between the points $\mathbf{r}_\epsilon(s)$ and $\mathbf{r}_\epsilon(t)$.

Let us consider the following integrals:

$$E_1(C_\epsilon) = \int_S \int_S \frac{1}{\|\mathbf{r}_\epsilon(s) - \mathbf{r}_\epsilon(t)\|^2} \|\mathbf{r}'(s)\| \, \|\mathbf{r}_\epsilon'(t)\| \, dsdt, \tag{2.323}$$

$$E_2(C_\epsilon) = \int_S \int_S \frac{1}{l_\epsilon(s,t)^2} \|\mathbf{r}_\epsilon'(s)\| \, \|\mathbf{r}'(t)\| \, dsdt. \tag{2.324}$$

We are going to check the variations δE_1 and δE_2. It holds

$$\begin{aligned}\|\mathbf{r}_\epsilon'(s)\| &= \|\mathbf{r}'(s) + \epsilon \mathbf{z}'(s)\| = (\|\mathbf{r}'(s) + \epsilon \mathbf{z}'(s)\|^2)^{1/2} \\ &= (\|\mathbf{r}'(s)\|^2 + 2\epsilon \mathbf{r}'(s) \cdot \mathbf{z}'(s) + \epsilon^2 \|\mathbf{z}'(s)\|^2)^{1/2}\end{aligned}$$

Using (2.24), (2.321), and Maclaurin formula, we obtain

$$\|\mathbf{r}'_\epsilon(s)\| = 1 + \frac{1}{2}\epsilon^2\|\mathbf{z}'(s)\|^2 - \frac{1}{8}\epsilon^4\|\mathbf{z}'(s)\|^4 + \cdots \tag{2.325}$$

Therefore, $\delta\|\mathbf{r}'(s)\| = 0$, i.e. $\|\mathbf{r}'_\epsilon(s)\| = \|\mathbf{r}'(s)\| = 1$ after neglecting the terms with ϵ^n, $n \geq 2$. Further,

$$\begin{aligned}
E_1(C_\epsilon) &= \int_S\int_S \frac{1}{\|\mathbf{r}_\epsilon(s) - \mathbf{r}_\epsilon(t)\|^2}\, dsdt \\
&= \int_S\int_S \frac{1}{\|(\mathbf{r}(s) - \mathbf{r}(t)) + \epsilon(\mathbf{z}(s) - \mathbf{z}(t))\|^2}\, dsdt \\
&= \int_S\int_S \frac{1}{\|\mathbf{r}(s) - \mathbf{r}(t)\|^2} \\
&\quad \cdot \left(1 - \frac{2\epsilon(\mathbf{r}(s) - \mathbf{r}(t)) \cdot (\mathbf{z}(s) - \mathbf{z}(t))}{\|\mathbf{r}(s) - \mathbf{r}(t)\|^2} + \epsilon^2 \ldots\right) dsdt
\end{aligned}$$

wherefrom, we obtain that

$$\delta E_1(C) = \frac{dE_1(C_\epsilon)}{ds}\bigg|_{\epsilon=0} = -2\int_S\int_S \frac{(\mathbf{r}(s) - \mathbf{r}(t)) \cdot (\mathbf{z}(s) - \mathbf{z}(t))}{\|\mathbf{r}(s) - \mathbf{r}(t)\|^4}\, dsdt \tag{2.326}$$

On the other hand, we have

$$\begin{aligned}
l_\epsilon(s,t) &= \int_s^t \|\dot{\mathbf{r}}_\epsilon(u)\|\, du = \int_s^t \|\dot{\mathbf{r}}(u) + \epsilon\dot{\mathbf{z}}(u)\|\, du \\
&= l(s,t) + \frac{1}{2}\epsilon^2\int_s^t \frac{\|\dot{\mathbf{z}}(u)\|^2}{\|\dot{\mathbf{r}}(u)\|}\, du - \frac{1}{8}\epsilon^4\int_s^t \frac{\|\dot{\mathbf{z}}(u)\|^4}{\|\dot{\mathbf{r}}(u)\|^3}\, du + \ldots,
\end{aligned}$$

i.e. $\delta l(s,t) = 0$ and, using (2.324) and (2.325), we get

$$\delta E_2(C) = \frac{dE_2(C_\epsilon)}{d\epsilon}\bigg|_{\epsilon=0} = 0. \tag{2.327}$$

According to (2.326) and (2.327), we obtain the next theorem.

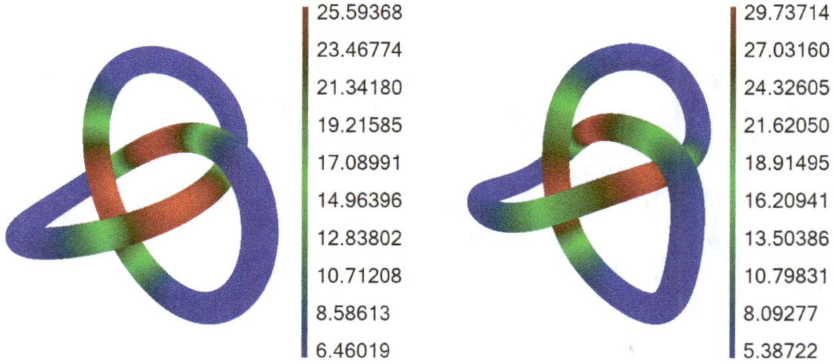

Fig. 2.61. Trefoil knot: basic and first order inf. bent with $\epsilon = 0.8$.

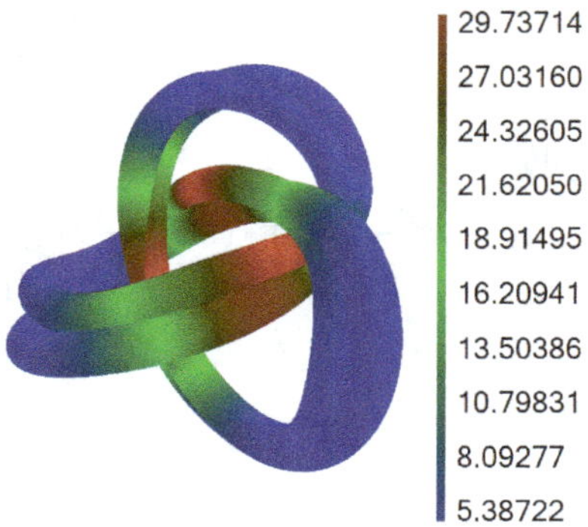

Fig. 2.62. Trefoil knot: both knots together.

Theorem 2.70 ([Kauffman *et al.* (2019)]). *Under infinitesimal bending of the knot C, variation of the Möbius energy is*

$$\delta E(C) = \frac{dE_1(C_\epsilon)}{ds}\bigg|_{\epsilon=0} = 2\int_S\int_S \frac{(\mathbf{r}(t)-\mathbf{r}(s))\cdot(\mathbf{z}(s)-\mathbf{z}(t))}{\|\mathbf{r}(s)-\mathbf{r}(t)\|^4}\,dsdt. \tag{2.328}$$

Möbius energy calculation on knots is already studied, also there are known efforts aimed to minimize Möbius energy on transformed knots. We will give some examples aimed to illustrate changes in Möbius energy on knots under infinitesimal bending. We will also use values of inner integral in Möbius energy definition for coloring bent knots.

The first example is a trefoil knot given by the parametric equations: $x = \sin(u) + 2\sin(2u)$, $y = \cos(u) - 2\cos(2u)$, $z = -\sin(3u)$.

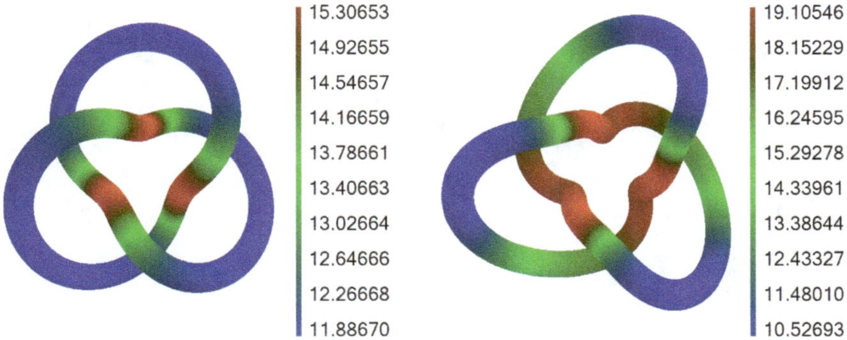

Fig. 2.63. $p = 2$, $q = 3$ torus knot: basic and first-order inf. bent with $\epsilon = 1.2$.

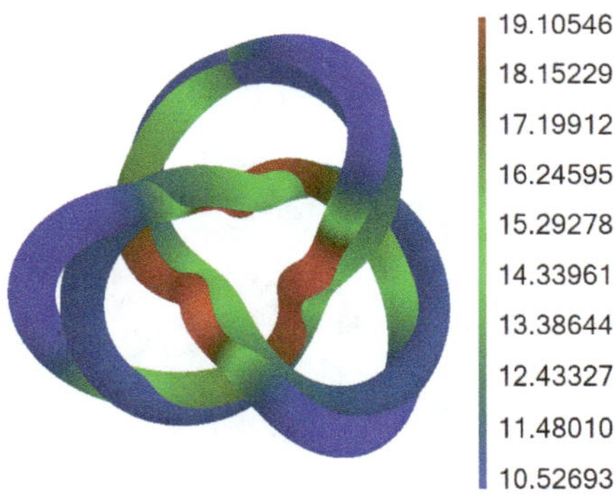

Fig. 2.64. $p = 2$, $q = 3$ torus knot: both knots together.

We will give trefoil knot bent by the first-order infinitesimal bending, given by $p(u) = \cos(3u)$ and $q(u) = \sin(6u)$, see Figs. 2.61 and 2.62.

The numerically calculated Möbius energy of the first-order infinitesimally bent trefoil knot is 75.9123 and 79.3857 for $\epsilon = 0$ and $\epsilon = 0.8$ respectively.

For $p2q3$ torus knot, see Figs. 2.63 and 2.64, given by the parametric equations: $x = (\cos(3u) + 2) \cdot \cos(2u)$, $y = (\cos(3u) + 2) \cdot \sin(2u)$, $z = -\sin(3u)$, bending fields are defined by: $p(u) = \cos(6u)$ and $q(u) = \sin(6u)$.

The numerically calculated Möbius energy of the first-order infinitesimally bent $p = 2$, $q = 3$ torus knot is 81.8068 and 95.1181 for $\epsilon = 0$ and $\epsilon = 1.2$ respectively.

Chapter 3

Surfaces

3.1 Surfaces

We are going to introduce some basic facts and definitions about surfaces. For more details, we refer to [Carmo (1976); Green and Zerna (1968); Struik (1961)].

Definition 3.1. A **parameterized surface** S is a injective map $\mathbf{r} : \mathcal{D} \to \mathcal{R}^3$ of class $C^k\,(k \geq 1)$, where $\mathcal{D}$ is an open set in $\mathcal{R}^2$. A surface is said to be **regular** if $\mathbf{r}_u \times \mathbf{r}_v \neq 0$ for all $(u, v) \in \mathcal{D}$.

Usually, we denote the parameterized surface as follows:

$$S : \mathbf{r} = \mathbf{r}(u, v), \quad (u, v) \in \mathcal{D} \subseteq \mathcal{R}^2. \tag{3.1}$$

Unless otherwise stated, we will assume all surfaces are regular. Also, we will not always specify the class of differentiability and will assume that k is a sufficiently large number as required by the consideration in question.

Further, we can consider a surface on a set $\mathcal{D}$ which is not open as the restriction of a surface defined on an open set $\mathcal{D}^* \supset \mathcal{D}$ in the sense of Definition 3.1.

Example 3.1. Let us consider a curve

$$C_0 \,:\, \boldsymbol{\rho} = \boldsymbol{\rho}(u)$$

and through each point of that curve, a unit vector $\mathbf{e} = \mathbf{e}(u)$. Then, the parameterized surface

$$S \,:\, \mathbf{r}(u, v) = \boldsymbol{\rho}(u) + v\mathbf{e}(u),$$

$v \in \mathcal{R}$, is said to be the **ruled surface**. The curve C_0 is called the **directrix** of the surface S. The straight lines in the direction $\mathbf{e}(u)$ are **generatrices** of that surface.

Example 3.2. Let

$$C : x = f(v), \quad z = g(v), \quad a < v < b, \quad f(v) > 0,$$

be a parameterization of a regular plane curve in the xOz plane. Let S be the surface obtained by rotation of C about the z axis. The vector equation of that surface is

$$S : \mathbf{r}(u, v) = (f(v) \cos u, f(v) \sin u, g(v)),$$

$0 < u < 2\pi$. Obviously, S is a regular parameterized surface which is called the **surface of revolution**. The curve C is called the **generating curve** of S, and the z axis is the **rotation axis** of S. The circles described by the points of C are called the **parallels** of S, and the various positions of C on S are called the **meridians** of S.

It can be shown that parts of a sphere can be parameterized, but there is no parameterization of the whole sphere. However, in accordance with the intuitive idea of the surface as a subset in $\mathcal{R}^3$, the notion of surface can be understood in a somewhat more complex way. In essence, a surface can be understood as a union of the images of parameterized surfaces $\mathbf{r} : \mathcal{D} \to \mathcal{R}^3$, so that the sets $\mathbf{r}(\mathcal{D}) = W$ overlap. At the overlap, we require that the parameterizations are related by coordinate transformations. Thus, the sphere is an example of surfaces. Furthermore, it is a **closed surface** (compact and without boundary).

A surface S is **orientable** if there is a unit normal vector field on S. A choice of unit normal vector field for a surface is called an **orientation of the surface**. An orientable surface with a prescribed unit normal vector field is called an **oriented surface**. Note that all the surfaces, we will study in this book are orientable.

Below, we will use the indices to indicate coordinate for simplicity. Let $S \subset \mathcal{R}^3$

$$S : \mathbf{r} = \mathbf{r}(u^1, u^2), \quad (u^1, u^2) \in D \subset \mathcal{R}^2, \quad (u^1 = u,\, u^2 = v), \tag{3.2}$$

be a surface in three-dimensional Euclidean space. The surface S is expressed in terms of the curvilinear surface coordinates u^i, and the relationship with the Cartesian coordinates x^i can be presented in the following way:

$$\begin{aligned}\mathbf{r} &= \mathbf{r}(u^1, u^2) \\ &= x^1(u^1, u^2)\mathbf{i}_1 + x^2(u^1, u^2)\mathbf{i}_2 + x^3(u^1, u^2)\mathbf{i}_3 \\ &= x^\alpha \mathbf{i}_\alpha, \end{aligned} \tag{3.3}$$

applying the **Einstein summation convention**, $\mathbf{i}_\alpha$ are unit vectors in the directions of the Cartesian axes. The index α by which the addition is performed is called **dumb index**.

Let us introduce the following **covariant base vectors**

$$\mathbf{r}_1 = \mathbf{r}_{,1}, \quad \mathbf{r}_2 = \mathbf{r}_{,2} \tag{3.4}$$

where the subscripts 1 and 2 mean partial differentiation with respect to u^1 and u^2, respectively. The vectors $\mathbf{r}_1$ and $\mathbf{r}_2$ lie in the tangent plane to the surface in the direction of the coordinate lines $u = u^1$ and $u = u^2$, respectively. The unit vector $\boldsymbol{\nu}$

$$\boldsymbol{\nu} = \frac{\mathbf{r}_1 \times \mathbf{r}_2}{\|\mathbf{r}_1 \times \mathbf{r}_2\|} \tag{3.5}$$

is the **unit normal** to the surface. Vectors $\mathbf{r}_1$, $\mathbf{r}_2$, $\boldsymbol{\nu}$ determine the **moving frame** at the point $M(\mathbf{r}(u^1, u^2))$ of the surface S.

The **contravariant base vectors**, $\mathbf{r}^j$, are defined by the scalar product

$$\mathbf{r}^j \cdot \mathbf{r}_i = \delta_i^j, \quad \mathbf{r}^j \cdot \boldsymbol{\nu} = 0, \tag{3.6}$$

where

$$\delta_i^j = \begin{cases} 1, & i = j \\ 0, & i \neq j \end{cases}$$

is the **Kronecker $\delta-$ symbol**. Thus $\mathbf{r}^1$ and $\mathbf{r}^2$ lie in the tangent plane to the surface perpendicular to $\mathbf{r}_2$ and $\mathbf{r}_1$, respectively. (Note that Latin indices i, j, k may have the values $1, 2$, unless otherwise stated.)

The **displacement** $d\mathbf{r}$ between two enough close points (u^1, u^2) and $(u^1 + du^1, u^2 + du^2)$ on the surface, is given by

$$d\mathbf{r} = \mathbf{r}_1 du^1 + \mathbf{r}_2 du^2 = \mathbf{r}_i du^i. \tag{3.7}$$

The **covariant and contravariant metric tensors** are defined by the scalar products respectively

$$g_{ij} = g_{ji} = \mathbf{r}_i \cdot \mathbf{r}_j \text{ i } g^{ij} = g^{ji} = \mathbf{r}^i \cdot \mathbf{r}^j. \tag{3.8}$$

Obviously, $g_{ip}g^{pj} = \delta_i^j$. Using the metric tensor, we perform the lowering and the rising of indices. For instance,

$$v^i = g^{ip} v_p, \quad v_i = g_{ip} v^p,$$

where v^i and v_i are contravariant and covariant coordinates, respectively, of the quantity v.

A quantity is a **tensor** if it satisfies the rule for changing its components under changing of surface coordinates. Thus, for instance, let u^i and $u^{i'}$ be two coordinate systems which can be used on the surface. Then, the components of a covariant tensor of the first order (a surface vector) satisfy

$$v_{i'} = \frac{\partial u^i}{\partial u^{i'}} v_i. \tag{3.9}$$

Similarly, the contravariant, the covariant and the mixed tensor of the second order respectively, satisfy

$$q^{i'j'} = \frac{\partial u^{i'}}{\partial u^i} \frac{\partial u^{j'}}{\partial u^j} q^{ij}, \quad q_{i'j'} = \frac{\partial u^i}{\partial u^{i'}} \frac{\partial u^j}{\partial u^{j'}} q_{ij}, \quad q^i_{j'} = \frac{\partial u^{i'}}{\partial u^i} \frac{\partial u^j}{\partial u^{j'}} q^i_j. \tag{3.10}$$

A quantity such as $q^i_i = q^1_1 + q^2_2$ is a scalar, and h^i_{ij} is a tensor of the first order and so on for higher order tensors. This process is known as **contraction**.

The square of the line element ds is determined by the equation

$$ds^2 = d\mathbf{r} \cdot d\mathbf{r} = g_{11}(du^1)^2 + 2g_{12} du^1 du^2 + g_{22}(du^2)^2, \tag{3.11}$$

which presents the **first fundamental form** of the surface. If

$$g = g_{11} g_{22} - (g_{12})^2 \tag{3.12}$$

is the determinant of the first fundamental form, then it will be valid

$$g^{11} = \frac{g_{22}}{g}, \quad g^{22} = \frac{g_{11}}{g}, \quad g^{12} = g^{21} = -\frac{g_{12}}{g}. \tag{3.13}$$

The **coefficients of the second fundamental form** are introduced in this way

$$b_{ij} = b_{ji} = \boldsymbol{\nu} \cdot \mathbf{r}_{i,j} = -\mathbf{r}_i \cdot \boldsymbol{\nu}_j, \tag{3.14}$$

$\boldsymbol{\nu}_j = \boldsymbol{\nu}_{,j}$, while the **second fundamental form** is given by the scalar product

$$d^2\mathbf{r} \cdot \boldsymbol{\nu} = b_{ij} du^i du^j. \tag{3.15}$$

The **shape operator** $\underline{S}$ is a linear map defined by

$$\underline{S}(\mathbf{r}_i) = -\boldsymbol{\nu}_i \tag{3.16}$$

and with the second fundamental form is connected by

$$d^2\mathbf{r} \cdot \boldsymbol{\nu} = \underline{S}(d\mathbf{r}) \cdot d\mathbf{r}.$$

The **normal curvature** in the direction $d\mathbf{r}$ is presented using the first and the second fundamental form by the equation

$$\bar{K} = \frac{d^2\mathbf{r} \cdot \boldsymbol{\nu}}{d\mathbf{r} \cdot d\mathbf{r}} = \frac{\underline{S}(d\mathbf{r}) \cdot d\mathbf{r}}{\|d\mathbf{r}\|^2}. \tag{3.17}$$

Let us consider the equation

$$k^2 - k(b_1^1 + b_2^2) + b_1^1 b_2^2 - b_1^2 b_2^1 = 0, \tag{3.18}$$

where $b_j^i = g^{ip} b_{pj}$. For each of two values of k obtained from the previous equation, there is a corresponding direction on the surface. These two directions are orthogonal, and that the two values of k are equal to the maximum and minimum values of normal curvature at that particular point on the surface and are known as the **principal curvatures**. The corresponding directions are said to be the **principal directions**. If the two values of k are equal at a point on the surface, then all directions are principal directions at that point.

The **mean curvature** H is the arithmetic mean of the two principal curvatures, i.e.

$$H = \frac{1}{2}\mathrm{tr}\underline{S} = \frac{k_{(1)} + k_{(2)}}{2} = \frac{b_1^1 + b_2^2}{2} = \frac{b_i^i}{2}$$
$$= \frac{g_{11}b_{22} - 2g_{12}b_{12} + g_{22}b_{11}}{2(g_{11}g_{22} - (g_{12})^2)}. \tag{3.19}$$

The **Gaussian curvature** K is the product of the two principal curvatures

$$K = \det\underline{S} = k_{(1)}k_{(2)} = b_1^1 b_2^2 - b_1^2 b_2^1 = \frac{1}{2}(b_i^i b_j^j - b_i^j b_j^i)$$
$$= \frac{b_{11}b_{22} - (b_{12})^2}{g_{11}g_{22} - (g_{12})^2}. \tag{3.20}$$

The equations

$$\Gamma_{i.jk} = \frac{1}{2}(g_{ji,k} - g_{jk,i} + g_{ik,j}), \tag{3.21}$$

$$\Gamma_{jk}^i = g^{ip}\Gamma_{p.jk} = \frac{1}{2}g^{ip}(g_{jp,k} - g_{jk,p} + g_{pk,j}) \tag{3.22}$$

define respectively the **Cristoffel symbols of the first and the second kind**. They are symmetric by the last two indices (the first kind) and by the lower indices (the second kind). The Cristoffel symbols satisfy

$$\Gamma_{pi}^i = \frac{\partial(\ln\sqrt{g})}{\partial u^p}. \tag{3.23}$$

Let $\mathbf{r}_{ij} = \mathbf{r}_{i,j}$. The **derivational formulas of the first kind** are

$$\mathbf{r}_{ij} = \Gamma_{ij}^p \mathbf{r}_p + b_{ij}\boldsymbol{\nu}, \tag{3.24}$$

while the equations

$$\boldsymbol{\nu}_i = -b_i^p \mathbf{r}_p \tag{3.25}$$

are the **derivational formulas of the second kind**. Using the derivational formulas, the vectors $\mathbf{r}_{jk}$ (i.e. $\mathbf{r}_{11}$, $\mathbf{r}_{12}$, and $\mathbf{r}_{22}$) and $\boldsymbol{\nu}_i$ can be expressed via the vectors of the moving frame $\mathbf{r}_1$, $\mathbf{r}_2$ and $\boldsymbol{\nu}$ of the surface.

Let $\mathbf{v} = v^i\mathbf{r}_i = v_i\mathbf{r}^i$ be the surface vector. The partial derivative of the vector $\mathbf{v}$ is given by

$$\begin{aligned}\frac{\partial \mathbf{v}}{\partial u^j} &= \mathbf{v}_{,j} = v^i_{,j}\mathbf{r}_i + v^i\mathbf{r}_{ij} \\ &= v_{i,j}\mathbf{r}^i + v_i\mathbf{r}^i_{,j}.\end{aligned} \tag{3.26}$$

Applying the derivative formulas of the first kind, we obtain

$$\mathbf{v}_{,j} = v^i_{;j}\mathbf{r}_i + v^i b_{ij}\boldsymbol{\nu}, \tag{3.27}$$

where the **covariant derivative**; is defined in the following way:

$$v^i_{;j} = v^i_{,j} + \Gamma^i_{pj}v^p, \tag{3.28}$$

The covariant derivative can be replaced by ordinary partial derivative:

$$v^i{}_{;i} = \frac{1}{g^{1/2}}(g^{1/2}v^i)_{,i}. \tag{3.29}$$

Equation (3.26) can be written as follows

$$\mathbf{v}_{,j} = v_{i;j}\mathbf{r}^i + v_i b^i_j\boldsymbol{\nu}, \tag{3.30}$$

where the covariant derivative is

$$v_{i;j} = v_{i,j} - \Gamma^p_{ij}v_p. \tag{3.31}$$

Similarly, the covariant derivative of the tensor of second order is given in the following way:

$$q^{ij}_{;k} = q^{ij}_{,k} + \Gamma^i_{kp}q^{pj} + \Gamma^j_{kp}q^{ip}, \tag{3.32}$$

$$q^i_{j;k} = q^i_{j,k} + \Gamma^i_{kp}q^p_j - \Gamma^p_{jk}q^i_p, \tag{3.33}$$

$$q_{ij;k} = q_{ij,k} - \Gamma^p_{ik}q_{pj} - \Gamma^p_{jk}q_{ip}. \tag{3.34}$$

The covariant derivative of a scalar is equal to its partial derivative. The covariant derivative of the metric tensor is equal to zero.

The covariant derivative has some properties that are consistent with the properties of ordinary derivatives. For instance,

- the covariant derivative of the sum (the difference) is equal to the sum (the difference) of the covariant derivatives,

- for the ordinary (external) product and the composition (internal product) the Leibniz rule holds.

The equations

$$b_{ip;j} = b_{ij;p}, \tag{3.35}$$

($j = 1$, $p = 2$ and $i = 1$ or 2) are said to be **Codazzi equations.** The equation

$$\frac{1}{2}(b^j_j b^p_p - b^p_j b^j_p) = \frac{1}{2}[g^{ij}(\Gamma^p_{ij})_{,p} - g^{ip}(\Gamma^j_{ij})_{,p} + g^{ij}\Gamma^q_{ij}\Gamma^p_{qp} - g^{ip}\Gamma^q_{ij}\Gamma^j_{qp}] \tag{3.36}$$

presents **Gauss' theorem**. As the left side of equality (3.36) is equal to the Gaussian curvature K, Gauss' theorem says that the Gaussian curvature can be expressed in terms of the metric tensor only, i.e. it belongs to the **intrinsic geometry of the surface**. Necessary conditions for existence of surfaces with given the first and the second fundamental form are the validity of Gauss–Coddazi equations.

The **integral theorem** says

$$\iint_S v^i{}_{;i}\, dA = \int_C c_{ij} v^i \frac{du^j}{ds}\, ds, \tag{3.37}$$

where dA is an element of the surface S and ds is an element of the boundary C of the surface. The boundary integral has to be performed in the direction that is clock-wise looking along the surface normal $\boldsymbol{\nu}$. $\mathbf{v} = v_i \mathbf{r}^i = v^i \mathbf{r}_i$ is a surface vector; c_{ij} is an anti-symmetric tensor defined in the following way

$$c_{11} = c_{22} = 0, \quad c_{12} = -c_{21} = \sqrt{g}. \tag{3.38}$$

Let be given a surface S (3.2) and a curve

$$C : \mathbf{r} = \mathbf{r}[u^1(s), u^2(s)], \tag{3.39}$$

on S. Let us observe the following decomposition of the vector $\mathbf{r}''$,

$$\mathbf{r}'' = k\mathbf{n} = \bar{K}\boldsymbol{\nu} + \lambda^1 \mathbf{r}_1 + \lambda^2 \mathbf{r}_2 = \bar{K}\boldsymbol{\nu} + \lambda^p \mathbf{r}_p.$$

A scalar $\bar{K}$ is the normal curvature of the curve C. The **geodesic curvature** of C is

$$\underset{g}{K} = \|\lambda^p \mathbf{r}_p\|. \tag{3.40}$$

In relation to the curvature (flexion) of C, we have

$$\bar{K} = k\cos\theta, \quad \underset{g}{K} = k\sin\theta,$$

where θ is the angle between the normal to the surface $\boldsymbol{\nu}$ and the normal to the curve $\mathbf{n}$. The geodesic curvature should be presented in the form

$$\underset{g}{K} = \sqrt{g}(u^{1'}\lambda^2 - u^{2'}\lambda^1),$$
$$\lambda^i = u^{i''} + \Gamma^i_{jk}u^{j'}u^{k'}, \quad u^{i'} = \frac{du^i}{ds}, \quad u^{i''} = \frac{d^2u^i}{ds^2}, \tag{3.41}$$

i.e. it is determined by the first fundamental form of the surface.

Curves of zero geodesic curvature are said to be **geodesics.**

3.2 Infinitesimal Deformations of Surfaces

Let us consider a surface S given in the vector form

$$S : \mathbf{r} = \mathbf{r}(u^1, u^2), \quad (u^1, u^2) \in \mathcal{D} \subset \mathcal{R}^2. \tag{3.42}$$

Definition 3.2. The surface S_ϵ is the **deformation** of the surface S if it is included in continuous family of surfaces

$$S_\epsilon : \tilde{\mathbf{r}}(u^1, u^2, \epsilon) = \mathbf{r}_\epsilon(u^1, u^2), \quad \epsilon \geq 0,\ \epsilon \to 0,$$

$$\mathbf{r}_\epsilon : \mathcal{D} \to \mathcal{R}^3,$$

and we get S for $\epsilon = 0$.

Definition 3.3. Let S_ϵ be a deformation of the surface S given with Eq. (3.42). If

$$S_\epsilon : \tilde{\mathbf{r}} = \mathbf{r}_\epsilon(u^1, u^2) = \mathbf{r}(u^1, u^2) + \epsilon \overset{(1)}{\mathbf{z}}(u^1, u^2) + \epsilon^2 \overset{(2)}{\mathbf{z}}(u^1, u^2) + \cdots$$
$$+ \epsilon^m \overset{(m)}{\mathbf{z}}(u^1, u^2), \quad m \geq 1, \tag{3.43}$$

where $\overset{(j)}{\mathbf{z}}(u, v) \in C^k (k \geq 1)$, $j = 1, \ldots, m$ are given fields, family S_ϵ is **infinitesimal deformation of the order m of the surface S.** The field $\overset{(j)}{\mathbf{z}}$ is said to be **infinitesimal deformation field of the order** j.

Giving different more special conditions we get different kinds of surface deformations.

3.3 Infinitesimal Bending of Surfaces

We are giving here some basic facts on infinitesimal bending of surfaces according to [Efimov (1948); Vekua (1959)].

Definition 3.4 ([Efimov (1948)]). Infinitesimal deformation of the order **m** of a surface S, is **infinitesimal bending of the order m** if

$$d\tilde{s}^2 - ds^2 = o(\epsilon^m), \tag{3.44}$$

i.e. if the difference of the squares of the first fundamental forms at the deformed and the initial surface is an infinitesimal value of the order higher then m with respect to ϵ. The field $\overset{(j)}{\mathbf{z}}(u^1, u^2)$ in (3.43) is **infinitesimal bending field of the order** j, $j = 1,\ldots, m$.

The equivalent to the definition of the infinitesimal bending is given with the next theorem.

Theorem 3.1. *During infinitesimal bending the length of each smooth curve stays unchanged with taken precision, i.e. the change of the arc length on the surface S is infinitesimal value of the higher order then m with respect to ϵ $(\epsilon \to 0)$.*

Based on the definition of the infinitesimal bending of the order m the next relation is valid

$$d\mathbf{r}_\epsilon^2 - d\mathbf{r}^2 = (d\mathbf{r}_\epsilon - d\mathbf{r}) \cdot (d\mathbf{r}_\epsilon + d\mathbf{r}) = o(\epsilon^m).$$

From here

$$d\left(\sum_{j=1}^{m} \epsilon^j \overset{(j)}{\mathbf{z}}\right) \cdot d\left(2\mathbf{r} + \sum_{j=1}^{m} \epsilon^j \overset{(j)}{\mathbf{z}}\right) = o(\epsilon^m),$$

or more precise

$$\left(\epsilon d\overset{(1)}{\mathbf{z}} + \epsilon^2 d\overset{(2)}{\mathbf{z}} + \epsilon^3 d\overset{(3)}{\mathbf{z}} + \epsilon^4 d\overset{(4)}{\mathbf{z}} + \cdots + \epsilon^m d\overset{(m)}{\mathbf{z}}\right)$$
$$\cdot \left(2d\mathbf{r} + \epsilon d\overset{(1)}{\mathbf{z}} + \epsilon^2 d\overset{(2)}{\mathbf{z}} + \epsilon^3 d\overset{(3)}{\mathbf{z}} + \cdots + \epsilon^m d\overset{(m)}{\mathbf{z}}\right) = o(\epsilon^m).$$

The necessary and sufficient condition for the left side to be infinitesimal value with respect to ε^m is to be

$$\begin{aligned}
&d\mathbf{r}\cdot d\overset{(1)}{\mathbf{z}}=0,\\
&2d\mathbf{r}\cdot d\overset{(2)}{\mathbf{z}}+(d\overset{(1)}{\mathbf{z}})^2=0,\\
&2d\mathbf{r}\cdot d\overset{(3)}{\mathbf{z}}+2d\overset{(1)}{\mathbf{z}}\cdot d\overset{(2)}{\mathbf{z}}=0,\\
&\dots\dots\dots\dots\dots\dots\dots\dots\\
&2d\mathbf{r}\cdot d\overset{(m)}{\mathbf{z}}+\sum_{l=1}^{m-1}d\overset{(l)}{\mathbf{z}}\cdot d\overset{(m-l)}{\mathbf{z}}=0,
\end{aligned} \tag{3.45}$$

or, shorter

$$d\mathbf{r}\cdot d\overset{(1)}{\mathbf{z}}=0,\ 2d\mathbf{r}\cdot d\overset{(j)}{\mathbf{z}}+\sum_{l=1}^{j-1}d\overset{(l)}{\mathbf{z}}\cdot d\overset{(j-l)}{\mathbf{z}}=0,\quad \text{for}\quad j=2,\dots,m. \tag{3.46}$$

According to that, the next theorem states.

Theorem 3.2. *Necessary and sufficient condition for the surface S_ϵ to be infinitesimal bending of the m-th order of the surface S is to be valid* (3.46).

The condition (3.46) given by Theorem 3.2 is equivalent to the system

$$\begin{aligned}
&\left(2\mathbf{r}_1\cdot\overset{(j)}{\mathbf{z}_1}+\sum_{l=1}^{j-1}\overset{(l)}{\mathbf{z}_1}\cdot\overset{(j-l)}{\mathbf{z}_1}\right)(du^1)^2\\
&+\left(\mathbf{r}_1\cdot\overset{(j)}{\mathbf{z}_2}+\mathbf{r}_2\cdot\overset{(j)}{\mathbf{z}_1}+\sum_{l=1}^{j-1}\overset{(l)}{\mathbf{z}_1}\cdot\overset{(j-l)}{\mathbf{z}_2}\right)du^1du^2\\
&+\left(2\mathbf{r}_2\cdot\overset{(j)}{\mathbf{z}_2}+\sum_{l=1}^{j-1}\overset{(l)}{\mathbf{z}_2}\cdot\overset{(j-l)}{\mathbf{z}_2}\right)(du^2)^2=0,\quad j=2,\dots,m,
\end{aligned}$$

i.e. to the system of the partial differential equations

$$
\begin{aligned}
&2\mathbf{r}_1 \cdot \overset{(j)}{\mathbf{z}_1} + \sum_{l=1}^{j-1} \overset{(l)}{\mathbf{z}_1} \cdot \overset{(j-l)}{\mathbf{z}_1} = 0,\\
&2\mathbf{r}_2 \cdot \overset{(j)}{\mathbf{z}_2} + \sum_{l=1}^{j-1} \overset{(l)}{\mathbf{z}_2} \cdot \overset{(j-l)}{\mathbf{z}_2} = 0, \qquad (3.47)\\
&\mathbf{r}_1 \cdot \overset{(j)}{\mathbf{z}_2} + \mathbf{r}_2 \cdot \overset{(j)}{\mathbf{z}_1} + \sum_{l=1}^{j-1} \overset{(l)}{\mathbf{z}_1} \cdot \overset{(j-l)}{\mathbf{z}_2} = 0, \quad \text{for } j = 2, \dots, m,
\end{aligned}
$$

$\overset{(j)}{\mathbf{z}_1} = \overset{(j)}{\mathbf{z}}_{,1} = \frac{\partial \overset{(j)}{\mathbf{z}}}{\partial u^1}$, $\overset{(j)}{\mathbf{z}_2} = \overset{(j)}{\mathbf{z}}_{,2} = \frac{\partial \overset{(j)}{\mathbf{z}}}{\partial u^2}$.

3.4 Infinitesimal Bending of the First Order

Let a surface S, given in the vector form (3.42), be included in the family of surfaces

$$S_\epsilon : \tilde{\mathbf{r}}(u^1, u^2, \epsilon) = \mathbf{r}_\epsilon(u^1, u^2) = \mathbf{r}(u^1, u^2) + \epsilon\mathbf{z}(u^1, u^2), \qquad (3.48)$$

which is infinitesimal bending of the first order, or shorter, infinitesimal bending. The vector field $\mathbf{z}(u^1, u^2)$ is infinitesimal bending field (of the first order), or **velocity field**. The necessary and sufficient condition for infinitesimal bending reduces to

$$d\mathbf{r} \cdot d\mathbf{z} = 0. \qquad (3.49)$$

Equation (3.49) is equivalent to the next three partial differential equations:

$$\mathbf{r}_1 \cdot \mathbf{z}_1 = 0, \quad \mathbf{r}_1 \cdot \mathbf{z}_2 + \mathbf{r}_2 \cdot \mathbf{z}_1 = 0, \quad \mathbf{r}_2 \cdot \mathbf{z}_2 = 0. \qquad (3.50)$$

From Eq. (3.48), we conclude

$$\frac{\partial \tilde{\mathbf{r}}(u^1, u^2, \epsilon)}{\partial \epsilon} = \mathbf{z}(u^1, u^2).$$

It expresses the fact that $\mathbf{z}$ is the velocity field under the moving of a surface point under bending taking that ϵ indicates time.

Theorem 3.3 ([Vekua (1959)]). *Under infinitesimal bending of the surfaces each line element gets non-zero addition, which is the infinitesimal value of the second order with respect to* ϵ, *i.e.*

$$ds_\epsilon - ds = o(\epsilon) \geq 0. \tag{3.51}$$

Proof. Based on (3.48) and (3.49), we have

$$(ds_\epsilon)^2 = (d\mathbf{r})^2 + \epsilon(d\mathbf{z})^2 = (ds)^2 + \epsilon^2(d\mathbf{z})^2 \Rightarrow (ds_\epsilon)^2$$
$$= (ds)^2 \left[1 + \epsilon^2 \left(\frac{d\mathbf{z}}{ds}\right)^2\right],$$

i.e.

$$ds_\epsilon = ds \left[1 + \epsilon^2 \left(\frac{d\mathbf{z}}{ds}\right)^2\right]^{\frac{1}{2}}.$$

After using the Maclaurin formula, we have

$$ds_\epsilon = ds \left[1 + \frac{1}{2}\epsilon^2 \left(\frac{d\mathbf{z}}{ds}\right)^2 - \frac{1}{8}\epsilon^4 \left(\frac{d\mathbf{z}}{ds}\right)^4 + \cdots\right],$$

i.e.

$$ds_\epsilon - ds = \frac{1}{2}\epsilon^2 \left(\frac{d\mathbf{z}}{ds}\right)^2 ds - \cdots$$

wherefrom, we obtain (3.51). □

Theorem 3.4 ([Vekua (1959)]). *Let* $s = s(\epsilon)$ *be the arc length of the curve* C_ϵ *on the surface* S_ϵ. *Necessary and sufficient condition for the infinitesimal bending of the initial surface* $S = S_0$ *is*

$$\left.\frac{\partial s_\epsilon}{\partial \epsilon}\right|_{\epsilon=0} = 0, \tag{3.52}$$

i.e. the velocity of the change of arc length at the initial moment to be zero.

Proof. If t is a parameter which determine the curve C_ϵ on the surface S_ϵ, then the position vector of an arbitrary point of that curve will be

$$\tilde{\mathbf{r}}(t,\epsilon) = \mathbf{r}_\epsilon(t) = \mathbf{r}(u^1(t), u^2(t)) + \epsilon\mathbf{z}(u^1(t), u^2(t)). \tag{3.53}$$

Then, we have

$$\frac{\partial}{\partial \epsilon}\left(\frac{\partial \tilde{\mathbf{r}}}{\partial t}\right) = \frac{\partial}{\partial t}\left(\frac{\partial \tilde{\mathbf{r}}}{\partial \epsilon}\right) = \frac{\partial \mathbf{z}}{\partial t} = \frac{d\mathbf{z}}{dt}.$$

For a fixed ϵ, arc length of the curve C_ϵ is

$$s_\epsilon(t) = \int_{t_o}^{t} \sqrt{\left(\frac{\partial \tilde{\mathbf{r}}}{\partial t}\right)^2}\, dt,$$

$$\frac{\partial s_\epsilon}{\partial \epsilon} = \int_{t_0}^{t} \frac{2\frac{\partial \tilde{\mathbf{r}}}{\partial t}\frac{\partial}{\partial \epsilon}\left(\frac{\partial \tilde{\mathbf{r}}}{\partial t}\right)}{2\sqrt{\left(\frac{\partial \tilde{\mathbf{r}}}{\partial t}\right)^2}}\, dt = \int_{t_0}^{t} \frac{\frac{\partial \tilde{\mathbf{r}}}{\partial t}\frac{d\mathbf{z}}{dt}}{\sqrt{\left(\frac{\partial \tilde{\mathbf{r}}}{\partial t}\right)^2}}\, dt.$$

For $\epsilon = 0$, according to (3.53), we have that

$$\tilde{\mathbf{r}}(t,0) = \mathbf{r}(t) = \mathbf{r}(u^1(t), u^2(t))$$

is a position vector of the point on the curve $C_0 = C$, so $\frac{\partial \tilde{\mathbf{r}}}{\partial t} = \frac{d\mathbf{r}}{dt}$. According to (3.49), we conclude that (3.52) is a required condition. □

If $\mathbf{a}$ and $\mathbf{b}$ are constant vectors, then the field

$$\mathbf{z} = \mathbf{a} \times \mathbf{r} + \mathbf{b} \tag{3.54}$$

always satisfies Eq. (3.49), because

$$d\mathbf{r} \cdot d\mathbf{z} = d\mathbf{r} \cdot (\mathbf{a} \times d\mathbf{r}) \equiv 0.$$

Definition 3.5. Bending field is **trivial**, i.e. it is a field of the rigid motion of the surface, if it can be given in the form (3.54).

Definition 3.6. A surface is **rigid** if it does not allow bending field other then trivial. Otherwise, a surface is **non-rigid** or **flexible**.

In the rest of this chapter, we will consider only infinitesimal bending of the first order.

3.4.1 *Infinitesimal Bending of the Surfaces at Descartes Coordinate System*

If the surface S is given in the Descartes coordinate system

$$S : \mathbf{r} = (x, y, z), \tag{3.55}$$

and the bending field is

$$\mathbf{z} = (\xi, \eta, \zeta), \tag{3.56}$$

the basic equation of infinitesimal bending (3.49) becomes

$$dx d\xi + dy d\eta + dz d\zeta = 0.$$

If we project on the $x0y$ plane, we can take x and y as parameters, and then

$$\mathbf{r} = (x, y, z(x, y)), \quad \mathbf{z} = (\xi(x, y), \eta(x, y), \zeta(x, y)).$$

Equation (3.49) is then equivalent to the system of the partial differential equations

$$\begin{aligned} &\xi_x + z_x \zeta_x = 0, \\ &\xi_y + \eta_x + z_x \zeta_y + z_y \zeta_x = 0, \\ &\eta_y + z_y \zeta_y = 0, \end{aligned}$$

wherefrom, we get the partial differential equation of the second order with respect to the unknown function $\zeta(x, y)$:

$$z_{xx}\zeta_{yy} - 2z_{xy}\zeta_{xy} + z_{yy}\zeta_{xx} = 0.$$

3.4.2 *Rotational Field*

We will here define, according to [Efimov (1948)], another field in connection to the bending field $\mathbf{z}(u^1, u^2)$.

Theorem 3.5 ([Efimov (1948)]). *Let* $\mathbf{z}$ *be the infinitesimal bending field of the surface* $S : \mathbf{r} = \mathbf{r}(u^1, u^2)$. *Then, there exists a unique vector field* $\mathbf{y}$ *so that*

$$d\mathbf{z} = \mathbf{y} \times d\mathbf{r}, \tag{3.57}$$

i.e.

$$\mathbf{z}_1 = \mathbf{y} \times \mathbf{r}_1, \quad \mathbf{z}_2 = \mathbf{y} \times \mathbf{r}_2. \tag{3.58}$$

Proof. If $\mathbf{z}$ is an infinitesimal bending field, then Eq. (3.50) are valid. It follows that there are vector fields $\mathbf{y}_{(1)}$ and $\mathbf{y}_{(2)}$ so that

$$\mathbf{z}_1 = \mathbf{y}_{(1)} \times \mathbf{r}_1, \quad \mathbf{z}_2 = \mathbf{y}_{(2)} \times \mathbf{r}_2. \tag{3.59}$$

Substituting it into the second equation in (3.50), we obtain

$$(\mathbf{y}_{(1)} - \mathbf{y}_{(2)}) \cdot (\mathbf{r}_1 \times \mathbf{r}_2) = 0$$

wherefrom, we conclude that the vectors in the mixed product are linearly dependent. As there is no linear dependence between $\mathbf{r}_1$ and $\mathbf{r}_2$, there are scalar functions $\varphi(u^1, u^2)$ and $\psi(u^1, u^2)$ so that

$$\mathbf{y}_{(1)} - \mathbf{y}_{(2)} = \varphi\mathbf{r}_1 + \psi\mathbf{r}_2 \text{ or } \mathbf{y}_{(1)} - \varphi\mathbf{r}_1 = \mathbf{y}_{(2)} + \mathbf{r}_2 \equiv \mathbf{y} \quad \text{wherefrom,}$$

$$\mathbf{y}_{(1)} = \mathbf{y} + \varphi\mathbf{r}_1, \quad \mathbf{y}_{(2)} = \mathbf{y} - \psi\mathbf{r}_2.$$

Putting that into Eq. (3.59), we obtain

$$\mathbf{z}_1 = (\mathbf{y} + \varphi\mathbf{r}_1) \times \mathbf{r}_1 = \mathbf{y} \times \mathbf{r}_1 \text{ and } \mathbf{z}_2 = (\mathbf{y} - \psi\mathbf{r}_2) \times \mathbf{r}_2 = \mathbf{y} \times \mathbf{r}_2.$$

This implies

$$d\mathbf{z} = \mathbf{z}_i du^i = \mathbf{z}_1 du^1 + \mathbf{z}_2 du^2 = (\mathbf{y} \times \mathbf{r}_1)du^1 + (\mathbf{y} \times \mathbf{r}_2)du^2 = \mathbf{y} \times d\mathbf{r}.$$

Let us prove the uniqueness of the field $\mathbf{y}$. If $\mathbf{y}$ and $\bar{\mathbf{y}}$ satisfy the conditions of the theorem, then

$$\mathbf{z}_1 = \mathbf{y} \times \mathbf{r}_1 = \bar{\mathbf{y}} \times \mathbf{r}_1 \Rightarrow (\mathbf{y} - \bar{\mathbf{y}}) \times \mathbf{r}_1 = 0,$$
$$\mathbf{z}_2 = \mathbf{y} \times \mathbf{r}_2 = \bar{\mathbf{y}} \times \mathbf{r}_2 \Rightarrow (\mathbf{y} - \bar{\mathbf{y}}) \times \mathbf{r}_2 = 0.$$

We conclude from above that $\mathbf{y} = \bar{\mathbf{y}}$, otherwise $\mathbf{y} - \bar{\mathbf{y}}$ would be collinear with $\mathbf{r}_1$ and $\mathbf{r}_2$, which is impossible. □

As a result of infinitesimal bending of surfaces, all its line elements are rotated with vector $\mathbf{y}$.

Definition 3.7. The field $\mathbf{y}$ determined by (3.57), (3.58) is **rotational field** of the surface S under infinitesimal bending with the bending field $\mathbf{z}(u^1, u^2)$.

In addition to the field $\mathbf{y}(u^1, u^2)$, we consider the field $\mathbf{s}(u^1, u^2)$ defined with the equation

$$\mathbf{s} = \mathbf{z} - \mathbf{y} \times \mathbf{r}, \quad \text{i.e.} \quad \mathbf{z} = \mathbf{y} \times \mathbf{r} + \mathbf{s}. \tag{3.60}$$

If we compare this equation with Eq. (3.54), we can see that $\mathbf{y}$ corresponds to the rotation vector $\mathbf{a}$, and $\mathbf{s}$ corresponds to the translation vector $\mathbf{b}$ during the rigid motion of the surface. This leads us to the following definition.

Definition 3.8. The field $\mathbf{s}(u^1, u^2)$ defined by Eq. (3.60) is **translation field** under infinitesimal bending generating by $\mathbf{z}(u^1, u^2)$.

If we imagine that vectors $\mathbf{z}, \mathbf{y}, \mathbf{s}, \mathbf{r}$ have the same starting point, then their ends generate the parameterized surfaces Z, Y, Σ, S. The surface Y generated by endpoints of rotational field $\mathbf{y}(u^1, u^2)$, taken with the same starting points at the origin, is the **indicatrix of rotation** for the surface S.

Let the vectors $\mathbf{r}, \mathbf{y}, \mathbf{z}, \mathbf{s}$ be the vectors of the surface S, rotational field, field of infinitesimal bending, field of translation, and generated by endpoints surfaces S, Y, Z, Σ with the same starting point O. Then, the next theorem is in the force:

Theorem 3.6 ([Efimov (1948)]). 1. *Corresponding the points of the surfaces S and Z one to another so that to the point $M(\mathbf{r}) \in S$ corresponds the point $M'(\mathbf{z}) \in Z$, then the line elements of this surfaces are perpendicular. The same is in the force for Y and Σ.*

2. *Surfaces $S(\mathbf{r}), Y(\mathbf{y}), Z(\mathbf{z}), \Sigma(\mathbf{s})$ correspond to the surfaces $Y(\mathbf{y})$, $S(\mathbf{r})$, $\Sigma(\mathbf{s}), Z(\mathbf{z})$ respectively. In other words, if we take Y for initial surface, then* $\mathbf{r}$ *is rotational field,* $\mathbf{s}$ *infinitesimal bending field, and* $\mathbf{z}$ *field of translations.*

3. *If Y is indicatrix of rotation for S, then S is indicatrix of rotation for Y.*

4. *In the corresponding points of the surface S and its indicatrix of rotation Y, tangent planes are parallel.*

Proof. 1. Orthogonality of the line elements of surfaces $S(\mathbf{r})$ and $Z(\mathbf{z})$ follows from $d\mathbf{r} \cdot d\mathbf{z} = 0$. For $Y(\mathbf{y})$ and $\Sigma(\mathbf{s})$ we have, according to (3.60) and (3.57),

$$d\mathbf{y} \cdot d\mathbf{s} = d\mathbf{y} \cdot (d\mathbf{z} - d\mathbf{y} \times \mathbf{r} - \mathbf{y} \times d\mathbf{r}) = 0.$$

2. If we start from the relations (3.60) and (3.57)

$$\mathbf{z} = \mathbf{y} \times \mathbf{r} + \mathbf{s}, \quad d\mathbf{z} = \mathbf{y} \times d\mathbf{r}$$

we have

$$\mathbf{s} = \mathbf{z} - \mathbf{y} \times \mathbf{r} = \mathbf{r} \times \mathbf{y} + \mathbf{z}, \quad d\mathbf{s} = \mathbf{r} \times d\mathbf{y}$$

wherefrom, we see that with the change of roles of $S(\mathbf{r})$ and $Y(\mathbf{y})$, the roles of $Z(\mathbf{z})$ and $\Sigma(\mathbf{s})$ change too.

3. The proof follows from the case 2.
4. From (3.58), we have

$$\mathbf{z}_1 = \mathbf{y} \times \mathbf{r}_1, \quad \mathbf{z}_2 = \mathbf{y} \times \mathbf{r}_2,$$

wherefrom,

$$\mathbf{z}_{12} = \mathbf{y}_2 \times \mathbf{r}_1 + \mathbf{y} \times \mathbf{r}_{12}, \quad \mathbf{z}_{21} = \mathbf{y}_1 \times \mathbf{r}_2 + \mathbf{y} \times \mathbf{r}_{21} = \mathbf{z}_{12},$$

$\mathbf{z}_{12} = \mathbf{z}_{1,2}$, $\mathbf{y}_1 = \mathbf{y}_{,1}$, etc. As $\mathbf{r}_{12} = \mathbf{r}_{21}$, from the previous equations, we have

$$\mathbf{y}_1 \times \mathbf{r}_2 = \mathbf{y}_2 \times \mathbf{r}_1. \tag{3.61}$$

If we mark the common value of the left and right sides with $\mathbf{p}$, we can see that all the vectors $\mathbf{r}_1$, $\mathbf{r}_2$, $\mathbf{y}_1$, $\mathbf{y}_2$ are perpendicular to $\mathbf{p}$, therefore, they are coplanar. As $\mathbf{r}_1$ and $\mathbf{r}_2$ determine tangent plane for S, and $\mathbf{y}_1$ and $\mathbf{y}_2$ for Y, we conclude that these planes are parallel, if they are determined. □

Theorem 3.7 ([Efimov (1948)]). *For the surface* $\mathbf{r} = \mathbf{r}(u^1, u^2)$ *derivative vectors* $\mathbf{y}_1 = \mathbf{y}_{,1}$, $\mathbf{y}_2 = \mathbf{y}_{,2}$ *are given by the equations*

$$\mathbf{y}_1 = \alpha \mathbf{r}_1 + \beta \mathbf{r}_2, \quad \mathbf{y}_2 = \gamma \mathbf{r}_1 - \alpha \mathbf{r}_2, \tag{3.62}$$

where scalar functions α, β, γ *satisfy the system of the equations*

$$\begin{aligned} b_{11}\gamma - 2b_{12}\alpha - b_{22}\beta &= 0, \\ \alpha_{,2} - \gamma_{,1} &= \Gamma^1_{11}\gamma - 2\Gamma^1_{12}\alpha - \Gamma^1_{22}\beta, \\ \alpha_{,1} + \beta_{,2} &= \Gamma^2_{11}\gamma - 2\Gamma^2_{12}\alpha - \Gamma^2_{22}\beta, \end{aligned} \tag{3.63}$$

where b_{ij}, $i, j = 1, 2$ *are the coefficients of the second fundamental form of the surface* S, *and* Γ^i_{jk}, $i, j, k = 1, 2$ *are Cristoffel symbols for* S.

Proof. As we obtained from Eq. (3.61), vectors $\mathbf{y}_1$, $\mathbf{y}_2$, $\mathbf{r}_1$, $\mathbf{r}_2$ are linearly dependent, so we can write

$$\mathbf{y}_1 = \alpha \mathbf{r}_1 + \beta \mathbf{r}_2, \tag{3.64}$$

$$\mathbf{y}_2 = \gamma \mathbf{r}_1 + \delta \mathbf{r}_2, \tag{3.65}$$

where α, β, γ, δ are scalar functions of u^1, u^2. Substituting into Eq. (3.61), we obtain

$$(\alpha\mathbf{r}_1 + \beta\mathbf{r}_2) \times \mathbf{r}_2 = (\gamma\mathbf{r}_1 + \delta\mathbf{r}_2) \times \mathbf{r}_1 \Rightarrow \alpha(\mathbf{r}_1 \times \mathbf{r}_2) = \delta(\mathbf{r}_2 \times \mathbf{r}_1)$$
$$\Rightarrow \delta = -\alpha,$$

so Eqs. (3.64) and (3.65) become (3.62).

Further, we have

$$\mathbf{y}_{12} = \mathbf{y}_{1,2} = \alpha_{,2}\mathbf{r}_1 + \alpha\mathbf{r}_{12} + \beta_{,2}\mathbf{r}_2 + \beta\mathbf{r}_{22},$$
$$\mathbf{y}_{21} = \mathbf{y}_{2,1} = \gamma_{,1}\mathbf{r}_1 + \gamma\mathbf{r}_{11} - \alpha_{,1}\mathbf{r}_2 - \alpha\mathbf{r}_{21}.$$

If we substitute $\mathbf{r}_{11}$, $\mathbf{r}_{12}$, $\mathbf{r}_{22}$ in the equation $\mathbf{y}_{12} = \mathbf{y}_{21}$ according to derivational formulas, then we obtain

$$\alpha_{,2}\mathbf{r}_1 + 2\alpha(\Gamma^1_{12}\mathbf{r}_1 + \Gamma^2_{12}\mathbf{r}_2 + b_{12}\boldsymbol{\nu}) + \beta_{,2}\mathbf{r}_2 + \beta(\Gamma^1_{22}\mathbf{r}_1 + \Gamma^2_{22}\mathbf{r}_2 + b_{22}\boldsymbol{\nu})$$
$$= \gamma_{,1}\mathbf{r}_1 + \gamma(\Gamma^1_{11}\mathbf{r}_1 + \Gamma^2_{11}\mathbf{r}_2 + b_{11}\boldsymbol{\nu}) - \alpha_{,1}\mathbf{r}_2.$$

As the vectors $\mathbf{r}_1\, \mathbf{r}_2$, $\boldsymbol{\nu}$ are linearly independent, we obtain

$$\begin{aligned}
&\alpha_{,2} + 2\alpha\Gamma^1_{12} + \beta\Gamma^1_{22} - \gamma\Gamma^1_{11} - \gamma_{,1} = 0,\\
&2\alpha\Gamma^2_{12} + \beta_{,2} + \beta\Gamma^2_{22} - \gamma\Gamma^2_{11} + \alpha_{,1} = 0,\\
&2\alpha b_{12} + \beta b_{22} - \gamma b_{11} = 0,
\end{aligned}$$

i.e. (3.63). □

By solving Eq. (3.63), we obtain α, β, γ. We can find the fields $\mathbf{y}$ and $\mathbf{z}$ in the following way. According to (3.62), we have

$$d\mathbf{y} = \mathbf{y}_1 du^1 + \mathbf{y}_2 du^2 = (\alpha\mathbf{r}_1 + \beta\mathbf{r}_2)du^1 + (\gamma\mathbf{r}_1 - \alpha\mathbf{r}_2)du^2.$$

The right side is total differential of the vector function, so by integration, we obtain the field $\mathbf{y}(u^1, u^2)$. Further, we have

$$d\mathbf{z} = \mathbf{y} \times d\mathbf{r} = (\mathbf{y} \times \mathbf{r}_1)du^1 + (\mathbf{y} \times \mathbf{r}_2)du^2. \tag{3.66}$$

According to

$$(\mathbf{y} \times \mathbf{r}_1)_{,2} = \mathbf{y}_2 \times \mathbf{r}_1 + \mathbf{y} \times \mathbf{r}_{12}, \quad (\mathbf{y} \times \mathbf{r}_2)_{,1} = \mathbf{y}_1 \times \mathbf{r}_2 + \mathbf{y} \times \mathbf{r}_{21}$$

as well as (3.61), we have $(\mathbf{y} \times \mathbf{r}_1)_{,2} = (\mathbf{y} \times \mathbf{r}_2)_{,1}$, so the right side in Eq. (3.66) is total differential. By integrating, we obtain $\mathbf{z}(u^1, u^2)$.

3.4.3 Kinematic System of Equations

Let us consider infinitesimal bending of the surface S, (3.42). We are going to determine infinitesimal bending field according to research by Vekua [Vekua (1959)].

Infinitesimal bending field is defined at the points of the surface S, so we can present it in the following way

$$\mathbf{z} = z^j \mathbf{r}_j + z\boldsymbol{\nu}, \tag{3.67}$$

where $z^j \mathbf{r}_j$ is tangent, and $z\boldsymbol{\nu}$ is normal component, z^j and z are functions of u^1, u^2. Therefore, the equation of the infinitesimal bending will be

$$S_\epsilon : \tilde{\mathbf{r}} = \mathbf{r} + \epsilon\mathbf{z} = \mathbf{r} + \epsilon(z^j \mathbf{r}_j + z\boldsymbol{\nu}). \tag{3.68}$$

We can also represent derivative vector of the field $\mathbf{z}$ by the vectors of the moving frame. Namely, if we differentiate Eq. (3.67) by u^i and denote $\frac{\partial \mathbf{z}}{\partial u^i} = \mathbf{z}_i$, $\frac{\partial z^j}{\partial u^i} = z^j{}_{,i}$, $\frac{\partial z}{\partial u^i} = z_{,i}$, we will have

$$\begin{aligned} \mathbf{z}_i &= z^j{}_{,i}\mathbf{r}_j + z^j \mathbf{r}_{j,i} + z_{,i}\boldsymbol{\nu} + z\boldsymbol{\nu}_i = z^j{}_{,i}\mathbf{r}_j + z^j(\Gamma^p_{ij}\mathbf{r}_p + b_{ij}\boldsymbol{\nu}) \\ &\quad + z_{,i}\boldsymbol{\nu} + z(-b^p_i \mathbf{r}_p), \end{aligned}$$

after using the derivative formulas of the first and the second kind. By making the necessary changes of dumb indices, we get

$$\mathbf{z}_i = (z^p{}_{,i} + z^j\Gamma^p_{ij} - zb^p_i)\mathbf{r}_p + (z^p b_{ip} + z_{,i})\boldsymbol{\nu}.$$

Using covariant derivative, we can present the previous equation in the following way

$$\mathbf{z}_i = (z^p{}_{;i} - zb^p_i)\mathbf{r}_p + (z^p b_{ip} + z_{,i})\boldsymbol{\nu}. \tag{3.69}$$

Let us now consider the basic equation of bending field (3.49). Since $d\mathbf{z} = \mathbf{z}_i du^i$, $d\mathbf{r} = \mathbf{r}_j du^j$, using (3.69), we obtain

$$\mathbf{r}_j du^j \cdot [(z^p{}_{;i} - b^p_i z)\mathbf{r}_p + (z^p b_{pi} + z_{,i})\boldsymbol{\nu}]du^i = 0.$$

We have also $\mathbf{r}_j \cdot \mathbf{r}_p = g_{jp}$ and $\mathbf{r}_j \cdot \boldsymbol{\nu} = 0$, so the previous equation becomes

$$(z^p{}_{;i} - b^p_i z)g_{jp}\, du^i du^j = 0. \tag{3.70}$$

This equation is valid for all du^i, du^j, so we obtain

$$z^p{}_{;i}g_{jp} - b_{ij}z = 0. \tag{3.71}$$

wherefrom, we obtain

$$z_{j;i} - b_{ij}z = 0, \tag{3.72}$$

where z_j are covariant coordinates of the vector $\mathbf{z}$. By changing the place of indices j and i, we obtain

$$z_{i;j} - b_{ij}z = 0, \tag{3.73}$$

considering the symmetry of the coefficients of the second fundamental form. By adding the previous two equations, we obtain

$$z_{i;j} + z_{j;i} - 2b_{ij}z = 0, \quad (i, j = 1, 2). \tag{3.74}$$

Therefore, the next theorem is valid.

Theorem 3.8 ([Vekua (1959)]). *Necessary and sufficient condition for the field* (3.67) *to be infinitesimal bending field of the surface* S (3.42), *is that the coordinates* z_i, z *satisfy the equations* (3.74).

System of equations (3.74), which is obviously Eq. (3.49) written by coordinates, is said to be **kinematic system of equations** or system of equations of infinitesimal bending field. It is a system of three equations of unknown functions z_1, z_2, z with arguments u^1, u^2.

However, z can be easily eliminated from this system. Namely, let us multiply Eq. (3.71) with g^{jk}. We obtain

$$z^k{}_{;i} - b_i^k z = 0,$$

then we perform a contraction by k, i

$$z^i{}_{;i} - b_i^i z = 0. \tag{3.75}$$

Since the mean curvature of the surface is given by Eq. (3.19), i.e. $H = \frac{1}{2}b_i^i$, comparing with (3.75), we obtain

$$z = \frac{1}{2H}z^i{}_{;i}. \tag{3.76}$$

Since $z^i{}_{;i} = z^i{}_{,i} + \Gamma^i_{pi} z^p = z^p{}_{,p} + \Gamma^i_{pi} z^p$, taking into account (3.23), Eq. (3.76) becomes

$$z = \frac{1}{2H}\left[\frac{\partial z^p}{\partial u^p} + \frac{\partial(\ln\sqrt{g})}{\partial u^p} z^p\right]. \tag{3.77}$$

Substituting (3.76) into (3.73), we obtain $z_{i;j} - b_{ij}\frac{1}{2H} z^p{}_{;p} = 0$, or

$$b_{ij} z^p{}_{;p} - 2H z_{i;j} = 0, \tag{3.78}$$

which we can write in the following form

$$b_{ij} z^p{}_{;p} - 2H g_{pi} z^p{}_{;j} = 0, \quad (i, j = 1, 2). \tag{3.79}$$

From here, we can find z^1 and z^2, and from (3.77), z. In this way, we can completely determine infinitesimal bending field $\mathbf{z}$ of the surface.

Let the rotation field $\mathbf{y}$ be given by the equation

$$\mathbf{y} = y^i \mathbf{r}_i + y\boldsymbol{\nu}, \tag{3.80}$$

where $\mathbf{z}$ is given in (3.67). Using $d\mathbf{z} = \mathbf{z}_j du^j$ and (3.69) in (3.57), we obtain

$$\begin{aligned}(z^p{}_{;j} - b^p_j z) du^j\, \mathbf{r}_p + (z^p b_{pj} + z_{,j}) du^j\, \boldsymbol{\nu} &= (y^i \mathbf{r}_i + y\boldsymbol{\nu}) \times \mathbf{r}_j du^j \\ &= y^i du^j (\mathbf{r}_i \times \mathbf{r}_j) + y du^j (\boldsymbol{\nu} \times \mathbf{r}_j).\end{aligned} \tag{3.81}$$

We can write

$$\mathbf{r}_i \times \mathbf{r}_j = c_{ij}\boldsymbol{\nu}, \tag{3.82}$$

where c_{ij} antisymmetric tensor defined in (3.38). It will be valid

$$\begin{aligned}\boldsymbol{\nu} \times \mathbf{r}_j &= \frac{\mathbf{r}_1 \times \mathbf{r}_2}{\sqrt{g}} \times \mathbf{r}_j = -\frac{1}{\sqrt{g}} \mathbf{r}_j \times (\mathbf{r}_1 \times \mathbf{r}_2) \\ &= -\frac{1}{\sqrt{g}}[(\mathbf{r}_j \cdot \mathbf{r}_2)\mathbf{r}_1 - (\mathbf{r}_j \cdot \mathbf{r}_1)\mathbf{r}_2] = -\frac{1}{\sqrt{g}}(g_{j2}\mathbf{r}_1 - g_{j1}\mathbf{r}_2),\end{aligned} \tag{3.83}$$

which can be written as follows

$$\boldsymbol{\nu} \times \mathbf{r}_j = c_{jq} g^{pq} \mathbf{r}_p. \tag{3.84}$$

Equation (3.81) becomes now

$$(z^p{}_{;j} - b^p_j z)du^j\,\mathbf{r}_p + (z^p b_{pj} + z_{,j})du^j\,\boldsymbol{\nu} = y^i du^j c_{ij}\,\boldsymbol{\nu} + y du^j c_{jq} g^{pq}\,\mathbf{r}_p,$$

wherefrom, we obtain

$$z^p{}_{;j} - b^p_j z = y c_{jq} g^{pq}, \tag{3.85}$$

$$z^p b_{pj} + z_{,j} \quad = c_{ij} y^i. \tag{3.86}$$

These two equations completely determine y^1, y^2, y, i.e. the rotation field $\mathbf{y}$.

3.4.4 *Gaudi's Surfaces*

One of the main tasks at infinitesimal bending problems is to find non trivial bending fields, provided the condition (3.49), for non-rigid or flexible surfaces. We are going to check the rigidity of Gaudi's surfaces (Antoni Gaudi, 1852–1926, a Catalan architect), i.e. to find the corresponding infinitesimal bending field, according to [Velimirović *et al.* (2012, 2013)]. Note that for sake of simplicity, we will not use here indices for coordinates, i.e. we are going to write $u = u^1$, $v = u^2$.

Gaudi's surfaces present a special class of ruled surfaces, so called sinusoidal conoids. They are used as elements of architecture and roof constructions.

Gaudi's surface is explicitly given by the scalar equation $z = kx \sin \frac{y}{c}$. The vector equation of this surface is

$$\mathbf{r}(u, v) = \left(u, v, ku \sin \frac{v}{c}\right), \quad (u = x, v = y) \tag{3.87}$$

i.e.

$$\mathbf{r} = u\mathbf{i} + v\mathbf{j} + ku \sin \frac{v}{c}\mathbf{k},$$

where $\mathbf{i}$, $\mathbf{j}$, $\mathbf{k}$ are mutually perpendicular unit vectors in the direction of Cartesian axes.

Examples of Gaudi's surface for parameters $c = 1$, $k = 0.5$ and $u \in (-\pi, \pi)$, $v \in (-2\pi, 2\pi)$, are given on Fig. 3.1:

Fig. 3.1. Gaudi's surface $c = 1, k = 0.5$ and $u \in (-\pi, \pi)$, $v \in (-2\pi, 2\pi)$.

The Cristoffel's symbols of this surface are

$$\Gamma^1_{11} = \Gamma^2_{11} = 0, \ \Gamma^1_{12} = \Gamma^1_{21} = \frac{1}{g}\frac{k^2}{c}\sin\frac{v}{c}\cos\frac{v}{c}, \ \Gamma^1_{22} = -\frac{1}{g}\frac{k^2}{c^2}u\sin^2\frac{v}{c}$$

$$\Gamma^2_{12} = \Gamma^2_{21} = \frac{1}{g}\frac{k^2}{c^2}u\cos^2\frac{v}{c}, \ \Gamma^2_{22} = -\frac{1}{g}\frac{k^2}{c^3}u^2\sin\frac{v}{c}\cos\frac{v}{c},$$

where $g = 1 + k^2\sin^2\frac{v}{c} + \frac{k^2}{c^2}u^2\cos^2\frac{v}{c}$ is a determinant of the coefficients of the first fundamental form of Gaudi's surface. The second fundamental form coefficients are:

$$b_{11} = 0, \quad b_{12} = b_{21} = \frac{1}{\sqrt{g}}\frac{k}{c}\cos\frac{v}{c}, \quad b_{22} = -\frac{1}{\sqrt{g}}\frac{k}{c^2}u\sin\frac{v}{c}.$$

The equations (3.63) become:

$$-2\frac{1}{\sqrt{g}}\frac{k}{c}\cos\frac{v}{c}\alpha + \frac{1}{\sqrt{g}}\frac{k}{c^2}u\sin\frac{v}{c}\beta = 0, \tag{3.88}$$

$$\alpha_v - \gamma_u = -2\frac{1}{g}\frac{k^2}{c}\sin\frac{v}{c}\cos\frac{v}{c}\alpha + \frac{1}{g}\frac{k^2}{c^2}u\sin^2\frac{v}{c}\beta, \tag{3.89}$$

$$\alpha_u + \beta_v = -2\frac{1}{g}\frac{k^2}{c^2}u\cos^2\frac{v}{c}\alpha + \frac{1}{g}\frac{k^2}{c^3}u^2\sin\frac{v}{c}\cos\frac{v}{c}\beta. \tag{3.90}$$

From (3.88), we have

$$\alpha = \frac{u}{2c}\tan\frac{v}{c}\beta, \quad v \neq \frac{c\pi}{2} + ck\pi. \tag{3.91}$$

If we put (3.91) into (3.90), then we get quasi-linear partial differential equation of unknown function β:

$$\frac{u}{2c}\tan\frac{v}{c}\beta_u + \beta_v = -\frac{1}{2c}\tan\frac{v}{c}\beta. \tag{3.92}$$

System of characteristics of this equation is

$$\frac{du}{\frac{u}{2c}\tan\frac{v}{c}} = \frac{dv}{1} = \frac{d\beta}{-\frac{1}{2c}\tan\frac{v}{c}\beta}. \tag{3.93}$$

From equations

$$\frac{du}{\frac{u}{2c}\tan\frac{v}{c}} = \frac{dv}{1}, \quad \frac{dv}{1} = \frac{d\beta}{-\frac{1}{2c}\tan\frac{v}{c}\beta}, \tag{3.94}$$

we obtain two integrals, respectively,

$$\psi_1 \equiv u^2\cos\frac{v}{c} = c_1, \quad \psi_2 \equiv \frac{\beta^2}{\cos\frac{v}{c}} = c_2, \tag{3.95}$$

c_1, c_2 are constants. Since Jacobian is

$$\frac{D(\psi_1,\psi_2)}{D(v,\beta)} = \begin{vmatrix} \frac{\partial\psi_1}{\partial v} & \frac{\partial\psi_1}{\partial\beta} \\ \frac{\partial\psi_2}{\partial v} & \frac{\partial\psi_2}{\partial\beta} \end{vmatrix} \neq 0, \tag{3.96}$$

the integrals ψ_1 and ψ_2 are independent, so the general solution of Eq. (3.92) is

$$V\left(u^2\cos\frac{v}{c}, \frac{\beta^2}{\cos\frac{v}{c}}\right) = 0, \tag{3.97}$$

where V is arbitrary continuous differentiable function.

The unknown function $\gamma(u,v)$ is got from (3.88), (3.91), and is the solution of the equation

$$\gamma_u = \frac{u}{2c^2}\frac{1}{\cos^2\frac{v}{c}}\beta + \frac{u}{2c}\tan\frac{v}{c}\beta_v$$

i.e.

$$\gamma = \int\left(\frac{u}{2c^2}\frac{1}{\cos^2\frac{v}{c}}\beta + \frac{u}{2c}\tan\frac{v}{c}\beta_v\right)du + \varphi(v), \tag{3.98}$$

where $\varphi(v)$ is an arbitrary function and β is from (3.97).

For instance, if we take $V(x,y) = xy - 1$, and $\varphi(v) = 0$, we get

$$\beta = \frac{1}{u}, \quad u \neq 0, \quad \alpha = \frac{1}{2c}\tan\frac{v}{c}, \quad \gamma = \frac{u}{2c^2\cos^2\frac{v}{c}}.$$

According to Eq. (3.62) and

$$d\mathbf{y} = \mathbf{y}_u du + \mathbf{y}_v dv = (dY_1, dY_2, dY_3),$$

by integration, we get

$$\mathbf{y} = (Y_1, Y_2, Y_3) = \left(\frac{u}{2c}\tan\frac{v}{c},\ \frac{1}{2}\ln\left(u^2\left|\cos\frac{v}{c}\right|\right),\right.$$
$$\left.\frac{ku}{2c}\tan\frac{v}{c}\sin\frac{v}{c} + \frac{ku}{c}\cos\frac{v}{c}\right),$$

assuming that the corresponding constants that occur during integration are equal to zero.

Applying (3.57), we will determine infinitesimal bending field. As Eq. (3.87) holds, for Gaudi's surface will be

$$d\mathbf{r} = \left(du, dv, k\sin\frac{v}{c}du + \frac{ku}{c}\cos\frac{v}{c}dv\right),$$

so

$$d\mathbf{z} = \mathbf{y}\times d\mathbf{r} = \begin{vmatrix} \mathbf{i} & \mathbf{j} & \mathbf{k} \\ \frac{u}{2c}\tan\frac{v}{c} & \frac{1}{2}\ln\left(u^2|\cos\frac{v}{c}|\right) & \frac{ku}{2c}\tan\frac{v}{c}\sin\frac{v}{c} + \frac{ku}{c}\cos\frac{v}{c} \\ du & dv & k\sin\frac{v}{c}du + \frac{ku}{c}\cos\frac{v}{c}dv \end{vmatrix}$$

Infinitesimal bending field of Gaudi's surface is obtained by integrating of this equation. Thus, for $u > 0$ and $v \in (0, c\pi/2)$, we have

$$\mathbf{z} = \left(k\sin\frac{v}{c}(u\ln u - u) + \frac{uk}{2}\sin\frac{v}{c}\ln\left(\cos\frac{v}{c}\right) + c_1,\right.$$
$$\left.\frac{ku^2}{2c}\cos\frac{v}{c} + c_2, -u\ln u + u - \frac{u}{2}\ln\left(\cos\frac{v}{c}\right) + c_3\right),$$

where c_1, c_2, c_3 are constants.

Example of Gaudi's surface for parameters $c = 1$, $k = 0.5$ and $u \in (0, \pi)$, $v \in (0.1, 1.472\pi)$, and its infinitesimal bending are given on Fig. 3.2 with different point of view:

Fig. 3.2. Gaudi's surface for $\epsilon = 0.00$ in the first row and $\epsilon = 0.35$ in the second row.

3.5 Variation of Geometric Magnitudes

We are going to define the variations of geometric magnitudes according to [Vekua (1959)].

Definition 3.9. Let $\mathcal{A} = \mathcal{A}(u^1, u^2)$ be the magnitude that characterizes a geometric property on the surface S and $\tilde{\mathcal{A}}(u^1, u^2, \epsilon) = \mathcal{A}_\epsilon(u^1, u^2)$ the corresponding magnitude on the surface S_ϵ being infinitesimal deformation of the surface S, and set

$$\Delta\mathcal{A} = \mathcal{A}_\epsilon - \mathcal{A} = \epsilon\delta\mathcal{A} + \epsilon^2\delta^2\mathcal{A} + \cdots + \epsilon^n\delta^n\mathcal{A} + \cdots \tag{3.99}$$

The coefficients $\delta\mathcal{A}, \delta^2\mathcal{A}, ..., \delta^n\mathcal{A}, ...$ are **the first, the second, ..., the n-th variation** of the geometric magnitude $\mathcal{A}$ respectively under infinitesimal deformation S_ϵ of the surface S.

Further, we will mainly consider the first variations under infinitesimal deformation of the first order so we will call them simply variations.

From (3.99), we have that

$$\frac{\Delta\mathcal{A}}{\epsilon} = \delta\mathcal{A} + \epsilon\delta^2\mathcal{A} + \cdots + \epsilon^{n-1}\delta^n\mathcal{A} + \cdots \tag{3.100}$$

According to that, variation of the geometric magnitude $\mathcal{A}$ is

$$\delta\mathcal{A} = \lim_{\epsilon\to 0}\frac{\Delta\mathcal{A}}{\epsilon} = \lim_{\epsilon\to 0}\frac{\mathcal{A}_\epsilon - \mathcal{A}}{\epsilon} \tag{3.101}$$

Then, according to (3.99), $\Delta\mathcal{A} = o(\epsilon)$ for $\delta\mathcal{A} = 0$. Equation (3.101) can be written also as

$$\delta\mathcal{A} = \left.\frac{\partial\mathcal{A}}{\partial\epsilon}\right|_{\epsilon=0} \tag{3.102}$$

Let us mark some properties of the variations:
I. $\delta(\mathcal{A}\mathcal{B}) = \mathcal{A}\delta\mathcal{B} + \mathcal{B}\delta\mathcal{A}$
Variation of the first order is derivative (for $\epsilon = 0$) and for the variation is effective

$$\delta(\mathcal{A}\mathcal{B}) = \left.\frac{\partial(\mathcal{A}\mathcal{B})}{\partial\epsilon}\right|_{\epsilon=0} = \mathcal{A}\delta\mathcal{B} + \mathcal{B}\delta\mathcal{A}.$$

II. Variation of the derivative is the derivative of the variation, i.e.

$$\delta\left(\frac{\partial\mathcal{A}}{\partial u^i}\right) = \frac{\partial(\delta\mathcal{A})}{\partial u^i} \tag{3.103}$$

As

$$\Delta\mathcal{A} = \mathcal{A}_\epsilon(u^1, u^2) - \mathcal{A}(u^1, u^2) \tag{3.104}$$

$$\Delta\mathcal{A} = \epsilon\delta\mathcal{A} + \epsilon^2\delta^2\mathcal{A} + \cdots \approx \epsilon\delta\mathcal{A} \tag{3.105}$$

for

$$\frac{\partial\mathcal{A}}{\partial u^i} = \mathcal{B} \tag{3.106}$$

we have using (3.105)

$$\Delta\frac{\partial\mathcal{A}}{\partial u^i} = \Delta\mathcal{B} = \epsilon\delta\mathcal{B} + \epsilon^2\delta^2\mathcal{B} + \cdots = \epsilon\delta\left(\frac{\partial\mathcal{A}}{\partial u^i}\right) + \epsilon^2\delta^2\left(\frac{\partial\mathcal{A}}{\partial u^i}\right) + \cdots. \tag{3.107}$$

It is also

$$\Delta\frac{\partial\mathcal{A}}{\partial u^i} = \Delta\mathcal{B} = \mathcal{B}_\epsilon(u^1,u^2) - \mathcal{B}(u^1,u^2)$$

$$= \frac{\partial\mathcal{A}_\epsilon(u^1,u^2)}{\partial u^i} - \frac{\partial\mathcal{A}(u^1,u^2)}{\partial u^i} = \frac{\partial}{\partial u^i}[\mathcal{A}_\epsilon(u^1,u^2) - \mathcal{A}(u^1,u^2)]$$

$$= \frac{\partial\Delta\mathcal{A}}{\partial u^i} = \frac{\partial(\epsilon\delta\mathcal{A} + \epsilon^2\delta^2\mathcal{A} + \cdots)}{\partial u^i} = \epsilon\frac{\partial(\delta\mathcal{A})}{\partial u^i} + \epsilon^2\frac{\partial(\delta^2\mathcal{A})}{\partial u^i} + \cdots \tag{3.108}$$

As the left sides at the last equation and (3.107) are equal, then (3.103) is effective.

The same is for the differential, i.e.

III. $\delta(d\mathcal{A}) = d(\delta\mathcal{A})$

On the surface $S_\epsilon : \tilde{\mathbf{r}} = \mathbf{r}_\epsilon(u^1, u^2)$ we can take the same coordinates for the point M_ϵ as well as for the point $M(u^1,u^2)$ and according to that $\Delta u^i = \epsilon\delta u^i = 0$, and from that $\delta u^i = 0$, $(i = 1, 2)$ i.e.

$$\delta u^1 = \delta u^2 = 0. \tag{3.109}$$

From here, the variation of the coordinate lines is zero, i.e. the coordinate lines stay the same.

Using **I** and (3.109), we have

$$\delta(d\mathcal{A}) = \delta\left(\frac{\partial\mathcal{A}}{\partial u^i}du^i\right) = \delta\left(\frac{\partial\mathcal{A}}{\partial u^i}\right)du^i$$

and

$$d(\delta\mathcal{A}) = \frac{\partial(\delta\mathcal{A})}{\partial u^i}du^i = \delta\left(\frac{\partial\mathcal{A}}{\partial u^i}\right)du^i.$$

As the right sides of the previous equations are equal, the left ones are equal too, i.e.

$$\delta(d\mathcal{A}) = d(\delta\mathcal{A}).$$

3.5.1 *Variations of Some Geometric Magnitudes under Infinitesimal Deformations*

We are going to examine variations of some geometric magnitudes of surfaces under infinitesimal deformations, according to [Velimirović *et al.* (2011); Williams (1987)].

Let us consider the surface S, (3.42), given by the equation

$$S : \mathbf{r} = \mathbf{r}(u^1, u^2), \quad (u^1, u^2) \in D \subset \mathcal{R}^2 \tag{3.110}$$

and an infinitesimal deformation of that surface (3.68),

$$S_\epsilon : \tilde{\mathbf{r}} = \mathbf{r} + \epsilon \mathbf{z} = \mathbf{r} + \epsilon \delta \mathbf{r} = \mathbf{r} + \epsilon (z^j \mathbf{r}_j + z \boldsymbol{\nu}). \tag{3.111}$$

The covariant base vectors of deformed surface will be, according to (3.69),

$$\begin{aligned}\tilde{\mathbf{r}}_i = \mathbf{r}_i + \epsilon \delta \mathbf{r}_i = (\mathbf{r} + \epsilon \delta \mathbf{r})_{,i} = (\mathbf{r} + \epsilon \mathbf{z})_{,i} = \mathbf{r}_i + \epsilon [(z^p{}_{;i} - z b_i^p) \mathbf{r}_p \\ + (z^p b_{ip} + z_{,i}) \boldsymbol{\nu}].\end{aligned} \tag{3.112}$$

Therefore, the next theorem holds.

Theorem 3.9. *Under infinitesimal deformation of the surface S, the variation of the covariant base vector is*

$$\delta \mathbf{r}_i = (z^p{}_{;i} - z b_i^p) \mathbf{r}_p + (z^p b_{ip} + z_{,i}) \boldsymbol{\nu}. \tag{3.113}$$

Theorem 3.10. *Under infinitesimal deformation of the surface S, the variation of the unit normal is*

$$\delta \boldsymbol{\nu} = -(z^p b_{pq} + z_{,q}) \mathbf{r}^q. \tag{3.114}$$

Proof. The unit normal of the deformed surface stays perpendicular to the surface, i.e. the following condition is valid

$$(\boldsymbol{\nu} + \epsilon \delta \boldsymbol{\nu}) \cdot (\mathbf{r}_i + \epsilon \delta \mathbf{r}_i) = 0$$

or, using (3.112)

$$(\boldsymbol{\nu} + \epsilon \delta \boldsymbol{\nu}) \cdot \{\mathbf{r}_i + \epsilon [(z^p{}_{;i} - z b_i^p) \mathbf{r}_p + (z^p b_{ip} + z_{,i}) \boldsymbol{\nu}]\} = 0,$$

i.e.

$$\begin{aligned}\boldsymbol{\nu} \cdot \mathbf{r}_i + \epsilon \{\delta \boldsymbol{\nu} \cdot \mathbf{r}_i + \boldsymbol{\nu} \cdot [(z^p{}_{;i} - z b_i^p) \mathbf{r}_p + (z^p b_{ip} + z_{,i}) \boldsymbol{\nu}]\} \\ + \epsilon^2 \delta \boldsymbol{\nu} \cdot [(z^p{}_{;i} - z b_i^p) \mathbf{r}_p + (z^p b_{ip} + z_{,i}) \boldsymbol{\nu}] = 0.\end{aligned}$$

As $\boldsymbol{\nu} \cdot \mathbf{r}_i = 0$, by neglecting the term with ϵ^2, we obtain

$$\delta \boldsymbol{\nu} \cdot \mathbf{r}_i + z^p b_{ip} + z_{,i} = 0. \tag{3.115}$$

On the other hand from the condition

$$(\boldsymbol{\nu} + \epsilon \delta \boldsymbol{\nu}) \cdot (\boldsymbol{\nu} + \epsilon \delta \boldsymbol{\nu}) = 1$$

we have

$$\boldsymbol{\nu} \cdot \delta\boldsymbol{\nu} = 0. \tag{3.116}$$

Comparing with (3.115), we obtain (3.114). □

Theorem 3.11. *Under infinitesimal deformation of the surface S, the variation of the contravariant base vector is*

$$\delta\mathbf{r}^j = -(z^j{}_{;p} - zb^j_p)\mathbf{r}^p + (z^p b^j_p + z_{,p}g^{pj})\boldsymbol{\nu}. \tag{3.117}$$

Proof. The contravariant base vectors of the deformed surface must lie in the plane of the surface and satisfy the condition (3.6), i.e.

$$(\mathbf{r}^j + \epsilon\delta\mathbf{r}^j) \cdot (\mathbf{r}_i + \epsilon\delta\mathbf{r}_i) = \delta^j_i \text{ and } (\mathbf{r}^j + \epsilon\delta\mathbf{r}^j) \cdot (\boldsymbol{\nu} + \epsilon\delta\boldsymbol{\nu}) = 0. \tag{3.118}$$

Using (3.112) and (3.114) in (3.118), we get

$$\mathbf{r}_i \cdot \delta\mathbf{r}^j = -(z^j_{;i} - zb^j_i), \quad \boldsymbol{\nu} \cdot \delta\mathbf{r}^j = z^p b^j_p + z_{,p}g^{pj},$$

i.e. (3.117). □

Theorem 3.12. *Under infinitesimal deformation of the surface S, the variation of the metric tensor is*

$$\delta g_{ij} = z_{i;j} + z_{j;i} - 2zb_{ij}, \tag{3.119}$$

$$\delta g^{ij} = z^j{}_{;p}g^{ip} + z^i{}_{;p}g^{jp} - 2zb^{ij}. \tag{3.120}$$

The variation of the determinant of the first fundamental form is

$$\delta g = 2g(z^i{}_{;i} - zb^i_i). \tag{3.121}$$

Proof. The coefficients of the first fundamental form (i.e. the metric tensor) can be obtained according to (3.8) in the following way

$$\tilde{g}_{ij} = g_{ij} + \epsilon\delta g_{ij} = (\mathbf{r}_i + \epsilon\delta\mathbf{r}_i) \cdot (\mathbf{r}_j + \epsilon\delta\mathbf{r}_j),$$

i.e. after using (3.112), we obtain (3.119). Similarly, we obtain (3.119).

The determinant of the first fundamental form of the surface $g = g_{11}g_{22} - (g_{12})^2$, after deformation, satisfies

$$\tilde{g} = g + \epsilon\delta g = (g_{11} + \epsilon\delta g_{11})(g_{22} + \epsilon\delta g_{22}) - (g_{12} + \epsilon\delta g_{12})^2.$$

Using (3.13), we obtain (3.121). □

Theorem 3.13. *Under infinitesimal deformation of the surface S, the variation of the coefficients of the second fundamental form is*

$$\delta b_{ij} = (z^p b_{pi} + z_{,i})_{;j} + (z^p{}_{;i} - z b_i^p) b_{pj}. \tag{3.122}$$

Proof. Using Eq. (3.30) which is valid for the partial differentiation of the vector $\mathbf{v} = v^i \mathbf{r}_i = v_i \mathbf{r}^i$, we directly get

$$\tilde{\boldsymbol{\nu}}_{,j} = (\boldsymbol{\nu} + \epsilon\delta\boldsymbol{\nu})_{,j} = \boldsymbol{\nu}_j - \epsilon[(z^q b_{qp} + z_{,p})_{;j}\mathbf{r}^p + (z^q b_{qp} + z_{,p}) b_j^p \boldsymbol{\nu}] \tag{3.123}$$

Applying Eqs. (3.14), (3.112), and (3.123), the coefficients of the second fundamental form become

$$\tilde{b}_{ij} = b_{ij} + \epsilon\delta b_{ij} = -(\mathbf{r}_i + \epsilon\delta\mathbf{r}_i) \cdot (\boldsymbol{\nu} + \epsilon\delta\boldsymbol{\nu})_{,j},$$

wherefrom, we obtain (3.122). □

Using Codazzi equations (3.35), we can put (3.122) into the symmetric form

$$\delta b_{ij} = z^p{}_{;j} b_{pi} + z^p{}_{;i} b_{pj} + z^p b_{ij;p} + z_{;ij} - z b_i^p b_{pj}. \tag{3.124}$$

This equation is symmetric by i and j due to the fact that the covariant derivative of the scalar equals the partial derivative and due to the definition of the covariant derivative, i.e.

$$z_{;ij} = z_{,i;j} = z_{,j;i} = z_{;ji}.$$

Equations (3.119) and (3.122) give

$$\begin{aligned}\tilde{b}_i^j &= b_i^j + \epsilon\delta b_i^j = (b_{ip} + \epsilon\delta b_{ip})(g^{pj} + \epsilon\delta g^{pj}) \\ &= b_{ip}g^{pj} + \epsilon(g^{pj}\delta b_{ip} + b_{ip}\delta g^{pj})\end{aligned}$$

$$= b_i^j + \epsilon\{g^{pj}[(z^q b_{pq} + z_{,p})_{;i} + (z^q_{;p} - z b^q_p) b_{iq}]$$
$$+ b_{ip}(-z^j_{;q} g^{pq} - z^p_{;q} g^{jq} + 2z b^{pj})\}$$
$$= b_i^j + \epsilon(z^q_{;i} b^j_q + z^q b^j_{q;i} + g^{pj} z_{;ip} + g^{pj} z^q_{;p} b_{iq} - z b^q_p b_{iq} g^{pj}$$
$$- z^j_{;q} b^q_i - b_{ip} g^{jq} z^p_{;q} + 2z b^{pj} b_{ip}) \tag{3.125}$$

after using the Codazzi equations (3.35), the change of dumb indices and the following equation

$$b^{jp} b_{pi} = b^j_q g^{qp} b^r_i g_{rp} = b^j_q b^r_i \delta^q_r = b^j_q b^q_i.$$

Therefore,

$$\delta b^j_i = z^p{}_{;i} b^j_p + z^p b^j_{p;i} + g^{pj} z_{;pi} - z^j{}_{;p} b^p_i + z b^j_p b^p_i. \tag{3.126}$$

Theorem 3.14. *Under infinitesimal deformation of the surface S, the variation of the mean curvature is*

$$\delta H = \frac{1}{2}(z^p b^i_{i;p} + g^{pi} z_{;pi} + z b^i_p b^p_i). \tag{3.127}$$

Proof. The mean curvature

$$H = \frac{1}{2} b^i_i \tag{3.128}$$

after the deformation satisfies $\tilde{H} = H + \epsilon \delta H = \frac{1}{2}(b^i_i + \epsilon \delta b^i_i)$, i.e. (3.127). We used Eq. (3.126) and

$$b^i_{p;i} = (b_{qp} g^{qi})_{;i} + b_{qp;i} g^{qi} + b_{qp} g^{qi}{}_{;i} = b_{qi;p} g^{qi} = b^i_{i;p},$$

due to Codazzi equations (3.35) and the fact that the covariant derivative of the metric tensor is equal to zero. □

Theorem 3.15. *Under infinitesimal deformation of the surface S, the variation of the Gaussian curvature is*

$$\delta K = z^p (b^i_i b^j_{j;p} - b^i_j b^j_{i;p}) + z_{;ij}(g^{ij} b^p_p - b^{ij}) + z(b^j_j b^i_p b^p_i - b^j_p b^p_i b^i_j). \tag{3.129}$$

Proof. The Gaussian curvature

$$K = \frac{1}{2}(b^i_i b^j_j - b^i_j b^j_i) \tag{3.130}$$

of the deformed surface will be

$$\tilde{K} = K + \epsilon\delta K = \frac{1}{2}[(b^i_i + \epsilon\delta b^i_i)(b^j_j + \epsilon\delta b^j_j) - (b^i_j + \epsilon\delta b^i_j)(b^j_i + \epsilon\delta b^j_i)]$$

i.e. using(3.35), (3.126) and changing the dumb indices, we obtain (3.129). □

Based on the theorems that precede, corresponding geometric magnitudes of deformed surfaces under infinitesimal deformation are:

$$\begin{aligned}
\tilde{\mathbf{r}}_i &= \mathbf{r}_i + \epsilon\delta\mathbf{r}_i = (\mathbf{r} + \epsilon\delta\mathbf{r})_{,i} = (\mathbf{r} + \epsilon\mathbf{z})_{,i} = \mathbf{r}_i + \epsilon[(z^p{}_{;i} - zb^p_i)\mathbf{r}_p \\
&\quad + (z^p b_{ip} + z_{,i})\boldsymbol{\nu}], \\
\tilde{\boldsymbol{\nu}} &= \boldsymbol{\nu} + \epsilon\delta\boldsymbol{\nu} = \boldsymbol{\nu} - \epsilon(z^p b_{pq} + z_{,q})\mathbf{r}^q, \\
\tilde{\mathbf{r}}^j &= \mathbf{r}^j + \epsilon\delta\mathbf{r}^j = \mathbf{r}^j + \epsilon[-(z^j{}_{;p} - zb^j_p)\mathbf{r}^p + (z^p b^j_p + z_{,p} g^{pj})\boldsymbol{\nu}], \\
\tilde{g}_{ij} &= g_{ij} + \epsilon\delta g_{ij} = g_{ij} + \epsilon(z_{i;j} + z_{j;i} - 2zb_{ij}) \\
\tilde{g}^{ij} &= g^{ij} + \epsilon\delta g^{ij} = g^{ij} - \epsilon(z^j{}_{;p} g^{ip} + z^i{}_{;p} g^{jp} - 2zb^{ij}), \\
\tilde{g} &= g + \epsilon\delta g = g + 2g\epsilon(z^i{}_{;i} - zb^i_i), \\
\tilde{b}_{ij} &= b_{ij} + \epsilon\delta b_{ij} = b_{ij} + \epsilon[(z^p b_{pi} + z_{,i})_{;j} + (z^p{}_{;i} - zb^p_i)b_{pj}] \\
\tilde{b}^j_i &= b^j_i + \epsilon\delta b^j_i = b^j_i + \epsilon(z^p{}_{;i} b^j_p + z^p b^j_{p;i} + g^{pj} z_{;pi} - z^j{}_{;p} b^p_i + zb^j_p b^p_i), \\
\tilde{H} &= H + \epsilon\delta H = H + \frac{1}{2}\epsilon(z^p b^i_{i;p} + g^{pi} z_{;pi} + zb^i_p b^p_i), \\
\tilde{K} &= K + \epsilon\delta K = K + \epsilon[z^p(b^i_i b^j_{j;p} - b^i_j b^j_{i;p}) + z_{;ij}(g^{ij} b^p_p - b^{ij}) \\
&\quad + z(b^j_j b^i_p b^p_i - b^j_p b^p_i b^i_j)].
\end{aligned} \tag{3.131}$$

3.5.2 *Variation of Some Geometric Magnitudes under Infinitesimal Bending*

Infinitesimal bending is a kind of deformation under which the metric tensor, i.e. the coefficients of the first fundamental form do not get the variations of the first order, i.e. these variations are zero. The magnitudes expressed by the coefficients of the first fundamental form and derivatives of these coefficients (e.g. Gaussian curvature, Cristofell's symbols and other) also have no variation of the first order. However, coefficients of the second fundamental form have, generally speaking, variations different from zero. More precisely, let us prove the following theorems.

Theorem 3.16. *Variation of the first fundamental form of the surface is zero under infinitesimal bending of the surface*

$$\delta I = 0, \quad \textit{where} \quad I = ds^2 = g_{ij}du^i du^j.$$

Also,

$$\delta(ds) = 0.$$

Variation of the coefficients of the first fundamental form (the metric tensor) is also zero:

$$\delta g_{ij} = 0, \quad i, j = 1, 2.$$

It is also

$$\delta g = 0, \quad g = g_{11}g_{22} - (g_{12})^2$$
$$\delta g^{ij} = 0, \quad i, j = 1, 2,$$
$$\delta \Gamma_{i.jk} = 0, \quad \delta \Gamma^i_{jk} = 0, \quad i, j, k = 1, 2.$$

Proof. For the surface S,

$$I = ds^2 = d\mathbf{r}^2 = g_{ij}du^i du^j,$$
$$\Delta I = \tilde{I}(u, v, \epsilon) - I(u, v) = ds_\epsilon^2 - ds^2 = d\mathbf{r}_\epsilon^2 - d\mathbf{r}^2$$
$$= (d\mathbf{r} + \epsilon d\mathbf{z})^2 - d\mathbf{r}^2,$$
$$\Delta I = 2\epsilon d\mathbf{r} \cdot d\mathbf{z} + \epsilon^2 d\mathbf{z}^2 = \epsilon \delta I + \epsilon^2 \delta^2 I. \tag{3.132}$$

According to the definition of the variation, in the case of infinitesimal bending of the first order $\Delta I = o(\epsilon)$, or equivalently $d\mathbf{r} \cdot d\mathbf{z} = 0$. According to (3.132) it is equivalent with the condition

$$\delta I = \delta(ds^2) = 0. \tag{3.133}$$

We also had that $ds_\varepsilon - ds = o(\epsilon)$ (Eq. (3.51)) and based on that

$$\Delta(ds) = 0 \cdot \epsilon + \epsilon^2 \delta^2(ds) + \cdots = \delta(ds)\epsilon + \delta^2(ds)\epsilon^2 + \cdots$$

and it is worth

$$\delta(ds) = 0. \tag{3.134}$$

Comparing (3.74) and (3.119), we conclude

$$\delta g_{ij} = 0. \tag{3.135}$$

Also,

$$\delta g^{ij} = 0, \tag{3.136}$$

where $g^{ij} = (g_{ij})^{-1}$. From Eqs. (3.75) and (3.121), we obtain

$$\delta g = 0. \tag{3.137}$$

For the Cristoffel symbols of the surface, we have: $\Gamma_{i.jk} = \frac{1}{2}(g_{ij,k} - g_{jk,i} + g_{ki,j})$, $\quad g_{ij,k} = \frac{\partial}{\partial u^k} g_{ij}$. Using (3.135) and the property II of the variation, we get

$$\delta \Gamma_{i.jk} = 0. \tag{3.138}$$

Based on the relation $\Gamma^i_{jk} = g^{ip}\Gamma_{p.jk}$, using property I of the variation, it is

$$\delta \Gamma^i_{jk} = 0. \tag{3.139}$$

□

Variation of the tensor field, given on the surface, is also tensor field of the same order. Relations connecting the components of the tensors of different types (covariant, contravariant and mixed), are effective also for their variations.

Theorem 3.17. *Variation of the determinant of the second fundamental form is zero.*

Proof. For the determinant of the coefficients of the second fundamental form of the surface

$$II = b_{ij} du^i du^j, \quad i, j = 1, 2$$

according to the Gaussian equation for the surface, we have

$$b = det(b_{ij}) = b_{11}b_{22} - (b_{12})^2$$

that it can be expressed in function of Cristoffel's symbols and their partial derivatives, and we have

$$\delta b = 0. \tag{3.140}$$

□

Variations of the coefficients of the second fundamental form b_{ij} are not zero. From (3.140), we have

$$\delta(b_{11}b_{22} - (b_{12})^2) = b_{11}\delta b_{22} + b_{22}\delta b_{11} - 2b_{12}\delta b_{12} = 0.$$

This equation presents the connection between coefficients of the second fundamental form and their variations.

Theorem 3.18. *Variation of the Gaussian curvature under infinitesimal bending of the surface is zero.*

Proof. For the Gaussian (total) curvature of the surface, we have:

$$K = \frac{b}{g} \Rightarrow \delta K = \frac{g\delta b - b\delta g}{g^2}$$

and according to that based on the previously mentioned

$$\delta K = 0. \tag{3.141}$$

□

Theorem 3.19. *Variation of the mean curvature under infinitesimal bending of the surface is*

$$\delta H = \frac{1}{2}(z^p b^i_p + z_{,p} g^{ip})_{;i}. \tag{3.142}$$

Proof. As $\delta H = \frac{1}{2}\delta b^i_i$, where δb^j_i is given in Eq. (3.126), we have

$$\begin{aligned} \delta H = \frac{1}{2}\delta b^i_i &= \frac{1}{2}(z^p_{;i} b^i_p + z^p b^i_{p;i} + g^{pi} z_{;pi} - z^i_{;p} b^p_i + z b^i_p b^p_i) \\ &= \frac{1}{2}[(z^p b^i_p)_{;i} + (z_{,p} g^{pi})_{;i} - z^i_{;p} b^p_i + z b^i_p b^p_i), \end{aligned} \tag{3.143}$$

having in mind that the covariant derivative of the metric tensor is zero. Also

$$\begin{aligned} z^i_{;p} b^p_i - z b^i_p b^p_i &= z^i_{;p} g_{ki} b^{kp} - z b^i_p b^p_i = b_{kp} b^{kp} z - z b^{ik} g_{kp} b_{iq} g^{pq} \\ &= b_{kp} b^{kp} z - b^{ik} b_{iq} \delta^q_k z = b_{kp} b^{kp} z - b^{ik} b_{ik} z \\ &= 0, \end{aligned} \tag{3.144}$$

due to $z^i_{;p} g_{ki} = b_{kp} z$, according to (3.71). Therefore, Eq. (3.142) is valid. □

For the normal curvature $\bar{K}$ (the curvature of the normal section) in the given direction $\frac{du^2}{du^1}$, we have:

Theorem 3.20. *Variation of the normal curvature is*

$$\delta\bar{K} = \delta k \cos\theta - k\sin\theta\delta\theta. \tag{3.145}$$

Proof. Having that

$$\bar{K} = k\cos\theta,$$

where k is the curvature (flexion) of the curve, having the same tangent as the considered normal section, θ the angle among the normal of the surface and the main normal of this, we can get (3.145). □

Let's remark that the variation of the geodesic curvature $\underset{g}{K}$ of the curve $u^i = u^i(s)$ on the surface is zero. As

$$\underset{g}{K} = k\sin\theta = \sqrt{g}(u^{1'}\lambda^2 - u^{2'}\lambda^1) \tag{3.146}$$

$$\lambda^i = u^{i''} + \Gamma^i_{jk}u^{j'}u^{k'}, \quad u^{i'} = \frac{du^i}{ds}, \quad u^{i''} = \frac{d^2u^i}{ds^2}$$

according to the previously presented, we have

$$\delta\lambda^i = 0$$

According to (3.146)

$$\delta\underset{g}{K} = 0. \tag{3.147}$$

Namely, it is known that geodesic curvature of the curve is the object of the intrinsic geometry (i.e. it can be expressed in function of the coefficients of the first fundamental form g_{ij} and by the reason of $\delta g_{ij} = 0$ it follows immediately (3.147). Namely,

Theorem 3.21. *Variation of geodesic curvature of the curve on the surface is zero.*

The next is the property of the geodesic line under infinitesimal bending of the surface.

Theorem 3.22. *Geodesics transform into geodesics under infinitesimal bending of the surface.*

Proof. Geodesics on the surface are the lines for which $u^{1'}\lambda^2 - u^{2'}\lambda^1 = 0$, and under infinitesimal bending this equation stays unchanged, i.e. geodesic line stays geodesic line. □

3.6 Willmore Energy

Let the surface S (3.42) be given. **The Willmore energy** of S is the surface integral

$$\mathcal{W} = \iint_S f(H, K)\, dA, \tag{3.148}$$

where $f(H, K) = H^2 - K$, H and K are mean and Gaussian curvature of the surface, respectively, and dA is an element of the surface area. For a closed surface, by the Gauss–Bonnet theorem, the integral of the Gaussian curvature may be computed in terms of the Euler characteristic $\chi(S)$ of the surface, so

$$\iint_S K\, dA = 2\pi\chi(S),$$

which is a topological invariant and thus independent of the particular embedding in $\mathcal{R}^3$ that was chosen. For this reason, the following functional is often observed

$$\mathcal{W}_1 = \iint_S H^2\, dA. \tag{3.149}$$

The Willmore energy has an interesting geometric property, which is its conformal invariance. This means that conformal transformations of the ambient metric do not change the value of the Willmore energy. According to [Spivak (1999)], a transformation F is conformal if and only if F is a composition of isometries, dilations, and spherical inversions.

The Willmore energy is named after an English mathematician Thomas Willmore, based on his work [Willmore (1965)]. He initiated the study of minimization problems on Willmore energy $\mathcal{W}_1$ by proving that round spheres in $\mathcal{R}^3$, having energy equal to 4π, are the only absolute minimizers of $\mathcal{W}_1$ among closed immersed surfaces in $\mathcal{R}^3$. Chen later expanded this result to higher dimensions

Chen (1971a, 1971b). The Li-Yau inequality further highlights the significance of the Willmore energy in variational problems by establishing 4π as the infimum of $\mathcal{W}_1$ among closed surface immersions and asserting that closed immersed surfaces in $\mathcal{R}^n$ with self-intersections have energy greater or equal than 8π.

In [Willmore (1965)], Willmore also proposed a conjecture regarding the minimization of Willmore energy among tori, suggesting that the infimum of $\mathcal{W}_1$ among immersions of tori equals $2\pi^2$, which can only be achieved by the torus of revolution given by a circle of radius 1 whose center is located at distance $\sqrt{2}$ from the axis of revolution, up to conformal transformation of the ambient. This conjecture spurred further research into minimization of $\mathcal{W}_1$ among surfaces of a fixed topology, that is closed surfaces of given genus in $\mathcal{R}^n$. We will mention the works of Simon [Simon (1993)] and Bauer–Kuwert [Bauer and Kuwert (2003)] on this topic.

The Willmore conjecture has been eventually proved by Marques-Neves in [Marques and Neves (2014)]. In this paper and in other recent works, the authors develop deep variational tool useful for the study of minimal surfaces in Riemannian manifolds. Their results are actually stronger than the statement of the Willmore conjecture.

Finding a surface that minimizes the energy entails finding its first variation with respect to perturbations of the surface. This surface is a critical point of the Willmore energy and is called Willmore surface. Many recent results related to the variation and minimization problems on the Willmore energy can be found for example in [Gruber *et al.* (2019); Pozzetta (2020); Weiner (1978)], as well as in the literature within them.

3.6.1 *Variation of the Willmore Energy under Infinitesimal Deformations*

Based on calculations in [Velimirović *et al.* (2011); Williams (1987)] using tensor calculus, we are going to present a concise derivation of the first variation of Willmore energy under infinitesimal deformation (3.68). Obtained expressions hold for surfaces with or without boundary and are based on the computation of surface fundamental forms and the knowledge of the infinitesimal deformation field of the first order.

Suppose that the parameterization of the surface (3.42), $S : \mathbf{r} = \mathbf{r}(u^1, u^2)$, is determined in the compact area D. An element of surface area on S is $g^{1/2}du^1du^2$, so

$$\mathcal{W} = \iint_D f(H,K)\, g^{1/2} du^1 du^2, \tag{3.150}$$

and analogously the Willmore energy of the deformed surface is

$$\tilde{\mathcal{W}} = \mathcal{W} + \epsilon\delta\mathcal{W} = \iint_D f(H+\epsilon\delta H, K+\epsilon\delta K)\,(g+\epsilon\delta g)^{1/2} du^1 du^2, \tag{3.151}$$

D is the area of integration.

Let us find $\delta\mathcal{W}$. Namely,

$$\begin{aligned}
\tilde{\mathcal{W}} &= \iint_D [(H+\epsilon\delta H)^2 - (K+\epsilon\delta K)](g+\epsilon\delta g)^{1/2}\, du^1 du^2 \\
&= \iint_D \left[f(H,K) + \epsilon\left(\frac{\partial f}{\partial H}\delta H + \frac{\partial f}{\partial K}\delta K\right)\right] g^{1/2}\left(1+\frac{\epsilon\delta g}{g}\right)^{1/2} \\
&= \iint_D \left[f(H,K) + \epsilon\left(\frac{\partial f}{\partial H}\delta H + \frac{\partial f}{\partial K}\delta K\right)\right] g^{1/2}\left(1+\frac{\epsilon\delta g}{2g}\right) \\
&= \iint_D \left[f(H,K) + \epsilon\left(\frac{\partial f}{\partial H}\delta H + \frac{\partial f}{\partial K}\delta K + f(H,K)\frac{\delta g}{2g}\right)\right] \\
&\quad \cdot g^{1/2} du^1 du^2,
\end{aligned} \tag{3.152}$$

after the development of $(1+\frac{\epsilon\delta g}{g})^{1/2}$ in a Maclaurin series up to the first degree (the other terms are neglected). Therefore,

$$\delta\mathcal{W} = \iint_D \left[\frac{\partial f}{\partial H}\delta H + \frac{\partial f}{\partial K}\delta K + f(H,K)\frac{\delta g}{2g}\right] g^{1/2} du^1 du^2. \tag{3.153}$$

Applying (3.121), (3.127), (3.128), (3.129), and (3.130), we get

$$\begin{aligned}
\delta\mathcal{W} &= \iint_D \left[2H\delta H - \delta K + (H^2-K)\frac{\delta g}{2g}\right] g^{1/2} du^1 du^2 \\
&= \iint_D \left\{\frac{1}{2}b_i^i(z^i b_{j;i}^j + g^{ij} z_{;ij} + z b_j^i b_i^j) - z^p(b_i^i b_{j;p}^j - b_j^i b_{i;p}^j)\right.
\end{aligned}$$

$$- z_{;ij}(g^{ij}b^p_p - g^{ip}b^j_p) - z(b^i_i b^j_p b^p_j - b^i_j b^j_p b^p_i)$$
$$+ \left[\frac{1}{4}b^i_i b^j_j - \frac{1}{2}(b^i_i b^j_j - b^j_i b^i_j)\right](z^i_{;i} - zb^i_i)\bigg\} g^{1/2}du^1du^2$$
$$= \iint_D \bigg\{\frac{1}{4}(b^j_j b^p_p z^i)_{;i} + \frac{1}{2}b^i_i g^{jp} z_{;jp} + \frac{1}{2}zb^i_i\left(b^j_p b^p_j - \frac{1}{2}b^j_j b^p_p\right)$$
$$- \frac{1}{2}[z^i(b^j_j b^p_p - b^j_p b^p_j)]_{;i} - z_{;ij}(g^{ij}b^p_p - g^{ip}b^j_p)$$
$$- \frac{z}{2}(3b^i_i b^j_p b^p_j - 2b^i_j b^j_p b^p_i - b^i_i b^j_j b^p_p)\bigg\} g^{1/2}du^1du^2 \quad (3.154)$$

i.e., after using

$$3b^i_i b^j_p b^p_j - 2b^i_j b^j_p b^p_i - b^i_i b^j_j b^p_p = 0$$

$$\delta\mathcal{W} = \iint_D \bigg\{\frac{1}{4}(b^j_j b^p_p z^i)_{;i} + \frac{1}{2}b^i_i g^{jp} z_{;jp} + \frac{1}{2}zb^i_i\left(b^j_p b^p_j - \frac{1}{2}b^j_j b^p_p\right)$$
$$- \frac{1}{2}[z^i(b^j_j b^p_p - b^j_p b^p_j)]_{;i} - z_{;ij}(g^{ij}b^p_p - g^{ip}b^j_p)\bigg\} g^{1/2}du^1du^2 \quad (3.155)$$

It holds

$$z_{;ij}(g^{ij}b^p_p - g^{ip}b^j_p) = (z_{;i})_{;j}(g^{ij}b^p_p - g^{ip}b^j_p)$$
$$= [z_{;i}(g^{ij}b^p_p - g^{ip}b^j_p)]_{;j} - z_{;i}(g^{ij}b^p_p - g^{ip}b^j_p)_{;j}$$
$$= [z_{;i}(g^{ij}b^p_p - g^{ip}b^j_p)]_{;j} \quad (3.156)$$

due to $(g^{ij}b^p_p - g^{ip}b^j_p)_{;j} = g^{ij}b^p_{p;j} - g^{ip}b^j_{p;j} = 0$, using Codazzi equations and the fact that the covariant derivative of the metric tensor is zero. Also,

$$b^i_i g^{jp} z_{;jp} = (b^i_i g^{jp} z_{;j})_{;p} - (b^i_i g^{jp})_{;p} z_{;j} = (b^i_i g^{jp} z_{;j})_{;p} - b^i_{i;p} g^{jp} z_{;j}$$
$$= (b^i_i g^{jp} z_{;j})_{;p} - (b^i_{i;p} g^{jp} z)_{;j} + b^i_{i;pj} g^{jp} z. \quad (3.157)$$

Substituting (3.156) and (3.157) into (3.155), we obtain

$$\delta\mathcal{W} = \iint_D \left\{ \frac{1}{4}(b^j_j b^p_p z^i)_{;i} + \frac{1}{2}(b^j_j g^{ip} z_{;p})_{;i} - \frac{1}{2}(b^p_{p;q} g^{iq} z)_{;i} + \frac{1}{2} b^p_{p;ij} g^{ij} z \right.$$
$$+ \frac{1}{2} z b^i_i \left(b^j_p b^p_j - \frac{1}{2} b^j_j b^p_p \right) - \frac{1}{2}[z^i (b^j_j b^p_p - b^j_p b^p_j)]_{;i}$$
$$\left. - [z_{;i}(g^{ij} b^p_p - g^{ip} b^j_p)]_{;j} \right\} g^{1/2} du^1 du^2. \quad (3.158)$$

Applying integral theorem in Eq. (3.158), we get

$$\delta\mathcal{W} = \iint_D \frac{z}{2} \left[b^p_{p;ij} g^{ij} + b^p_p (b^i_j b^j_i - \frac{1}{2} b^i_i b^j_j) \right] g^{1/2} du^1 du^2$$
$$+ \int_C \frac{c_{ij}}{2} \left[\frac{1}{2} z^i b^p_p b^q_q + b^p_p g^{iq} z_{;q} - b^p_{p;q} g^{iq} z - z^i (b^p_p b^q_q - b^p_q b^q_p) \right.$$
$$\left. - 2 z_{;q} (g^{ip} b^q_q - g^{iq} b^p_p) \right] \left(\frac{du^j}{ds} \right) ds. \quad (3.159)$$

Therefore, we have proved the next theorem.

Theorem 3.23. *The variation of the Willmore energy of the surface* (3.42) *under infinitesimal deformation* (3.68) *is* (3.159).

Obviously, we have the following corollary.

Corollary 3.1. *The variation of the Willmore energy of a closed surface* (3.42) *under infinitesimal deformation* (3.68) *is*

$$\delta\mathcal{W} = \iint_D \frac{z}{2} \left[b^p_{p;ij} g^{ij} + b^p_p \left(b^i_j b^j_i - \frac{1}{2} b^i_i b^j_j \right) \right] g^{1/2} du^1 du^2. \quad (3.160)$$

Note that the surface integral (3.159) will be zero for each $\mathbf{z}$ with support in the interior of D if

$$b^p_{p;ij} g^{ij} + b^p_p \left(b^i_j b^j_i - \frac{1}{2} b^i_i b^j_j \right) = 0. \quad (3.161)$$

This is the equation of the surfaces which are critical points of the Willmore energy and represents a differential equation of the fourth order. We can rewrite it in the terms of the mean and the Gaussian curvature as

$$H_{;ij}g^{ij} + 2H(H^2 - K) = 0. \tag{3.162}$$

The covariant differentiation can be changed into the partial differentiation using Eqs. (3.29) and (3.162) becomes

$$\frac{1}{g^{1/2}}(g^{1/2}g^{ij}H_{,i})_{,j} + 2H(H^2 - K) = 0. \tag{3.163}$$

In particular, let us emphasize the following theorem.

Theorem 3.24. *The variation of the Willmore energy of a closed surface whose mean and Gaussian curvatures, H and K, satisfy Eq.* (3.163), *is equal to zero under infinitesimal deformation.*

3.6.2 *Variation of the Willmore Energy under Infinitesimal Bending*

Let us examine the change of the Willmore energy under infinitesimal bending, i.e. let us determine the first variation of the Willmore energy, having in mind the kinematic system of equations that represent the equations of the bending field.

Obviously,

$$\delta\mathcal{W} = \iint_D 2H\delta H g^{1/2} du^1 du^2, \tag{3.164}$$

according to Eq. (3.153) and $\delta K = \delta g = 0$.

Applying (3.19) and (3.142) in (3.164), we obtain

$$\delta\mathcal{W} = \frac{1}{2}\iint_D b^q_q[(z^p b^i_p)_{;i} + (z_{,p}g^{ip})_{;i}]g^{1/2}du^1du^2, \tag{3.165}$$

i.e.

$$\delta\mathcal{W} = \frac{1}{2}\iint_D [(z^p b^i_p b^q_q)_{;i} - z^p b^i_p b^q_{q;i} + (z_{,p}g^{ip}b^q_q)_{;i} - z_{,p}g^{ip}b^q_{q;i}]g^{1/2}du^1du^2. \tag{3.166}$$

Using the integral theorem, we have

$$\delta\mathcal{W} = -\frac{1}{2}\iint_D (z^p b_p^i + z_{,p} g^{ip}) b_{q;i}^q g^{1/2} du^1 du^2$$
$$+\frac{1}{2}\int_C c_{ij}(z^p b_p^i + z_{,p} g^{ip}) b_q^q \frac{du^j}{ds} ds. \tag{3.167}$$

Multiplying Eq. (3.86) with g^{ij} and changing corresponding dumb indices we get

$$z^p b_p^i + z_{,p} g^{ip} = c_{pq} y^p g^{iq}. \tag{3.168}$$

If we put (3.168) into (3.167), then we have

$$\delta\mathcal{W} = -\frac{1}{2}\iint_D c_{pq} y^p g^{iq} b_{r;i}^r g^{1/2} du^i du^j + \frac{1}{2}\int_C c_{ij} c_{pq} y^p g^{iq} b_r^r \frac{du^j}{ds} ds. \tag{3.169}$$

Let us define a contravariant tensor c^{ij} in the following way

$$c^{ij} = g^{iq} g^{jp} c_{qp}. \tag{3.170}$$

It satisfies

$$c^{11} = c^{22} = 0, \quad c^{12} = -c^{21} = \frac{1}{\sqrt{g}}. \tag{3.171}$$

We have obtained that directly from Eqs. (3.170), (3.13), and (3.38). Therefore, it will be valid

$$g_{jp} = -g^{iq} c_{pq} c_{ij}. \tag{3.172}$$

As

$$\mathbf{y} \cdot d\mathbf{r} = (y^p \mathbf{r}_p + y\boldsymbol{\nu}) \cdot \mathbf{r}_j du^j = y^p g_{jp} du^j, \tag{3.173}$$

we obtain

$$\delta\mathcal{W} = -\iint_D c_{pq} y^p g^{iq} H_{,i}\, g^{1/2} du^i du^j - \int_C H(\mathbf{y} \cdot d\mathbf{r}). \tag{3.174}$$

Also, we have

$$c_{pq} y^p g^{iq} H_{,i} = \sqrt{g}(y^1 g^{2i} - y^2 g^{1i}) H_{,i},$$

so

$$\delta\mathcal{W} = \iint_D (y^2 g^{1i} - y^1 g^{2i}) H_{,i}\, g\, du^i du^j - \int_C H(\mathbf{y} \cdot d\mathbf{r}). \tag{3.175}$$

On the other hand, according to (3.84), we have

$$\boldsymbol{\nu} \times \mathbf{y} = \boldsymbol{\nu} \times (y^j \mathbf{r}_j + y\boldsymbol{\nu}) = y^j (\boldsymbol{\nu} \times \mathbf{r}_j) = y^j c_{jq} g^{pq} \mathbf{r}_p = y^j c_{jq} \mathbf{r}^q, \tag{3.176}$$

and

$$c_{pq} y^p g^{iq} H_{,i} = c_{pq} y^p \mathbf{r}^q \cdot \mathbf{r}^i H_{,i} = (\boldsymbol{\nu} \times \mathbf{y}) \cdot \mathbf{r}^i H_{,i} = [\boldsymbol{\nu}, \mathbf{y}, \mathbf{r}^i] H_{,i},$$

where $[,,]$ denotes scalar triple product of vectors. Therefore,

$$\delta\mathcal{W} = -\iint_D [\boldsymbol{\nu}, \mathbf{y}, \mathbf{r}^i] H_{,i}\, g^{1/2} du^i du^j - \int_C H(\mathbf{y} \cdot d\mathbf{r}). \tag{3.177}$$

Thus, we have proved the following theorem.

Theorem 3.25 ([Velimirović *et al.* (2011)]). *The variation of the Willmore energy of the surface S under infinitesimal bending is given by any of three equivalent equations:* (3.174), (3.175), *and* (3.177).

As we can see, the variation of the Willmore energy is expressed as a function of the rotational field $\mathbf{y}$.

The following equations are equivalent

$$c_{pq} y^p g^{iq} H_{,i} = 0, \quad (y^1 g^{2i} - y^2 g^{1i}) H_{,i} = 0, \quad [\boldsymbol{\nu}, \mathbf{y}, \mathbf{r}^i] H_{,i} = 0, \tag{3.178}$$

so we obtain.

Corollary 3.2. *Willmore energy of the closed surface S is stationary under infinitesimal bending whose rotational field satisfies one of three equivalent equations* (3.178).

3.6.3 *CMC Surfaces: Minimal Surfaces*

A surface is said to be **constant mean curvature surface** or **CMC surface**, if its mean curvature is constant and different from zero. If the mean curvature of a surface is identically zero, one speaks about **minimal surface**.

The spheres and round cylinders are the first few examples of CMC surfaces. In 1841, Delaunay classified all revolutional CMC surfaces which are called Delaunay surfaces [Eells (1987)]. For some time, no new CMC surfaces were found. The question was opened whether there are any compact CMC surfaces other then the spheres. In 1986, Wente was proved that must exist the CMC tori and furthermore there exist infinitely many constant mean curvature tori [Wente (1986)]. Moreover, for each integer $g = 2$, there is a compact constant mean curvature surface of genus g. In 1992, Bobenko managed to derive formulas that describe all CMC tori.

The CMC surfaces can be regarded as interface surfaces in nature. Particularly, the interface shape is described by a constant mean curvature surface that satisfies some particular conditions. The interface shape separates a gas layer within a superhydrophic surface consisting of a square lattice of posts from a pressurized liquid above the surface.

Equivalent definition of minimal surfaces is that they are the surfaces of the smallest area spanned by a given boundary. Minimal area property makes this surface suitable for application in architecture. The main reasons for application are: weight and amount of material are reduced on minimum. Famous architects like Otto Frei created this new trend in architecture.

The Plateau's problem is the problem in calculus of variations to find the minimal surface for a boundary with specified constraints (having no singularities on the surface). In 1873, a physicist named Joseph Plateau observed that soap film bounded by wire appeared to form minimal surfaces. The problem named after him, Belgian physicist experimentally solved for some special cases. Jess Douglas 1931 solved this problem. In general, there may be one, multiple, or no minimal surfaces spanning a given closed curve in a space.

Soap film must go to the state at which the surface area is minimized in order to minimize surface tension and reach equilibrium.

Because of the exceptional application in architecture and the significance of minimal and CMC surfaces, as well as the place of Wilmore energy in the theory of shells, we single out the following theorems that represent direct consequences of Theorem 3.25.

Corollary 3.3. *The variation of the Willmore energy of minimal surface under infinitesimal bending equals zero.*

Corollary 3.4. *The variation of the Willmore energy of CMC surface under infinitesimal bending is*

$$\delta W = -H \int_C \mathbf{y} \cdot d\mathbf{r}. \tag{3.179}$$

Corollary 3.5. *The variation of the Willmore energy of closed CMC surface under infinitesimal bending equals zero.*

3.7 Total Mean Curvature

The total mean curvature of a surface S is the surface integral

$$\mathcal{H}(S) = \iint_S H\, dA = \iint_D H g^{1/2} du^1 du^2, \tag{3.180}$$

H is mean curvature of the surface, dA is an element of the surface area.

Theorem 3.26 ([Velimirović *et al.* (2011)]). *The variation of the total mean curvature of a surface S under infinitesimal bending equals*

$$\delta\mathcal{H}(S) = -\frac{1}{2} \int_C \mathbf{y} \cdot d\mathbf{r}. \tag{3.181}$$

Proof. The total mean curvature of deformed surface is

$$\tilde{\mathcal{H}}(S) = \mathcal{H}(S) + \epsilon\, \delta\mathcal{H}(S) = \iint_S (H + \epsilon\, \delta H)(g + \epsilon\, \delta g)^{1/2} du^1 du^2.$$

As $\delta g = 0$ under infinitesimal bending, we have

$$\begin{aligned} \tilde{\mathcal{H}}(S) = \mathcal{H}(S) + \epsilon\, \delta\mathcal{H}(S) &= \iint_S (H + \epsilon\, \delta H) g^{1/2} du^1 du^2 \\ &= \mathcal{H}(S) + \epsilon \iint_S \delta H g^{1/2} du^1 du^2, \end{aligned}$$

i.e. the variation of the total mean curvature under infinitesimal bending is

$$\delta\mathcal{H}(S) = \iint_D \delta H g^{1/2} du^1 du^2. \tag{3.182}$$

Using Eq. (3.142) and the integral theorem, we get

$$\delta\mathcal{H}(S) = \frac{1}{2}\int_C c_{ij}(z^p b^i_p + z_{,p} g^{ip})\frac{du^j}{ds}ds. \tag{3.183}$$

Due to Eqs. (3.168) and (3.172), Eq. (3.183) reduces to

$$\delta\mathcal{H}(S) = -\frac{1}{2}\int_C g_{jp} y^p \frac{du^j}{ds}ds. \tag{3.184}$$

Comparing the previous equation with Eq. (3.173), we obtain (3.181). □

The following corollary immediately follows.

Corollary 3.6. *The variation of the total mean curvature of closed surface under infinitesimal bending equals zero.*

3.8 Volume of Generalized Cone

Let us introduce a scalar quantity $J = \mathbf{r} \cdot \boldsymbol{\nu}$, which is the scalar product of the position vector of a point of S, and the unit normal to the surface. Clearly, J changes if the origin moves in 3-dimensional space, but J remains unchanged at a particular point on the surface if only the surface coordinate system is modified, and thus J is a scalar. The change in J is required in calculating the volume swept as the surface moves. The surface integral of J is equal to three times the volume bounded by the surface S and the cone joining the origin to the boundary of the surface, i.e.

$$V = \frac{1}{3}\iint_S J\,dA, \tag{3.185}$$

dA is an element of surface area.

Theorem 3.27 ([Najdanović (2014)]). *Variation of the volume bounded by the surface S and the cone joining the origin to the boundary of the surface, under infinitesimal bending of S with the field of* *translation* $\mathbf{s}$, *is equal to one third of the flux of the field* $\mathbf{s}$ *through the given surface* S, i.e.

$$\delta V = \frac{1}{3}\iint_S \mathbf{s}\cdot\boldsymbol{\nu}\,dA, \tag{3.186}$$

where $\boldsymbol{\nu}$ *is the unit normal to the surface* S, $\mathbf{s}\cdot\boldsymbol{\nu}$ *is the scalar product of the corresponding vectors,* dA *is an element of surface area.*

Proof. The scalar quantity J after deformation satisfies

$$\tilde{J} = J + \epsilon\delta J = (\mathbf{r} + \epsilon\delta\mathbf{r})\cdot(\boldsymbol{\nu} + \epsilon\delta\boldsymbol{\nu}).$$

Using (3.68) and (3.114), we obtain

$$\tilde{J} = J + \epsilon\delta J = \mathbf{r}\cdot\boldsymbol{\nu} + \epsilon(z - \mathbf{r}\cdot\mathbf{r}^p(z^q b_{pq} + z_{,p})).$$

Therefore,

$$\delta J = z - \mathbf{r}\cdot\mathbf{r}^p(z^q b_{pq} + z_{,p}). \tag{3.187}$$

According to Eq. (3.86), we have

$$\delta J = z - \mathbf{r}\cdot\mathbf{r}^p c_{qp} y^q,$$

or, due to (3.176):

$$\delta J = z - \mathbf{r}\cdot(\boldsymbol{\nu}\times\mathbf{y}) = z - [\boldsymbol{\nu},\mathbf{y},\mathbf{r}]. \tag{3.188}$$

Let us multiply Eq. (3.67) scalarly with $\boldsymbol{\nu}$, we get

$$\mathbf{z}\cdot\boldsymbol{\nu} = z. \tag{3.189}$$

Put (3.189) into (3.188) and use the vectors' cyclic permutation in the mixed product $[\boldsymbol{\nu},\mathbf{y},\mathbf{r}] = [\mathbf{y},\mathbf{r},\boldsymbol{\nu}]$. We obtain

$$\delta J = \mathbf{z}\cdot\boldsymbol{\nu} - (\mathbf{y}\times\mathbf{r})\cdot\boldsymbol{\nu} = (\mathbf{z} - \mathbf{y}\times\mathbf{r})\cdot\boldsymbol{\nu} = \mathbf{s}\cdot\boldsymbol{\nu}, \tag{3.190}$$

where $\mathbf{s}$ is the translation field under infinitesimal bending. Therefore,

$$\delta V = \frac{1}{3}\iint_D \delta J\sqrt{g}\,du^1du^2 = \frac{1}{3}\iint_D \mathbf{s}\cdot\boldsymbol{\nu}\sqrt{g}\,du^1du^2,$$

i.e. we proved (3.186). □

3.8.1 *An Example of Volume Variation Calculation*

We shall examine variation of the volume bounded by the elliptic paraboloid $x^2 + y^2 = 1 - 2z$, $0 \le x^2 + y^2 \le 1/3$, and the cone joining the origin to the boundary of the mentioned surface, under its infinitesimal bending. The vector parametric equation of this surface is

$$\mathbf{r}(u,v) = (u, v, (1-u^2-v^2)/2) \quad D: 0 \le u^2+v^2 \le 1/3. \tag{3.191}$$

In the polar coordinates (ρ, φ), the equation has the next form:

$$\mathbf{r}(\rho,\varphi) = (\rho\cos\varphi, \rho\sin\varphi, (1-\rho^2)/2), \quad 0 \le \rho \le \sqrt{3}/3, \\ 0 \le \varphi \le 2\pi. \tag{3.192}$$

According to (3.185), it is easy obtained that the volume is

$$V = \frac{1}{3}\iint_S \mathbf{r}\cdot\boldsymbol{\nu}\, dA = \frac{1}{3}\iint_D \mathbf{r}\cdot\boldsymbol{\nu}\sqrt{g}\, dudv = \frac{1}{6}\iint_D (1+u^2+v^2)\, dudv$$

$$= \frac{1}{6}\int_0^{2\pi} d\varphi \int_0^{\sqrt{3}/3} (1+\rho^2)\rho\, d\rho = \frac{14\pi}{216}, \tag{3.193}$$

where $\boldsymbol{\nu} = \frac{1}{\sqrt{1+u^2+v^2}}(u, v, 1)$ is unit normal vector and $g = 1+u^2+v^2$ is the determinant of the metric tensor.

The Cristoffel symbols of the surface (3.191) are

$$\Gamma^1_{12} = \Gamma^2_{12} = 0,\ \Gamma^1_{11} = \Gamma^1_{22} = \frac{u}{1+u^2+v^2},\ \Gamma^2_{11} = \Gamma^2_{22} = \frac{v}{1+u^2+v^2}.$$

The second fundamental form coefficients are:

$$b_{12} = b_{21} = 0, b_{11} = b_{22} = \frac{-1}{\sqrt{1+u^2+v^2}}.$$

The equations (3.63) become:

$$\beta = \gamma, \quad \alpha_v - \beta_u = 0, \quad \alpha_u + \beta_v = 0,$$

or

$$\beta(u,v) = \gamma(u,v), \quad \alpha(u,v) = \int \beta_u\, dv - \beta_v\, du. \tag{3.194}$$

Let be, for instance, $\beta(u,v)=u$. Then, $\gamma(u,v)=u$ and $\alpha(u,v)=v$. Using these facts in (3.62), we obtain that the rotational field has the form

$$\mathbf{y}=\left(uv,\frac{u^2}{2}-\frac{v^2}{2},-u^2v+\frac{v^3}{3}\right). \tag{3.195}$$

From (3.57), we have

$$d\mathbf{z}=\mathbf{y}\times d\mathbf{r}=\begin{vmatrix}\mathbf{i} & \mathbf{j} & \mathbf{k}\\ uv & \frac{u^2}{2}-\frac{v^2}{2} & -u^2v+\frac{v^3}{3}\\ du & dv & -udu-vdv\end{vmatrix}$$

wherefrom, we get infinitesimal bending field applying one more integration

$$\mathbf{z}=\left(\frac{u^2v^2}{4}-\frac{u^4}{8}+\frac{v^4}{24},\frac{uv^3}{3},\frac{uv^2}{2}-\frac{u^3}{6}\right), \tag{3.196}$$

(assuming that the appropriate constants that occur during integration equal to zero).

In Fig. 3.3, we have infinitesimal bending of the elliptic paraboloid under mentioned infinitesimal bending field. Furthermore on example surfaces are colored by local value of subintegral function of Willmore energy scaled on color range.

The field of translation is, according to (3.60),

$$\mathbf{s}=\left(-\frac{3u^2v^2}{4}-\frac{u^2}{4}+\frac{v^2}{4}+\frac{u^4}{8}+\frac{v^4}{8},\frac{u^3v}{2}-\frac{uv^3}{2}+\frac{uv}{2},-uv^2+\frac{u^3}{3}\right). \tag{3.197}$$

Now we can calculate the variation of the volume under mentioned infinitesimal bending according to Theorem 3.27:

$$\begin{aligned}\delta V&=\frac{1}{3}\iint_S \mathbf{s}\cdot\boldsymbol{\nu}\,dA=\frac{1}{3}\iint_D \mathbf{s}\cdot\boldsymbol{\nu}\sqrt{g}\,dudv\\ &=\frac{1}{3}\iint_D\Big(-\frac{3}{8}uv^4-\frac{uv^2}{4}+\frac{u^3}{12}+\frac{u^5}{8}-\frac{u^3v^2}{4}\Big)\,dudv.\end{aligned} \tag{3.198}$$

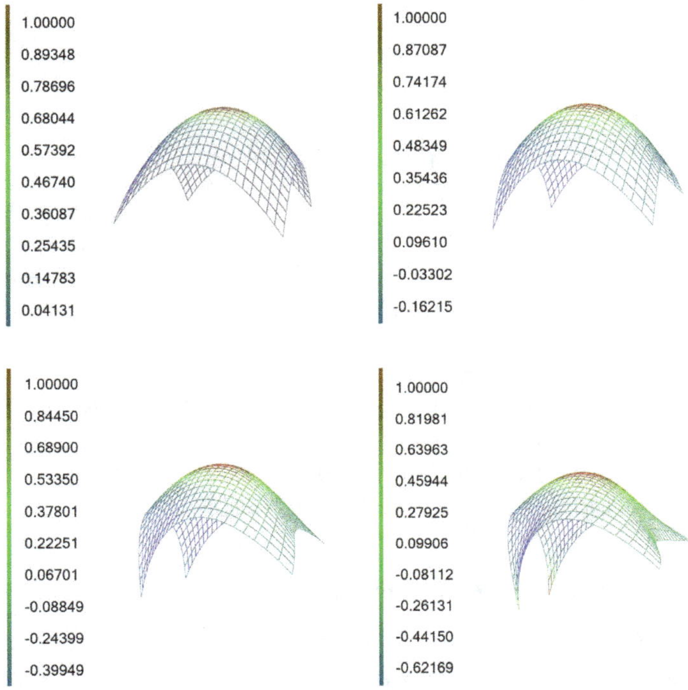

Fig. 3.3. Infinitesimal bending of elliptical paraboloid for $\epsilon = 0.0, 0.2, 0.4, 0.6$.

After moving to the polar coordinates, we have

$$\delta V = \frac{1}{3}\int_0^{2\pi} d\varphi \int_0^{\sqrt{3}/3}\Big(-\frac{3\rho^5\cos\varphi\sin^4\varphi}{8} + \frac{\rho^5\cos^5\varphi}{8} + \frac{\rho^3\cos^3\varphi}{12}$$

$$-\frac{\rho^3\cos\varphi\sin^2\varphi}{4} - \frac{1}{4}\rho^5\cos^3\varphi\sin^2\varphi\Big)\,\rho\,d\rho,$$

i.e.

$$\delta V = \frac{1}{3}\left[\frac{-3}{56}\left(\frac{\sqrt{3}}{3}\right)^7 \int_0^{2\pi}\cos\varphi\sin^4\varphi\,d\varphi + \frac{1}{56}\left(\frac{\sqrt{3}}{3}\right)^7\right.$$

$$\cdot \int_0^{2\pi}\cos^5\varphi\,d\varphi + \frac{1}{60}\left(\frac{\sqrt{3}}{3}\right)^5\int_0^{2\pi}\cos^3\varphi\,d\varphi$$

$$
\begin{aligned}
&-\frac{1}{20}\left(\frac{\sqrt{3}}{3}\right)^{5}\int_{0}^{2\pi}\cos\varphi\sin^{2}\varphi\,d\varphi \\
&\left.-\frac{1}{28}\left(\frac{\sqrt{3}}{3}\right)^{7}\int_{0}^{2\pi}\cos^{3}\varphi\sin^{2}\varphi\,d\varphi\right]. \qquad (3.199)
\end{aligned}
$$

Since all integrals in the last equation equal to zero, we obtain

$$\delta V = 0. \qquad (3.200)$$

Note that if we use at (3.194) $\beta(u,v) = u^2$, and as a consequence $\gamma(u,v) = u^2$ and $\alpha(u,v) = 2uv$, after calculations similar to those that preceded, we get such infinitesimal bending field under which the variation of the volume is not zero, i.e. $\delta V \neq 0$.

3.8.2 *Deformed Swept Volume*

A swept volume is the volume generated by the motion of an arbitrary object along an arbitrary path (or even a surface) possibly with arbitrary rotations [Abdel-Malek *et al.* (2004)]. The problem of swept volume determination is concerned with the calculation of properties associated with the resulting sweep, including the delineation of the boundary for visualization and computer graphics, formulations for modeling the sweep procedure, and the calculation of the volume's mass properties for solid modeling.

Let us describe the main idea of the sweep operations. For precise definition and basic facts about the concept on swept volumes, we refer to [Abdel-Malek *et al.* (2004)].

For example, consider the extrusion of a curve $\mathbf{c}(t)$ along axis x of coordinate system $Oxyz$ to generate a surface $\mathbf{s}(t,u)$. This surface is then revolved along a curve (about y axis) to generate volume $\mathbf{v}(t,u,v)$ parameterized in three variables. The volume is then swept again along a curve (about z axis) to generate a more complex shape $\boldsymbol{\xi}(t,u,v,w)$, etc. Note that $\boldsymbol{\xi}(t,u,v,w)$ is the totality of points in the swept volume representing every point in the solid model.

Because this formulation yields closed form surface equations describing the boundary of the solid model, it is possible to apply the divergence theorem to compute the volume. The contribution of a variety $\boldsymbol{\xi}^{(i)}(\mathbf{u})$, $\mathbf{u} = (u^1, u^2)$, to the volume of a solid model with unit normal $\boldsymbol{\nu}^{(i)}$ using divergence theorem is given by

$$3V = \iint_S \boldsymbol{\xi}^{(i)}(\mathbf{u}) \cdot \boldsymbol{\nu}^{(i)}\, dA.$$

Note that the varieties are expressed by substituting the singularities $\mathbf{p}_i$ into $\boldsymbol{\xi}(\mathbf{u})$ such that the equation of the variety is

$$\boldsymbol{\xi}(\mathbf{u}) = \boldsymbol{\xi}(\mathbf{u}, \mathbf{p_i}),$$

(for more precise explanations see [Abdel-Malek and Othman (1999)]). The area is given by $dA = \|\boldsymbol{\xi}_1^{(i)} \times \boldsymbol{\xi}_2^{(i)}\|\, du^1 du^2$ and for a variety which is a parametric entity, the normal vector is written as

$$\boldsymbol{\nu}^{(i)} = \frac{\boldsymbol{\xi}_1^{(i)} \times \boldsymbol{\xi}_2^{(i)}}{\|\boldsymbol{\xi}_1^{(i)} \times \boldsymbol{\xi}_2^{(i)}\|}.$$

The volume V of a solid model enclosed by varieties is the summation of the contribution of all boundary varieties as (note that $\tilde{\boldsymbol{\nu}}^{(i)}$ is not a unit vector)

$$V = \frac{1}{3}\sum_i^n \iint_S \boldsymbol{\xi}^{(i)}(\mathbf{u}) \cdot \boldsymbol{\nu}^{(i)}\, dA = \frac{1}{3}\sum_i^n \iint_D \boldsymbol{\xi}^{(i)}(\mathbf{u}) \cdot \tilde{\boldsymbol{\nu}}^{(i)}\, du^1 du^2.$$

It is very interesting to observe deformed swept volume, especially in modeling membranes and deformable objects. Blackmore *et al.* in their paper [Blackmore *et al.* (1994)] observed objects experiencing general smooth deformations as opposed to the rigid sweeps.

Here, we point to the possibility of studying infinitesimal bending of swept volume and opens new questions in the theory of sweeping. According to Theorem 3.27, we obtain that under infinitesimal bending, the variation of the volume of a solid model is equal

$$\delta V = \frac{1}{3}\sum_i^n \iint_S \mathbf{s}^{(i)}(\mathbf{u}) \cdot \boldsymbol{\nu}^{(i)}\, dA = \frac{1}{3}\sum_i^n \iint_D \mathbf{s}^{(i)}(\mathbf{u}) \cdot \tilde{\boldsymbol{\nu}}^{(i)}\, du^1 du^2,$$

where $\mathbf{s}^{(i)}$ is the field of translation of the corresponding variety.

This is a new application of the translation field. Some other interesting results and manifestations of the field $\mathbf{s}$ can be found in [Alexandrov (2010)] (in particular Theorems 4 and 5).

Chapter 4

Surfaces of Revolution

4.1 Infinitesimal Bending of the First Order of Surfaces of Revolution

Many works at the theory of deformation of surfaces are dedicated to infinitesimal bending problem of surfaces of revolution. The problem of generating the field of infinitesimal bending is in this case reduced to the problem of solving the system of ordinary differential equations.

We shall consider infinitesimal bending of surfaces of revolution using Cohn-Vossen's method (Cohn-Vossen, 1959), which is the basic for this class of surfaces.

The special coordinate system with the orthonormal base $\mathbf{e}$, $\mathbf{a}(v)$, $\mathbf{a}'(v)$ is introduced in the following way.

The axe of rotation has the orth $\mathbf{e}$ and orth $\mathbf{a}(v) \perp \mathbf{e}$ at the point O. At the coordinate plane generated by the orths $\mathbf{e}$, $\mathbf{a}(v)$, we consider the curve C which is the meridian of the surface of revolution

$$C : \rho = \rho(u),$$

where u-axe is the axe of rotation. The equation of the surface of revolution generated by the curve C as meridian is

$$S : \mathbf{r}(u, v) = u\mathbf{e} + \rho(u)\mathbf{a}(v). \tag{4.1}$$

Infinitesimal bending field $\mathbf{z}(u,v)$ can be presented in this coordinate system as

$$\mathbf{z}(u,v) = \alpha(u,v)\mathbf{e} + \beta(u,v)\mathbf{a}(v) + \gamma(u,v)\mathbf{a}'(v). \tag{4.2}$$

Then, the following theorem is in force (Cohn-Vossen, 1959).

Theorem 4.1. *Necessary and sufficient condition for the field $\mathbf{z}(u,v)$ to be a field of infinitesimal bending of the surface of revolution S is that the functions $\alpha(u,v)$, $\beta(u,v)$, $\gamma(u,v)$ satisfy the system of partial differential equations*

$$\begin{aligned} &\alpha_u + \rho'(u)\beta_u = 0,\\ &\beta + \gamma_v = 0,\\ &\alpha_v + \rho'(u)(\beta_v - \gamma) + \rho(u)\gamma_u = 0. \end{aligned} \tag{4.3}$$

The functions $\alpha(u,v), \beta(u,v), \gamma(u,v)$ are periodical functions with respect to v with period 2π. They can be developed in Fourier series:

$$\begin{aligned} &\alpha(u,v) = \sum_{k=-\infty}^{+\infty} \varphi_k(u)e^{ikv}, \quad \beta(u,v) = \sum_{k=-\infty}^{+\infty} \psi_k(u)e^{ikv},\\ &\gamma(u,v) = \sum_{k=-\infty}^{+\infty} \chi_k(u)e^{ikv}, \end{aligned} \tag{4.4}$$

where $\varphi_k(u)$, $\psi_k(u)$, $\chi_k(u)$, $k \in \mathbb{Z}$, are the complex functions.

The field $\mathbf{z}(u,v)$ can be written as

$$\begin{aligned} \mathbf{z}(u,v) &= \sum_{k=-\infty}^{+\infty} [\varphi_k(u)\mathbf{e} + \psi_k(u)\mathbf{a}(v) + \chi_k(u)\mathbf{a}'(v)]e^{ikv}\\ &= \sum_{k=-\infty}^{+\infty} \mathbf{z}_k(u,v), \end{aligned} \tag{4.5}$$

where

$$\mathbf{z}_k(u,v) = [\varphi_k(u)\mathbf{e} + \psi_k(u)\mathbf{a}(v) + \chi_k(u)\mathbf{a}'(v)]e^{ikv}, \quad k \in \mathbb{Z}. \tag{4.6}$$

The following theorem is valid too (Cohn-Vossen, 1959).

Theorem 4.2. *The functions* $\varphi_k(u)$, $\psi_k(u)$, $\chi_k(u)$, $k \in \mathbb{Z}$, *given with* (4.4) *satisfy the next system of differential equations*

$$\begin{aligned} \varphi_k'(u) + \rho'(u)\psi_k'(u) &= 0, \\ \psi_k(u) + ik\chi_k(u) &= 0, \\ ik\varphi_k(u) + \rho'(u)\left[ik\psi_k(u) - \chi_k(u)\right] + \rho(u)\chi_k'(u) &= 0, \end{aligned} \tag{4.7}$$

and the functions $\psi_k(u)$, $\chi_k(u)$, *satisfy the same ordinary differential equation of the second order*

$$\rho(u)\psi_k''(u) + (k^2 - 1)\rho''(u)\psi_k(u) = 0 \tag{4.8}$$

or

$$\rho(u)\chi_k''(u) + (k^2 - 1)\rho''(u)\chi_k(u) = 0. \tag{4.9}$$

According to Cohn-Vossen (1959), the following theorem is in the force.

Theorem 4.3. *Each solution* $\psi_k(u)$ *of Eq.* (4.8), *which is not identically equal to zero for each* $k \in \{\pm 2, \pm 3, \ldots\}$, *corresponds to non-trivial field* $\mathbf{z}_k(u, v)$ (4.6) *of the infinitesimal bending of the considered part of the surface of revolution.*

As it is well known (Efimov, 1948), for each $k \in \{2, 3, \ldots\}$, there exist a field of infinitesimal bending

$$\begin{aligned} \mathbf{z}(u, v) = \mathbf{z}_k(u, v) = &\left[\varphi_k(u)e^{ikv} + \tilde{\varphi}_k(u)e^{-ikv}\right]\mathbf{e} \\ &+ [\psi_k(u)e^{ikv} + \tilde{\psi}_k(u)e^{-ikv}]\mathbf{a}(v) \\ &+ \left[\chi_k(u)e^{ikv} + \tilde{\chi}_k(u)e^{-ikv}\right]\mathbf{a}'(v), \end{aligned} \tag{4.10}$$

where for instance, $\tilde{\varphi}_k(u)$ is the conjugated value for $\varphi_k(u)$.

Equation (4.10) can in other way be written as follows:

$$\begin{aligned} \mathbf{z}(u, v) &= \mathbf{z}_k(u, v) \\ &= [\varphi_k(u)\cos kv + \tilde{\varphi}_k(u)\sin kv]\,\mathbf{e} \\ &\quad + [\psi_k(u)\cos kv + \tilde{\psi}_k(u)\sin kv]\mathbf{a}(v) \\ &\quad + [\chi_k(u)\cos kv + \tilde{\chi}_k(u)\sin kv]\,\mathbf{a}'(v). \end{aligned} \tag{4.11}$$

For the consideration of infinitesimal bending of a surface of revolution generated by meridian being continuous but not regular curve, we use the consideration of Cohn-Vossen (1959).

We start with theorem of Cohn-Vossen (1959).

Theorem 4.4. *If the functions $\varphi_k(u)$, $\chi_k(u)$ are continuous at the breaking points $u = \sigma$ of the meridian $\rho = \rho(u)$,* i.e. *at the points where $\rho'(\sigma - 0) \neq \rho'(\sigma + 0)$, then the functions $\psi_k(u)$ at this points satisfy the equation*

$$[\rho(u)\psi_k'(u) + (k^2 - 1)\rho'(u)\psi_k(u)] \,|_{\sigma-0}^{\sigma+0} = 0, \tag{4.12}$$

i.e.

$$\begin{aligned}\rho(\sigma)[\psi_k'(\sigma + 0) - \psi_k'(\sigma - 0)]&\\ + (k^2 - 1)[\rho'(\sigma + 0) - \rho'(\sigma - 0)]\psi_k(\sigma) &= 0.\end{aligned} \tag{4.13}$$

4.2 Classification of Toroids Generated by Quadrangle with Respect to Flexibility

It is known that the circular torus is rigid. Among the surfaces that are topologically equivalent to the torus, Belov has pointed to the class of non-rigid toroids. At a work, Belov (1968) has considered infinitesimal bending of the toroid surface of revolution with a meridian in a shape of special quadrangle and among them pointed to a class of non-rigid surfaces with a convex meridian.

From further consideration, we exclude the toroids containing plane part generated by a side of quadrangle perpendicular on the axis of revolution. Based on the work of Efimov (1948), this surface is always non-rigid, and for bending field, we can get

$$\mathbf{z}(M) = \begin{cases} \varphi(M)\mathbf{e}, & M \in \omega, \\ 0, & M \notin \omega, \end{cases} \tag{4.14}$$

where $\varphi(M)$ is a scalar function of a point from a plane part ω of a surface S and $\mathbf{e}$ is the unit normal vector.

If the side of the meridian is parallel to the axis of revolution, then the generated part of toroid is cylinder. The examples of surfaces containing cylindrical parts were studied in the work of Efimov (1948).

For considering toroid surface of revolution with any quadrangular meridian, we use Cohn-Vossen's method (Cohn-Vossen, 1959).

First, we present Belov's theorem given in the work of Belov (1968), and we here give further generalization, considering toroid with any quadrangular meridian.

Theorem 4.5 (Belov's Theorem, Belov, 1968). *Suppose that quadrangle $\mathcal{B}$ with apices*

$$A(-1,b), \quad B(0,b+c_1), \quad C(1,b), \quad D(0,b-c_2) \tag{4.15}$$

rotates around u-axis at $u0\rho$ plane. Toroid surface of revolution generated by meridian $\mathcal{B}$ is non-rigid if and only if

$$\frac{1}{c_2} - \frac{1}{c_1} = \frac{k^2}{b}, \tag{4.16}$$

where $k \geq 2$ is an integer number.

Theorem 4.6. *Let $\mathcal{Q} = A_1A_2A_3A_4$ be a convex quadrangle with apices*

$$\begin{aligned} &A_1(a - p\cos\alpha, b - p\sin\alpha), A_2(a + s\cos\beta, b + s\sin\beta), \\ &A_3(a + q\cos\alpha, b + q\sin\alpha), A_4(a - r\cos\beta, b - r\sin\beta), \end{aligned} \tag{4.17}$$

where the point $O(a,b)$ is an intersection of diagonals, p, q, r, s are the lengths of the segments on the diagonals, and α, $\beta \in [0,\pi]$ are the angles between diagonals and the positive part of x-axis (Fig. 4.1).

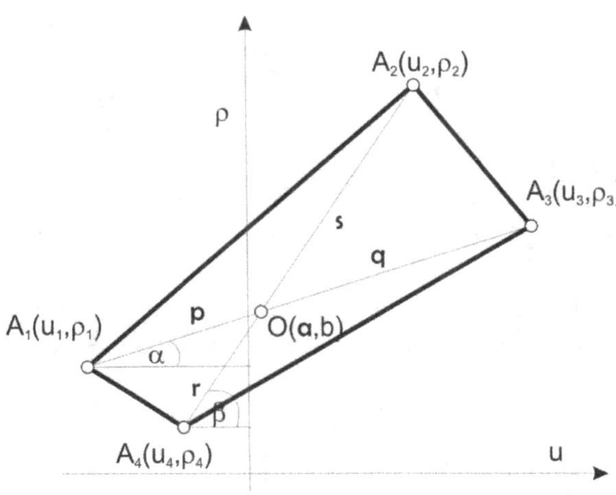

Fig. 4.1. A convex quadrangle.

Toroid surface of revolution $\mathcal{S}$, generated by meridian $\mathcal{Q}$, is non-rigid if and only if

$$\frac{k^2\sin(\alpha-\beta)}{b}=\left(\frac{1}{s}-\frac{1}{r}\right)\cos\alpha+\left(\frac{1}{p}-\frac{1}{q}\right)\cos\beta \tag{4.18}$$

or

$$bpq(s-r)\cos\alpha+brs(p-q)\cos\beta+k^2rspq\sin(\alpha-\beta)=0, \tag{4.19}$$

where $k\geq 2$ is an integer number.

Proof. Based on the fact that the bending field is continuous at the vertices of the meridian and using the procedure presented in the work of Velimirović and Ran̆cıć (2009), we get the system of algebraic homogenous linear equations.

Necessary and sufficient condition for the existence of the field of infinitesimal bending of a toroid rotational surface with quadrangular meridian is that the determinant of this system is zero, i.e. Eq. (4.19) is valid. □

The question about the number of non-rigid surfaces among all such generated toroids arises. The answer offers the following theorem.

Theorem 4.7. *There exist countably many values of the parameter s*

$$s=\frac{bpqr\cos\alpha}{bpq\cos\alpha+brp\cos\beta-brq\cos\beta+k^2rpq\sin(\alpha-\beta)},$$

yielding to non-rigid cases ($k\geq 2,\ k\in\mathbb{Z}$). In the limit for $k\to\infty$, we have $s=0$, i.e. *we have triangular meridian (rigid case in the work of Velimirović,* 1994*).*

Proof. Starting from the condition (4.19), it is easy to get value for s. In this case, s is the function of k, where $k\in\mathbb{Z}$. □

Theorem 4.8. *The infinitesimal bending field of the surface of revolution $\mathcal{S}$ generated by convex meridian $\mathcal{Q}$, given by* (4.17), *satisfying condition* (4.18), *is unique.*

Proof. From (4.18), $k \geq 2,\ k \in \mathbb{Z}$, we get

$$k = \sqrt{\frac{bpq(r-s)\cos\alpha + rsb(q-p)\cos\beta}{rspq\sin(\alpha-\beta)}}.$$

Using Cohn-Vossen's method (Cohn-Vossen, 1959) and remark that for positive and negative values of k we get the same equation for ordering bending field. □

It is easy to prove that the angle between diagonals of quadrangle $\mathcal{Q}$ given by (4.17) is $|\beta - \alpha|$.

4.2.1 *Geometrical Meaning of the Non-Rigidity Condition*

In the previous section, we got the non-rigidity condition (4.18) for toroid with quadrangular meridian. We try to find the geometrical meaning of this condition and classify toroid surfaces with quadrangular meridian as rigid or non-rigid. Based on previous considerations, we have the following theorem.

Theorem 4.9 (Velimirović and Rančić, 2007). *Toroid surface of revolution generated by a meridian in a shape of parallelogram, containing conical parts, is rigid if and only if one of the meridian sides is not orthogonal to the axis of revolution.*

A. We discuss the condition (4.18) based on the position of quadrangular's diagonals.

A.1. If one diagonal (e.g. A_1A_3) is parallel to the axis of revolution, i.e. $\alpha = 0$, the non-rigidity condition (4.18) becomes

$$\frac{k^2}{b}\sin\beta = \frac{1}{r} - \frac{1}{s} + \left(\frac{1}{q} - \frac{1}{p}\right)\cos\beta. \tag{4.20}$$

This formula yields us to the rigid and non-rigid surfaces. We discuss (4.20).

A.1.1. In the case of orthogonality of the other diagonal (A_2A_4) to the axis of revolution, i.e. $\beta = \pi/2$, we get Belov's condition

$$\frac{k^2}{b} = \frac{1}{r} - \frac{1}{s}. \tag{4.21}$$

Corollary 4.1. *In the case of orthogonality of the diagonals of meridian $\mathcal{Q}$ with apices given by* (4.17) *of toroid surface $\mathcal{S}$, where one diagonal is parallel to the axis of revolution, the non-rigidity condition is* (4.21)*. For $s \leq r$, and $r > b/4$, the surface is always rigid.*

A.1.2. If one diagonal (A_2A_4) divides the other (A_1A_3) into equal parts ($p = q$, $r \neq s$, $\beta \neq \pi/2$), the condition (4.20) becomes

$$\frac{k^2}{b}\sin\beta = \frac{1}{r} - \frac{1}{s}. \tag{4.22}$$

Corollary 4.2. *In the case where one diagonal of meridian (e.g. A_1A_3) is parallel to the axis of revolution and if the other diagonal (A_2A_4) divides the first (A_1A_3) into equal parts ($p = q$, $r \neq s$), the non-rigidity condition is* (4.22)*. For $s \leq r$, and $r > b/4\sin\beta$, the surface is always rigid.*

A.1.3. If diagonal (A_1A_3) divides the other (A_2A_4) into equal parts ($p \neq q$, $r = s$, $\beta \neq \pi/2$), the condition (4.20) becomes

$$\frac{k^2}{b}\tan\beta = \frac{1}{q} - \frac{1}{p}. \tag{4.23}$$

Corollary 4.3. *In the case of one diagonal (e.g. A_1A_3) being parallel to the axis of revolution and if it divides the other (A_2A_4) into equal parts ($p \neq q$, $r = s$), then the non-rigidity condition is* (4.23)*. For $p\tan\beta \leq q\tan\beta$ and $q > b/4\tan\beta$, the surface is always rigid.*

A.1.4. If the diagonal divides the other in halves to each devices, i.e. $p = q$, $r = s$, we get a class of parallelograms. The surfaces generated by parallelogram are rigid (see the work of Velimirović and Rančić, 2007).

A.2. If one diagonal (e.g. A_1A_3) is orthogonal to the axis of revolution $\alpha = \pi/2$, the non-rigidity condition (4.18) becomes

$$b(p-q) + k^2pq = 0 \quad \text{or} \quad \frac{k^2}{b} = \frac{1}{p} - \frac{1}{q}. \tag{4.24}$$

This condition is equivalent to Belov's condition.

Corollary 4.4. *If one diagonal is orthogonal to the axis of revolution of meridian $\mathcal{Q}$ with apices given by (4.17) of toroid surface $\mathcal{S}$, the non-rigidity condition is (4.24). For $q \leq p$ and $p > b/4$, the surface is always rigid.*

In the case of $\alpha = \pi$, discussion is the same as in the case **A**.
We discuss here the special classes of quadrangles:

If the meridian is trapezium, the non-rigidity condition (4.18) becomes

$$\frac{k^2}{b}\sin(\alpha-\beta) = \left(\frac{1}{s} - \frac{1}{r}\right)\cos\alpha + \left(\frac{1}{p} - \frac{p}{rs}\right)\cos\beta. \tag{4.25}$$

If $p = r$ and $q = s$, $s \neq r$, the meridian is isosceles trapezium. The non-rigidity condition (4.18) reduces to

$$\frac{k^2\cos\frac{\alpha-\beta}{2}}{b\sin\frac{\alpha+\beta}{2}} = \frac{1}{r} - \frac{1}{s}, \quad \alpha-\beta \notin \{0,\pi\},\ \alpha+\beta \notin \{0,2\pi\}. \tag{4.26}$$

If one diagonal of meridian divides the other diagonal into two equal parts:

In the case $r = s$, $p \neq q$, then (4.18) becomes

$$\frac{k^2\sin(\alpha-\beta)}{b\cos\beta} = \frac{1}{p} - \frac{1}{q}, \quad \alpha-\beta \notin \{0,\pi\}. \tag{4.27}$$

The same in case of $r \neq s, p = q$, the condition (4.18) becomes

$$\frac{k^2\sin(\alpha-\beta)}{b\cos\alpha} = \frac{1}{s} - \frac{1}{r}, \quad \alpha-\beta \notin \{0,\pi\}. \tag{4.28}$$

If $r = s, p = q$, the quadrangle is parallelogram, the condition (4.18) reduces to

$$k^2pqrs\sin(\alpha-\beta) = 0, \quad \alpha-\beta \notin \{0,\pi\}, \tag{4.29}$$

i.e. the toroid with meridian in shape of parallelogram is rigid.

In the case of the quadrangle with mutually orthogonal diagonals $|\alpha - \beta| = \pi/2$, the non-rigidity condition (4.18) reduces to

$$\frac{k^2}{b} = \pm\left(\left(\frac{1}{p} - \frac{1}{q}\right)\cos\beta + \left(\frac{1}{s} - \frac{1}{r}\right)\sin\beta\right), \tag{4.30}$$

where for $\beta = 0$ or $\beta = \pi/2$, we get Belov's condition (4.16) as a special case. In this class of quadrangle are deltoids.

Corollary 4.5. *The toroids generated by meridian in a shape of a deltoid (kite) with diagonal parallel to the axis of revolution are rigid.*

The following theorem holds.

Theorem 4.10. *Let* $\mathcal{Q} = A_1A_2A_3A_4$ *be a non-convex quadrangle with apices*

$$\begin{aligned} &A_1(a - p\cos\alpha, b - p\sin\alpha), \quad A_2(a + s\cos\beta, b + s\sin\beta), \\ &A_3(a - q\cos\alpha, b - q\sin\alpha), \quad A_4(a - r\cos\beta, b - r\sin\beta), \end{aligned} \tag{4.31}$$

where the point $O(a,b)$ is an intersection of diagonals, p, q, r, s are the lengths of the segments on the diagonals, and $\alpha < \pi/2$, $\beta > \pi/2$ or $\alpha < \pi/2$, $\beta < \pi/2$ are the angles between diagonals and the positive part of x-axis (Fig. 4.2). The toroid surface of revolution $\mathcal{S}$, generated by meridian $\mathcal{Q}$, is non-rigid if and only if

$$\frac{k^2\sin(\alpha - \beta)}{b} = \left(\frac{1}{s} - \frac{1}{r}\right)\cos\alpha + \left(\frac{1}{p} + \frac{1}{q}\right)\cos\beta, \tag{4.32}$$

where $k \geq 2$ is an integer number.

Proof. Proof of this theorem is similar like the proof of Theorem 4.6. □

We also consider the number of non-rigid toroid surfaces with non-convex quadrangular meridian. Starting from (4.32), we have the following conclusion.

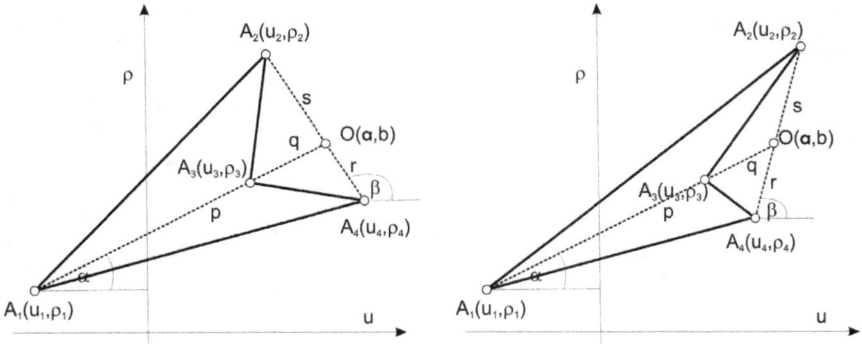

Fig. 4.2. Non-convex quadrangles.

Theorem 4.11. *There exist countably many values of the parameter s*

$$s = \frac{bpqr\cos\alpha}{bpq\cos\alpha - brp\cos\beta - brq\cos\beta + k^2 rpq\sin(\alpha-\beta)},$$

yielding to non-rigid cases ($k \geq 2,\ k \in \mathbb{Z}$). *In the limit for* $k \to \infty$, *we have* $s = 0$, *we have triangular meridian (rigid case in the work of Velimirović,* 1994*).*

Proof. Starting from the condition (4.32), it is easy to get value for s. In this case, s is the function of k, where $k \in \mathbb{Z}$. □

Theorem 4.12. *The infinitesimal bending field of the surface of revolution* $\mathcal{S}$ *generated by non-convex meridian* $\mathcal{Q}$ *given by* (4.31) *is unique.*

Proof. From (4.32), $k \geq 2$, $k \in \mathbb{Z}$, we get

$$k = \sqrt{\frac{bpq(r-s)\cos\alpha + rsb(q+p)\cos\beta}{rspq\sin(\alpha-\beta)}}. \tag{4.33}$$

Using Cohn-Vossen's method (Cohn-Vossen, 1959) and remark that for positive and negative values of k we get the same equation. □

Previously, we got the non-rigidity condition (4.32) for toroid with non-convex quadrangular meridian. We try to find the geometrical meaning of this condition and classify toroid surfaces with quadrangular meridian as rigid or non-rigid.

B. We discuss the condition (4.32) based on the position of it's diagonals.

B.1. If diagonal A_1A_3 is parallel to the axis of revolution $\alpha = 0$, the non-rigidity condition (4.32) becomes

$$\frac{k^2}{b}\sin\beta = \left(\frac{1}{r} - \frac{1}{s}\right) - \left(\frac{1}{p} + \frac{1}{q}\right)\cos\beta. \tag{4.34}$$

This formula yields us to the rigid and non-rigid surfaces. We discuss (4.34).

B.1.1. In the case of orthogonality of the other diagonal A_2A_4 to the axis of revolution $\beta = \pi/2$, we get

$$\frac{k^2}{b} = \frac{1}{r} - \frac{1}{s}. \tag{4.35}$$

This condition is equivalent to Belov's condition.

Corollary 4.6. *In the case of orthogonality of the diagonals of meridian $\mathcal{Q}$ with apices given by* (4.31) *of toroid surface $\mathcal{S}$, where one diagonal is parallel to the axis of revolution, the non-rigidity condition is* (4.35). *For $r \leq s$ and $s > b/4$, the surface is always rigid.*

B.2. If diagonal A_2A_4 is parallel to the axis of revolution $\beta = 0$, the non-rigidity condition (4.32) becomes

$$\frac{k^2}{b}\sin\alpha = \left(\frac{1}{s} - \frac{1}{r}\right)\cos\alpha + \left(\frac{1}{p} + \frac{1}{q}\right). \tag{4.36}$$

This formula yields us to the rigid and non-rigid surfaces. We discuss (4.36).

B.2.1. In the case of orthogonality of the other diagonal A_1A_3 to the axis of revolution $\alpha = \pi/2$, we get

$$\frac{k^2}{b} = \frac{1}{p} + \frac{1}{q}. \tag{4.37}$$

Corollary 4.7. *In the case of orthogonality of the diagonals of meridian $\mathcal{Q}$ with apices given by* (4.31) *of toroid surface $\mathcal{S}$, where one diagonal is parallel to the axis of revolution, the non-rigidity condition is* (4.37).

B.3. If the diagonal A_1A_3 is orthogonal to the axis of revolution $\alpha = \pi/2$, the non-rigidity condition (4.32) becomes

$$\frac{k^2}{b} = \frac{1}{p} + \frac{1}{q}. \tag{4.38}$$

Corollary 4.8. *If one diagonal is orthogonal to the axis of revolution of meridian $\mathcal{Q}$ with apices given by* (4.31) *of toroid surface $\mathcal{S}$, the non-rigidity condition is* (4.38).

B.4. If the diagonal A_2A_4 is orthogonal to the axis of revolution $\beta = \pi/2$, the non-rigidity condition (4.32) becomes

$$\frac{k^2}{b} = \frac{1}{r} - \frac{1}{s}. \tag{4.39}$$

This condition is equivalent to Belov's condition.

Corollary 4.9. *In the case of orthogonality of the diagonals of meridian $\mathcal{Q}$ with apices given by* (4.31) *of toroid surface $\mathcal{S}$, where one diagonal is parallel to the axis of revolution, the non-rigidity condition is* (4.39). *For $s \leq r$ and $r > b/4$, the surface is always rigid.*

In the case of $\alpha = \pi$, the discussion is the same as in the case **B.**

In the following pictures, we have the basic and deformed toroid of the class defined by Theorem 4.6 in Fig. 4.3 and of the class given by Theorem 4.10 in Fig. 4.4.

In Fig. 4.3, we have an example with $p \neq q$ i $r \neq s$ and $k = 3$, $\beta = \frac{5\pi}{6}, \alpha = \frac{\pi}{3}, a = 1, b = 6, p = 4, q = 3, r = 5, s = \frac{60}{(192-5\sqrt{3})}$. Then,

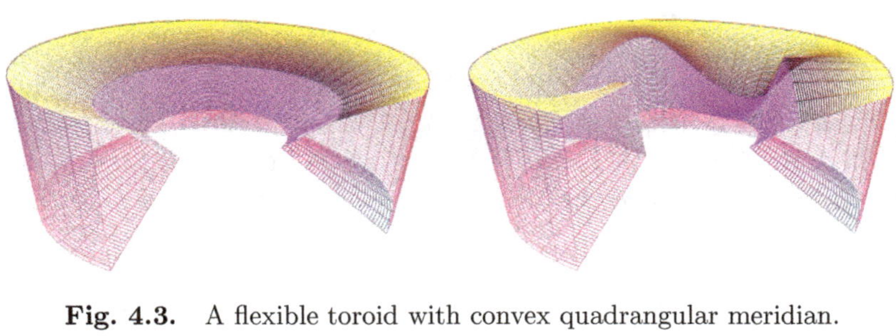

Fig. 4.3. A flexible toroid with convex quadrangular meridian.

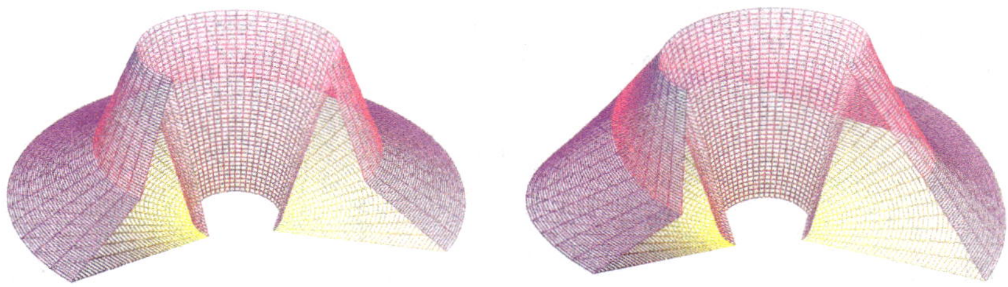

Fig. 4.4. A flexible toroid with non-convex quadrangular meridian.

we have vertices

$$\left\{(-1,2(3+\sqrt{3})),\left(7+\frac{1152}{-192+5\sqrt{3}},6-\frac{30}{192-5\sqrt{3}}\right),\right.$$
$$\left.\left(\frac{5}{2},\frac{3}{2}(4-\sqrt{3})\right),\left(1+\frac{5\sqrt{3}}{2},\frac{17}{2}\right)\right\}.$$

For $\varepsilon = 0.2$, the basic and the transformed toroids are given in Fig. 4.3.

On Fig. 4.4, we have an example with $p \neq q$ i $r \neq s$ and $k = 2$, $\alpha = \frac{\pi}{4}$, $\beta = \frac{5\pi}{6}$, $a = \frac{3}{2}, b = 3 + \frac{\sqrt{2}}{2}$, $p = \frac{5\sqrt{2}}{4}$, $q = \frac{\sqrt{2}}{2}$, $r = \frac{3\sqrt{2}}{4}$, $s = (\frac{3}{2} - \frac{9}{41})\frac{\sqrt{2}}{2}$, then we have vertices

$$\left\{(-1,1),\left(\frac{9}{41},5-\frac{9}{41}\right),(1,3),(3,2)\right\}.$$

For $\varepsilon = 0.05$, the basic and the transformed toroids are given in Fig. 4.4.

4.3 Variation of the Volume of Surface of Revolution

Sabitov (Connelly, 1980) made conjecture that for any infinitesimal C^1 smooth bending of a closed surface, the volume, bounded by it, will be stationary, i.e. that the variation of this volume is zero. Aleksandrov has proved in the work of Alexandrov (1989) that the mentioned conjecture is true for the surface of revolution of type 0 or 1 with C^1-smooth meridian not containing a segment perpendicular to the axis of rotation.

Namely, at the starting point, Aleksandrov made a connection between the variation of volume and flow. He has proved the following lemma.

Lemma 4.1 (Alexandrov, 1989). *The variation of the volume of domain D_0 under the action of the field $\mathbf{z}$ is equal to the flux of the field $\mathbf{z}$ through the outer side of the surface S_0,* i.e.

$$\delta V(D_0, \mathbf{z}) = \iint_{S_0} \mathbf{z} \cdot \boldsymbol{\nu} dS, \tag{4.40}$$

where $\boldsymbol{\nu}$ is the outer normal to the surface S_0 and $\mathbf{z} \cdot \boldsymbol{\nu}$ is the scalar product of the vectors $\mathbf{z}$ and $\boldsymbol{\nu}$.

Using this lemma, Aleksandrov has proved the following theorem.

Theorem 4.13 (Alexandrov, 1989). *Let the meridian of a connected closed surface of rotation be a C^1-smooth curve that does not contain a segment perpendicular to the axis of rotation. Then, the flux through the given surface of any C^1-smooth bending field is equal to zero.*

The hypotheses of this theorem satisfy non-rigid surfaces topologically equivalent to a sphere and to a torus.

Aleksandrov (1989) finished his work with the following open question: Is the volume bounded by a piecewise-smooth surface of revolution always stationary under infinitesimal bending?

We are proving the following theorem.

Theorem 4.14 (Velimirović, 1998b). *Let the meridian of the closed connected surface of revolution be a piecewise smooth curve, not containing a segment perpendicular to the axis of rotation. Then, the variation of the volume, bounded by this surface, is equal to zero, under the C^1-smooth field of infinitesimal bending.*

Proof. Let $\rho = \rho(u)$ be a meridian, rotating around u-axis at Cartesian coordinate system $uO\rho$. The equation of the surface of revolution S_0, at special coordinate system with trihedron $\mathbf{e}$, $\mathbf{a}(v)$, $\mathbf{a}'(v)$, is

$$S_0 : \mathbf{r}(u, v) = u\mathbf{e} + \rho(u)\mathbf{a}(v). \tag{4.41}$$

Let us suppose that the meridian generating surface S_0 is a piecewise smooth curve

$$C_0 = \bigcup_{j=1}^{n} C_j, \tag{4.42}$$

where

$$C_j : \rho = \rho_{(j)}(u), \quad j = 1, \ldots, n; \quad u \in [u_j, u_{j+1}] \subset [\alpha, \beta]. \tag{4.43}$$

At the breaking points $u = \sigma$ of the meridian $\rho = \rho(u)$, i.e. at the points where meridian is continuous, but not smooth curve, we have

$$\rho_{(j)}(\sigma) = \rho_{(j+1)}(\sigma), \quad \rho'_{(j)}(\sigma) \neq \rho'_{(j+1)}(\sigma). \tag{4.44}$$

Infinitesimal bending field $\mathbf{z}(u, v)$ can be considered as

$$\mathbf{z} = \sum_{k=2}^{m} a_k \mathbf{z}_k, \quad a_k = \text{const.}, \quad m = 2, 3, \ldots, \tag{4.45}$$

where

$$\begin{aligned} \mathbf{z}_k(u, v) = {} & [\varphi_k(u)e^{ikv} + \widetilde{\varphi}_k(u)e^{-ikv}]\mathbf{e} + [\psi_k(u)e^{ikv} + \widetilde{\psi}_k(u)e^{-ikv}]\mathbf{a}(v) \\ & + [\chi_k(u)e^{ikv} + \widetilde{\chi}_k(u)e^{-ikv}]\mathbf{a}'(v), \end{aligned} \tag{4.46}$$

and $\widetilde{\chi}_k(u)$ is the conjugated value for $\chi_k(u)$, etc.

Infinitesimal bending field $\mathbf{z}(u, v)$ must be a continuous vector field on the whole surface, including the circles described by the breaking points $u = \sigma$ of the meridian. From (4.45) and (4.46), we have

$$\begin{aligned} & \varphi_{k(j)}(\sigma) = \varphi_{k(j+1)}(\sigma), \quad \psi_{k(j)}(\sigma) = \psi_{k(j+1)}(\sigma), \\ & \chi_{k(j)}(\sigma) = \chi_{k(j+1)}(\sigma). \end{aligned} \tag{4.47}$$

Surface S_0 consists of smooth parts $S_o^{(j)}$:

$$S_0 = \bigcup_{j=1}^{n} S_0^{(j)}. \tag{4.48}$$

According to the lemma of Aleksandrov, the variation of the volume is flux, calculated as surface integral. Based on the characteristics of the surface integral, according to (4.40), we have

$$\delta V = \iint_{S_0} \mathbf{z} \cdot \boldsymbol{\nu} dS = \sum_{j=1}^{n} \iint_{S_0^{(j)}} \mathbf{z} \cdot \boldsymbol{\nu} dS. \tag{4.49}$$

We transform surface integrals to doubles:

$$\boldsymbol{\nu} = \frac{\mathbf{r}_u \times \mathbf{r}_v}{\sqrt{EG - F^2}}, \quad dS = \sqrt{EG - F^2} du dv,$$

$$\delta V = \sum_{j=1}^{n} \iint\limits_{A_0^{(j)}} [\mathbf{z}, \mathbf{r}_u, \mathbf{r}_v] du dv, \tag{4.50}$$

where $[\mathbf{z}, \mathbf{r}_u, \mathbf{r}_v]$ is the scalar triple product of the vectors $\mathbf{z}$, $\mathbf{r}_u$, $\mathbf{r}_v$:

$$A_0^{(j)} = [u_j, u_{j+1}] \times [0, 2\pi] \subset [\alpha, \beta] \times [0, 2\pi], \tag{4.51}$$

$A_0^{(j)}$ is a part of u, v-plane, corresponding to the part $S_o^{(j)}$ of the surface. Using (4.45), we have

$$\delta V = \sum_{j=1}^{n} \sum_{k=2}^{m} a_k \iint\limits_{A_0^{(j)}} [\mathbf{z}_k, \mathbf{r}_u, \mathbf{r}_v] du dv. \tag{4.52}$$

According to (4.46) and (4.41), we have

$$\begin{aligned}[\mathbf{z}_k, \mathbf{r}_u, \mathbf{r}_v] &= e^{ikv}[\rho(u)\rho'(u)\varphi_k(u) - \rho(u)\psi_k(u)] \\ &\quad + e^{-ikv}[\rho(u)\rho'(u)\widetilde{\varphi}_k(u) - \rho(u)\widetilde{\psi}_k(u)]\end{aligned}$$

and then

$$\begin{aligned}\iint\limits_{A_0^{(j)}} [\mathbf{z}_k, \mathbf{r}_u, \mathbf{r}_v] du dv &= \int_0^{2\pi} dv \int_{u_j}^{u_{j+1}} \{e^{ikv} \rho(u)[\rho'(u)\varphi_k(u) - \psi_k(u)] \\ &\quad + e^{-ikv} \rho(u)[\rho'(u)\widetilde{\varphi}_k(u) - \widetilde{\psi}_k(u)]\} du \\ &= \int_0^{2\pi} e^{ikv} dv \int_{u_j}^{u_{j+1}} \rho(u)[\rho'(u)\varphi_k(u) - \psi_k(u)] du \\ &\quad + \int_0^{2\pi} e^{-ikv} dv \int_{u_j}^{u_{j+1}} \rho(u)[\rho'(u)\widetilde{\varphi}_k(u) \\ &\quad - \widetilde{\psi}_k(u)] du = 0.\end{aligned}$$

The last equality holds as

$$\int_o^{2\pi} e^{ikv} dv = \int_o^{2\pi} e^{-ikv} dv = 0.$$

Substituting at (4.52), we get

$$\delta V(D_0, \mathbf{z}) = 0$$

and the theorem is proved. □

4.3.1 *An Example of Direct Calculation of Variation of the Volume*

Belov (1968) considered toroid surface of revolution generated by quadrangular meridian $\mathcal{B}$ at $uO\rho$-plane:

$$\mathcal{B}: \quad A(-1, b), \quad B(0, b + c_1), \quad C(1, b), \quad D(0, b - c_2). \tag{4.53}$$

Necessary and sufficient condition for this surface to be non-rigid is

$$1/c_2 - 1/c_1 = k^2/b, \quad (k \in N, k \geq 2). \tag{4.54}$$

According to Theorem 4.14, variation of the volume bounded by Belov's surface is zero. We here prove this fact in an another way, by direct calculation of the variation of the volume. For that purpose, we find the equation of deformed surface under the infinitesimal bending. Let us consider the surface of revolution S at coordinate system with base $\mathbf{e}$, $\mathbf{a}(v)$, $\mathbf{a}'(v)$. The equation of the surface S is given by (4.41), and deformed surface S_ϵ, with infinitesimal bending field $\mathbf{z}(u, v)$:

$$S_\epsilon : \tilde{\mathbf{r}}(u, v, \epsilon) = \mathbf{r}(u, v) + \epsilon \mathbf{z}(u, v). \tag{4.55}$$

The field $\mathbf{z}_k(u, v)$ (4.46), with every $k \in \{2, 3, \ldots\}$, can be considered as a bending field. Equation (4.46) can be written as

$$\begin{aligned}\mathbf{z}(u, v) = {} & \{\Re e[\varphi(u)] \cos kv - \Im m[\varphi(u)] \sin kv\}\mathbf{e} \\ & + \{\Re e[\psi(u)] \cos kv - \Im m[\psi(u)] \sin kv\}\mathbf{a}(v) \\ & + \{\Re e[\chi(u)] \cos kv - \Im m[\chi(u)] \sin kv\}\mathbf{a}'(v). \end{aligned} \tag{4.56}$$

According to Velimirović (1998a), we have

$$\begin{aligned}\psi_{(1)}(u) &= (u-P), \quad \psi_{(2)}(u) = -(u+P),\\ \psi_{(3)}(u) &= \left[\frac{c_1}{c_2}(u-1)-(P+1)\right],\\ \psi_{(4)}(u) &= -\left[\frac{c_1}{c_2}(u+1)+P+1\right],\end{aligned} \tag{4.57}$$

$$\begin{aligned}\chi_{(1)}(u) &= \frac{i}{k}(u-P), \quad \chi_{(2)}(u) = -\frac{i}{k}(u+P),\\ \chi_{(3)}(u) &= \frac{i}{k}\left[\frac{c_1}{c_2}(u-1)-(P+1)\right],\\ \chi_{(4)}(u) &= -\frac{i}{k}\left[\frac{c_1}{c_2}(u+1)+P+1\right],\end{aligned} \tag{4.58}$$

$$\begin{aligned}\varphi_{(1)}(u) &= \varphi_{(2)}(u) = \varphi_{(3)}(u) = \varphi_{(4)}(u) = -c_1 u,\\ P &= \frac{b+c_1}{c_1(k^2-1)} = \frac{bc_2+c_1c_2}{bc_1-bc_2-c_1c_2},\\ Q &= \frac{bc_1-c_1c_2}{bc_1-bc_2-c_1c_2},\end{aligned} \tag{4.59}$$

where $\varphi_{(i)}(u)$, $\psi_{(i)}(u)$, $\chi_{(i)}(u)$ are the functions $\varphi(u)$, $\psi(u)$, $\chi(u)$, $i=1,\ldots,4$, on the parts generated by the sides AB, BC, CD, DA, respectively.

According to (4.55) and (4.41) for the parts of deformed surface S_ϵ, corresponding to the sides of the meridian, we have

$$\tilde{\mathbf{r}}_{(i)}(u,v,\epsilon) = u\mathbf{e} + \rho_{(i)}(u)\mathbf{a}(v) + \epsilon\mathbf{z}_{(i)}(u,v), \quad i=1,2,3,4, \tag{4.60}$$

where $\rho_{(i)}(u)$ are the values of $\rho=\rho(u)$ on the corresponding sides of meridian. Using (4.56)–(4.59), we have

$$\begin{aligned}\tilde{\mathbf{r}}_{(1)}(u,v,\epsilon) = {}& u(1-\epsilon c_1\cos kv)\mathbf{e}\\ &+ [c_1u+b+c_1+\epsilon(u-P)\cos kv]\mathbf{a}(v)\\ &+ \frac{\epsilon}{k}(P-u)\sin kv\mathbf{a}'(v),\end{aligned} \tag{4.61}$$

$$\begin{aligned}\tilde{\mathbf{r}}_{(2)}(u,v,\epsilon) &= u(1-\epsilon c_1\cos kv)\mathbf{e}\\ &\quad+[-c_1u+b+c_1-\epsilon(u+P)\cos kv]\mathbf{a}(v)\\ &\quad+\frac{\epsilon}{k}(P+u)\sin kv\mathbf{a}'(v), \qquad (4.62)\\ \tilde{\mathbf{r}}_{(3)}(u,v,\epsilon) &= u(1-\epsilon c_1\cos kv)\mathbf{e}\\ &\quad+\left[c_2u+b-c_2+\epsilon\frac{c_1}{c_2}(u-Q)\cos kv\right]\mathbf{a}(v)\\ &\quad+\epsilon\frac{c_1}{kc_2}(Q-u)\sin kv\mathbf{a}'(v), \qquad (4.63)\\ \tilde{\mathbf{r}}_{(4)}(u,v,\epsilon) &= u(1-\epsilon c_1\cos kv)\mathbf{e}\\ &\quad+[c_1u+b+c_1+\epsilon(u-P)\cos kv]\mathbf{a}(v)\\ &\quad+\frac{1}{k}\epsilon(P-u)\sin kv\mathbf{a}'(v). \qquad (4.64)\end{aligned}$$

At the plane $(\mathbf{a},\mathbf{a}')$, we introduce xOy coordinate system so that for $v=0$, we obtain the positive part of x-axis, and u-axis is identical to z-axis. Then,

$$\mathbf{a}(v)=\cos v\mathbf{i}+\sin v\mathbf{j}, \quad \mathbf{a}'(v)=-\sin v\mathbf{i}+\cos v\mathbf{j}, \quad \mathbf{e}=\mathbf{k}, \qquad (4.65)$$

and from (4.61)–(4.64), we have

$$\begin{aligned}\tilde{x}_{(1)}(u,v,\epsilon) &= (c_1u+b+c_1)\cos v\\ &\quad+\epsilon(u-P)\left(\cos kv\cos v+\frac{1}{k}\sin kv\sin v\right), \qquad (4.66)\\ \tilde{y}_{(1)}(u,v,\epsilon) &= (c_1u+b+c_1)\sin v\\ &\quad+\epsilon(u-P)\left(\cos kv\sin v+\frac{1}{k}\sin kv\cos v\right),\\ \tilde{z}_{(1)}(u,v,\epsilon) &= u(1-\epsilon c_1\cos kv),\\ &\qquad (u\in[-1,0], v\in[0,2\pi]),\\ \tilde{x}_{(2)}(u,v,\epsilon) &= (-c_1u+b+c_1)\cos v\\ &\quad-\epsilon(u+P)\left(\cos kv\cos v+\frac{1}{k}\sin kv\sin v\right),\end{aligned}$$

$$\tilde{y}_{(2)}(u,v,\epsilon) = (-c_1 u + b + c_1)\sin v - \epsilon(u+P)\left(\cos kv \sin v - \frac{1}{k}\sin kv \cos v\right), \tag{4.67}$$

$$\tilde{z}_{(2)}(u,v,\epsilon) = u(1-\epsilon c_1 \cos kv), \quad (u \in [0,1], v \in [0,2\pi]),$$

and, analogously, for the remaining two parts. From these equations, we can see that deformed surface is symmetrical related to the plane xOy (i.e. $(\mathbf{a}, \mathbf{a}')$).

From the previous equations, we have the following:

1. For $u = 0$,

$$\begin{aligned}
\tilde{x}_{(1)}(0,v,\epsilon) = \tilde{x}_{(2)}(0,v,\epsilon) &= (b+c_1)\cos v \\
&\quad - \epsilon P\left(\cos kv \cos v - \frac{1}{k}\sin kv \sin v\right), \\
\tilde{y}_{(1)}(0,v,\epsilon) = \tilde{y}_{(2)}(0,v,\epsilon) &= (b+c_1)\sin v \\
&\quad -\epsilon P\left(\cos kv \sin v - \frac{1}{k}\sin kv \cos v\right), \\
\tilde{z}_{(1)}(0,v,\epsilon) = \tilde{z}_{(2)}(0,v,\epsilon) &= 0,
\end{aligned} \tag{4.68}$$

i.e. the circle described by the apex B is deformed at the plane curve ($z = 0$) that is not a circle.

2. For $u = -1$ and $u = 1$,

$$\begin{aligned}
\tilde{x}_{(1)}(-1,v,\epsilon) = \tilde{x}_{(2)}(1,v,\epsilon) &= b\cos v \\
&\quad - \epsilon(1+P)\left(\cos kv \cos v + \frac{1}{k}\sin kv \sin v\right), \\
\tilde{y}_{(1)}(-1,v,\epsilon) = \tilde{y}_{(2)}(1,v,\epsilon) &= b\sin v \\
&\quad - \epsilon(1+P)\left(\cos kv \sin v - \frac{1}{k}\sin kv \cos v\right), \\
-\tilde{z}_{(1)}(-1,v,\epsilon) = \tilde{z}_{(2)}(1,v,\epsilon) &= 1 - \epsilon c_1 \cos kv.
\end{aligned} \tag{4.69}$$

Hence, we conclude that the circles described by the apices A, C are deformed to the plane curves that are symmetrical related to the xOy-plane and have the same projection.

The first two equations at (4.68) determine closed curve L, and at (4.69), closed curve L_1 at xOy-plane. Let us prove that L_1 is inside L. It is sufficient to prove that $d^2 - d_1^2 > 0$, where d is the distance from $M \in L$ to O, and d_1 is the distance from $M_1 \in L_1$ to O. From (4.68) and (4.69), we have

$$\begin{aligned} d^2 - d_1^2 &= (\tilde{x}_{(2)}(0, v, \epsilon))^2 + (\hat{y}_{(2)}(0, v, \epsilon))^2 - (\tilde{x}_{(2)}(1, v, \epsilon))^2 \\ &\quad - (\tilde{y}_{(2)}(1, v, \epsilon))^2 \\ &= 2bc_1 + c_1^2 - (1 + 2P)\epsilon^2 \left(\cos kv^2 + \frac{1}{k^2}\sin^2 kv\right) \\ &\quad + 2\epsilon(b - Pc_1)\cos kv > 0 \end{aligned}$$

for sufficiently small ϵ. Let us observe the parts of the volume of toroid of Belov with $V_{(1)}$, $V_{(2)}$, $V_{(3)}$, $V_{(4)}$, for the parts that are between xOy-plane and the parts that are generated by AB, BC, CD, DA, respectively, and with $\tilde{V}_{(j)}(\epsilon)$, $j = 1, \ldots, 4$, the volume of corresponding deformed part. If $\tilde{V}(\epsilon)$ denotes the volume of deformed toroid, we will have

$$\tilde{V}(\epsilon) = 2\tilde{V}_{(2)}(\epsilon) + 2\tilde{V}_{(3)}(\epsilon). \tag{4.70}$$

The part that has volume $\tilde{V}_{(2)}(\epsilon)$ is projected at the "ring" between L and L_1, i.e. on the area denoted as $\tilde{R}(\epsilon)$. Then,

$$\tilde{V}_{(2)}(\epsilon) = \iint\limits_{\tilde{R}(\epsilon)} \tilde{z}_{(2)}(u, v, \epsilon)|J|dudv = \int_0^{2\pi} dv \int_0^1 \tilde{z}_{(2)}(u, v, \epsilon)|J|du, \tag{4.71}$$

where $\tilde{z}_{(2)}(u, v, \epsilon)$ is given at (4.67). From

$$J = \frac{\partial(\tilde{x}_{(2)}, \tilde{y}_{(2)})}{\partial(u, v)} = (c_1u - b - c_1)(c_1 + \epsilon\cos kv) - \frac{k^2 - 1}{k^2}\epsilon^2(u + P)\sin^2 kv,$$

for sufficiently small ϵ, $J \approx c_1(c_1u - b - c_1)$. As $u \in [0, 1]$, we have that $J < 0$. From (4.71) and (4.67), we get

$$\tilde{V}_{(2)}(\epsilon) = \int_0^{2\pi} dv \int_0^1 u(1 - \epsilon c_1 \cos kv)$$

$$\left[(b+c_1-c_1u)(c_1+\epsilon\cos kv)+\frac{k^2-1}{k^2}\epsilon^2(u+P)\sin^2 kv\right]du$$

$$=\pi\left(\frac{1}{3}c_1^2+bc_1\right)+\pi\epsilon^2\lambda(b,c_1,c_2).$$

In the same manner,

$$\tilde{V}_{(3)}(\epsilon)=\pi\left(-\frac{1}{3}c_1^2+bc_2\right)+\pi\epsilon^2\mu(b,c_1,c_2),$$

where λ,μ are the functions of their arguments. Substituting at (4.70), we have

$$\tilde{V}(\epsilon)=\frac{2\pi}{3}(c_1+c_2)(3b+c_1-c_2)+2\pi\epsilon^2(\lambda+\mu)$$

and

$$\delta V(\epsilon)=\left(\frac{\partial\tilde{V}}{\partial\epsilon}\right)_{\epsilon=0}=0.$$

4.4 Variation of the Volume of a Surface of Revolution Generated by a Meridian Being Infinitesimally Bent

In this part, we confront following question:

How does the volume, bounded by a surface of revolution, generated by a meridian being infinitesimally bent, change?

For infinitesimal bending field of a plane curve, which is a meridian of the surface of revolution, we choose infinitesimal bending field that meridian makes staying in the plane.

We consider plane, closed curve at xOy-plane, containing the pole O in the interior.

If ρ and θ are polar coordinates, the vector form of this curve is

$$K:\mathbf{r}(\theta)=\rho(\theta)\cos\theta\mathbf{i}+\rho(\theta)\sin\theta\mathbf{j},\quad \theta\in[0,2\pi]. \tag{4.72}$$

The equation of a curve that remains at xOy-plane according to Theorem 2.9 and Corollary 2.1 is

$$\tilde{K}:\tilde{\mathbf{r}}(\theta)=\rho(\theta)(\cos\theta+\epsilon\sin\theta)\mathbf{i}+\rho(\theta)(\sin\theta-\epsilon\cos\theta)\mathbf{j},\quad \theta\in[0,2\pi].$$

According to (4.72), the parametric form of the equation of the curve K is

$$K : x = \rho(\theta)\cos\theta, \quad y = \rho(\theta)\sin\theta. \tag{4.73}$$

Let us introduce a new coordinate system XO_1Y, where $OO_1 = h$, $Y = y + h$, $X = x$. Parametric equations of the curve K are

$$K : X = \rho(\theta)\cos\theta, \quad Y = \rho(\theta)\sin\theta + h. \tag{4.74}$$

Let us divide the curve K by end points A and B in the curves K_1 and K_2. Then, for the volume generated by the curve K, we have

$$V = \pi \int_a^b (Y_2^2 - Y_1^2)dX, \quad dX = dx = (\dot{\rho}\cos\theta - \rho\sin\theta)d\theta, \tag{4.75}$$

where

$$\begin{aligned} Y &= Y_2, \quad \theta \in [\theta_1, \theta_2] \quad \text{for } K_2 \\ Y &= Y_1, \quad \theta \in [\theta_2, 2\pi + \theta_1] \quad \text{for } K_1. \end{aligned} \tag{4.76}$$

Further, we have

$$V = \pi \int_{\theta_1}^{\theta_2} \varphi(\theta)d\theta - \pi \int_{\theta_2}^{2\pi+\theta_1} \varphi(\theta)d\theta,$$

where

$$\varphi(\theta) = (\rho\sin\theta + h)^2(\dot{\rho}\cos\theta - \rho\sin\theta).$$

Deformed curve $\tilde{K}$ has the equations:

$$\tilde{x} = \rho(\cos\theta + \epsilon\sin\theta), \quad \tilde{y} = \rho(\sin\theta - \epsilon\cos\theta).$$

At the new coordinate system, the equations of the curve $\tilde{K}$ are

$$\tilde{X} = \rho(\cos\theta + \epsilon\sin\theta), \quad \tilde{Y} = \rho(\sin\theta - \epsilon\cos\theta) + h.$$

The volume bounded by the surface generated by rotation of deformed curve $\tilde{K}$ is

$$\tilde{V} = \pi \int_{\tilde{a}}^{\tilde{b}} (\tilde{Y}_2^2 - \tilde{Y}_1^2)d\tilde{X},$$

i.e.

$$\tilde{V} = \pi \int_{\tilde{\theta}_1}^{\tilde{\theta}_2} \tilde{\varphi}(\theta) d\theta - \pi \int_{\tilde{\theta}_2}^{2\pi+\tilde{\theta}_1} \tilde{\varphi}(\theta) d\theta,$$

where

$$\tilde{\varphi}(\theta) = (\rho \sin\theta - \epsilon\rho\cos\theta + h)^2(\dot{\rho}\cos\theta - \rho\sin\theta + \epsilon\dot{\rho}\sin\theta + \epsilon\rho\cos\theta).$$

If we take that $\theta_1 < \theta_1 + \epsilon_1 = \tilde{\theta}_1 < \tilde{\theta}_2 = \theta_2 - \epsilon_2 < \theta_2$, where ϵ_1, $\epsilon_2 > 0$, ϵ_1, $\epsilon_2 \to 0$, when $\epsilon \to 0$ (in this case, we take ϵ_1, $\epsilon_2 > 0$, but we can take ϵ_1, ϵ_2 grater or less then 0). We can express $\tilde{V}$ in the following way:

$$\begin{aligned} \tilde{V} = & \left[\pi \int_{\theta_1}^{\theta_2} \tilde{\varphi}(\theta) d\theta - \pi \int_{\theta_1}^{\theta_1+\epsilon_1} \tilde{\varphi}(\theta) d\theta - \pi \int_{\theta_2-\epsilon_2}^{\theta_2} \tilde{\varphi}(\theta) d\theta\right] \\ & - \left[\pi \int_{\theta_2}^{2\pi+\theta_1} \tilde{\varphi}(\theta) d\theta + \pi \int_{\theta_2-\epsilon_2}^{\theta_2} \tilde{\varphi}(\theta) d\theta + \pi \int_{\theta_1}^{\theta_1+\epsilon_1} \tilde{\varphi}(\theta) d\theta\right], \end{aligned} \tag{4.77}$$

where

$$\tilde{\varphi}(\theta) = \varphi(\theta) + \epsilon\lambda(\theta) + \epsilon^2\mu(\theta) + \epsilon^3\nu(\theta). \tag{4.78}$$

Substituting at (4.77), we get

$$\begin{aligned} \tilde{V} = V - 2\pi & \left[\int_{\theta_1}^{\theta_1+\epsilon_1} \varphi(\theta) d\theta + \int_{\theta_2-\epsilon_2}^{\theta_2} \varphi(\theta) d\theta\right] \\ & + \epsilon\pi \left[\int_{\theta_1}^{\theta_2} \lambda(\theta) d\theta - \int_{\theta_2}^{2\pi+\theta_1} \lambda(\theta) d\theta \right. \\ & \left. - 2\int_{\theta_1}^{\theta_1+\epsilon_1} \lambda(\theta) d\theta - 2\int_{\theta_2-\epsilon_2}^{\theta_2} \lambda(\theta) d\theta\right] \\ & + \epsilon^2\pi \left[\int_{\theta_1}^{\theta_2} \mu(\theta) d\theta - \int_{\theta_2}^{2\pi+\theta_1} \mu(\theta) d\theta \right. \\ & \left. -2\int_{\theta_1}^{\theta_1+\epsilon_1} \mu(\theta) d\theta - 2\int_{\theta_2-\epsilon_2}^{\theta_2} \mu(\theta) d\theta\right] \end{aligned}$$

$$+\epsilon^3\pi\left[\int_{\theta_1}^{\theta_2}\nu(\theta)d\theta-\int_{\theta_2}^{2\pi+\theta_1}\nu(\theta)d\theta\right.$$
$$\left.-2\int_{\theta_1}^{\theta_1+\epsilon_1}\nu(\theta)d\theta-2\int_{\theta_2-\epsilon_2}^{\theta_2}\nu(\theta)d\theta\right].$$

The variation of the volume in this case is

$$\delta V=\pi\left[\int_{\theta_1}^{\theta_2}\lambda(\theta)d\theta-\int_{\theta_2}^{2\pi+\theta_1}\lambda(\theta)d\theta\right], \tag{4.79}$$

where

$$\lambda(\theta)=(\rho\sin\theta+h)$$
$$(-2\rho\dot{\rho}\cos^2\theta+3\rho^2\sin\theta\cos\theta+h\dot{\rho}\sin\theta+h\rho\cos\theta+\rho\dot{\rho}\sin^2\theta). \tag{4.80}$$

We have the following theorem.

Theorem 4.15. *The variation of the volume, bounded by surface of revolution, under infinitesimal bending of the meridian does not have to be zero and it is given by* (4.79) *and* (4.80).

4.5 Infinitesimal Bending of Higher Order of Surfaces of Revolution

Let S be a piecewise regular surface, given by the equation

$$S:\mathbf{r}=\mathbf{r}(u,v), \tag{4.81}$$

included in a family of surfaces being infinitesimal bending of the mth order of the surface S

$$S_\epsilon:\mathbf{r}_\epsilon=\mathbf{r}(u,v)+\sum_{j=1}^{m}\epsilon^j\overset{(j)}{\mathbf{z}}(u,v), \tag{4.82}$$

where $\epsilon\geq 0$ is a small parameter and $\overset{(j)}{\mathbf{z}}(u,v)$ are continuous vector functions defined at the points of S.

The equation of a surface of revolution, in the coordinate system with the base **e**, **a**, **a**′, is

$$S : \mathbf{r}(u, v) = u\mathbf{e} + \rho(u)\mathbf{a}(v), \tag{4.83}$$

where $\rho = \rho(u)$ is the equation of the meridian. The fields of infinitesimal bending $\overset{(j)}{\mathbf{z}}(u, v)$, $j = 1, \ldots, m$, can be represented by the system with mutual perpendicular unit vectors **e**, $\mathbf{a}(v)$, $\mathbf{a}'(v)$ in the form

$$\overset{(j)}{\mathbf{z}}(u, v) = \overset{(j)}{\alpha}(u, v)\mathbf{e} + \overset{(j)}{\beta}(u, v)\mathbf{a} + \overset{(j)}{\gamma}(u, v)\mathbf{a}'. \tag{4.84}$$

The coefficients $\overset{(j)}{\alpha}(u, v)$, $\overset{(j)}{\beta}(u, v)$, $\overset{(j)}{\gamma}(u, v)$, $j = 1, 2, \ldots, m$, are periodical functions with respect to v, with a period 2π, and according to Cohn-Vossen (1959), they can be presented for $j = 1$ in the form

$$\begin{aligned} \overset{(1)}{\alpha}(u, v) &= \overset{(1)}{\alpha}_k(u, v) = \overset{(1)}{\varphi}_k(u)e^{ikv} + \overset{(1)}{\varphi}_{-k}(u)e^{-ikv}, \\ \overset{(1)}{\beta}(u, v) &= \overset{(1)}{\beta}_k(u, v) = \overset{(1)}{\psi}_k(u)e^{ikv} + \overset{(1)}{\psi}_{-k}(u)e^{-ikv}, \\ \overset{(1)}{\gamma}(u, v) &= \overset{(1)}{\gamma}_k(u, v) = \overset{(1)}{\chi}_k(u)e^{ikv} + \overset{(1)}{\chi}_{-k}(u)e^{-ikv}, \end{aligned} \tag{4.85}$$

where $k \geq 2$, $k \in Z$, and $\overset{(1)}{\varphi}_{-k}(u)$ are the conjugate values of $\overset{(1)}{\varphi}_k(u)$ and so on. The functions $\overset{(1)}{\varphi}_k(u)$, $\overset{(1)}{\psi}_k(u)$, $\overset{(1)}{\chi}_k(u)$ satisfy the following system of differential equations:

$$\begin{aligned} &\overset{(1)}{\varphi}{}_k' + \rho' \overset{(1)}{\psi}{}_k' = 0, \quad ik\overset{(1)}{\chi}_k + \overset{(1)}{\psi}_k = 0, \\ &ik\overset{(1)}{\varphi}_k + \rho'(k\overset{(1)}{\psi}_k - \overset{(1)}{\chi}_k) + \rho\overset{(1)}{\chi}{}_k' = 0, \end{aligned} \tag{4.86}$$

where from one obtains a differential equation of the second order with respect to $\overset{(1)}{\psi}_k$

$$\rho\overset{(1)}{\psi}{}_k'' + (k^2 - 1)\rho''\overset{(1)}{\psi}_k = 0. \tag{4.87}$$

We first solve Eq. (4.87) and then determine $\overset{(1)}{\varphi}_k(u)$, $\overset{(1)}{\chi}_k(u)$ from (4.86). Thus, we obtain

$$\begin{aligned}\overset{(1)}{\mathbf{z}}(u,v) &= \overset{(1)}{\mathbf{z}}_k(u,v)\\ &= e^{ikv}[\overset{(1)}{\varphi}_k(u)\mathbf{e} + \overset{(1)}{\psi}_k(u)\mathbf{a}(v) + \overset{(1)}{\chi}_k(u)\mathbf{a}'(v)]\\ &\quad + e^{-ikv}[\overset{(1)}{\varphi}_{-k}(u)\mathbf{e} + \overset{(1)}{\psi}_{-k}(u)\mathbf{a}(v) + \overset{(1)}{\chi}_{-k}(u)\mathbf{a}'(v)]. \end{aligned} \tag{4.88}$$

For $j > 1$, the functions $\overset{(j)}{\alpha}_k(u,v)$, $\overset{(j)}{\beta}_k(u,v)$, $\overset{(j)}{\gamma}_k(u,v)$, $j = 2,\dots,m$, have

(1) for even j, only even powers $e^{\pm ikv}$ and e^0,
(2) for odd j, only odd powers $e^{\pm ikv}$, i.e.

$$\begin{aligned}\overset{(j)}{\alpha}_k(u,v) &= \sum_{(p)}\left[\overset{(j)}{\varphi}_{pk}(u)e^{ipkv} + \overset{(j)}{\varphi}_{-pk}(u)e^{-ipkv}\right],\\ \overset{(j)}{\beta}_k(u,v) &= \sum_{(p)}\left[\overset{(j)}{\psi}_{pk}(u)e^{ipkv} + \overset{(j)}{\psi}_{-pk}(u)e^{-ipkv}\right],\\ \overset{(j)}{\gamma}_k(u,v) &= \sum_{(p)}\left[\overset{(j)}{\chi}_{pk}(u)e^{ipkv} + \overset{(j)}{\chi}_{-pk}(u)e^{-ipkv}\right],\end{aligned} \tag{4.89}$$

where

(1) $p = 0, 2, 4, \dots, j$ for even j,
(2) $p = 1, 3, 5, \dots, j$ for odd j.

The fundamental field of infinitesimal bending of order j is

$$\begin{aligned}\overset{(j)}{\mathbf{z}}_k(u,v) &= \left[\sum_{(p)}(\overset{(j)}{\varphi}_{pk}(u)e^{ipkv} + \overset{(j)}{\varphi}_{-pk}(u)e^{-ipkv})\right]\mathbf{e}\\ &\quad + \left[\sum_{(p)}(\overset{(j)}{\psi}_{pk}(u)e^{ipkv} + \overset{(j)}{\psi}_{-pk}(u)e^{-ipkv})\right]\mathbf{a}(v)\\ &\quad + \left[\sum_{(p)}(\overset{(j)}{\chi}_{pk}(u)e^{ipkv} + \overset{(j)}{\chi}_{-pk}(u)e^{-ipkv})\right]\mathbf{a}'(v).\end{aligned}$$

The functions $\overset{(j)}{\varphi}_{pk}(u)$, $\overset{(j)}{\psi}_{pk}(u)$, $\overset{(j)}{\chi}_{pk}(u)$ satisfy the following system of differential equations:

$$\begin{aligned}
&\overset{(j)}{\varphi}{}'_{pk}(u) + \rho'(u)\overset{(j)}{\psi}{}'_{pk}(u) = \overset{(j)}{A}_{pk}(u),\\
&ipk\overset{(j)}{\chi}_{pk}(u) + \overset{(j)}{\psi}_{pk}(u) = \overset{(j)}{B}_{pk}(u), \qquad (4.90)\\
&ipk\overset{(j)}{\varphi}_{pk}(u) + \rho'(u)(ipk\overset{(j)}{\psi}_{pk}(u) - \overset{(j)}{\chi}_{pk}(u)) + \rho(u)\overset{(j)}{\chi}{}'_{pk}(u) = \overset{(j)}{C}_{pk}(u).
\end{aligned}$$

This system is equivalent to a differential equation of the second order with respect to $\overset{(j)}{\psi}_{pk}$

$$\rho\overset{(j)}{\psi}{}''_{pk} + (p^2k^2 - 1)\rho''\overset{(j)}{\psi}_{pk} = \overset{(j)}{R}_{pk}(u), \qquad (4.91)$$

where

$$\overset{(j)}{R}_{pk}(u) = -\rho''\overset{(j)}{B}_{pk}(u) + \rho\overset{(j)}{B}{}''_{pk}(u) - p^2k^2\overset{(j)}{A}_{pk}(u) - ipk\overset{(j)}{C}{}'_{pk}(u). \qquad (4.92)$$

The right sides of the system (4.90) can be written in more detail in the well-known way according to Ivanova-Karatopraklieva (1992)

$$\begin{aligned}
\overset{(j)}{A}_{pk}(u) &= \overset{(j)}{A}_{(j-2h_j)k}(u)\\
&= -\frac{1}{2}\sum_{s=1}^{j-1}\sum_{r_s}\Big(\overset{(s)}{\varphi}{}'_{r_sk}\overset{(j-s)}{\varphi}{}'_{(j-2h_j-r_s)k}\\
&\quad + \overset{(s)}{\psi}{}'_{r_sk}\overset{(j-s)}{\psi}{}'_{(j-2h_j-r_s)k} + \overset{(s)}{\chi}{}'_{r_sk}\overset{(j-s)}{\chi}{}'_{(j-2h_j-r_s)k}\Big), \qquad (4.93)
\end{aligned}$$

$$\begin{aligned}
\overset{(j)}{B}_{pk}(u) &= \overset{(j)}{B}_{(j-2h_j)k}(u)\\
&= -\frac{1}{2\rho}\sum_{s=1}^{j-1}\sum_{(r_s)}\Big\{-k^2r_s(j-2h_j-r_s)\Big(\overset{(s)}{\varphi}_{r_sk}\overset{(j-s)}{\varphi}_{(j-2h_j-r_s)k}\\
&\quad + \Big(ir_{sk}\overset{(s)}{\psi}_{r_sk} - \overset{(s)}{\chi}_{r_sk}\Big)\Big[i(j-2h_j-r_s)k\overset{(j-s)}{\psi}{}'_{(j-2h_j-r_s)k}\\
&\quad - \overset{(j-s)}{\chi}_{(j-2h_j-r_s)k}\Big]\\
&\quad + \Big(ir_{sk}\overset{(s)}{\chi}_{r_sk} + \overset{(s)}{\psi}_{r_sk}\Big)\Big[i(j-2h_j-r_s)k\overset{(j-s)}{\chi}_{(j-2h_j-r_s)k}\\
&\quad + \overset{(j-s)}{\psi}_{(j-2h_j-r_s)k}\Big]\Big\}, \qquad (4.94)
\end{aligned}$$

$$
\begin{aligned}
\overset{(j)}{C}_{pk}(u) &= \overset{(j)}{C}_{(j-2h_j)k}(u) \\
&= -\sum_{s=1}^{j-1}\sum_{(r_s)} \Big\{ ik(j-2h_j-r_s)\Big(\overset{(s)}{\varphi}{}'_{r_sk}\overset{(j-s)}{\varphi}_{(j-2h_j-r_s)k} \\
&\quad + \overset{(s)}{\psi}{}'_{r_sk}\Big[i(j-2h_j-r_s)k\overset{(j-s)}{\psi}_{(j-2h_j-r_s)k} - \overset{(j-s)}{\chi}_{(j-2h_j-r_s)k}\Big] \\
&\quad + \overset{(s)}{\chi}{}'_{r_sk}\Big[i(j-2h_j-r_s)k\overset{(j-s)}{\chi}_{(j-2h_j-r_s)k} + \overset{(j-s)}{\psi}_{(j-2h_j-r_s)k}\Big]\Big\},
\end{aligned}
\tag{4.95}
$$

where

(1) $j = 2, \ldots, m$;
(2) $p_j = \frac{j}{2}$ for even j and $p_j = \frac{j-1}{2}$ for odd j;
(3) $h_j = 0, \ldots, p_j, j = 2, \ldots, m$;
(4) the summation index r_s in (4.93)–(4.95) takes values in the set $\{\pm(s-2h_s), h_s = 0, 1, \ldots, p_s\}$ so that the numbers $j - 2h_j - r_s$ belong to the set $\{\pm(j-s-2h_{(j-s)}), h_{(j-s)} = 0, 1, \ldots, p_{(j-s)}\}$;
(5) to every number j, $2 \le j \le m$, there correspond $p_j + 1$ systems of equations (4.90).

Cohn-Vossen (1959) proved that at the breaking points of the meridian, Theorem 4.4 is valid.

We can also prove the following theorem for the infinitesimal bending of higher order.

Theorem 4.16. *Suppose that at the breaking points $u = \sigma$ of the meridian $\rho = \rho(u)$, $(\rho'(\sigma+0) \ne \rho'(\sigma-0))$, the functions $\overset{(j)}{\varphi}_{pk}(u)$ and $\overset{(j)}{\chi}_{pk}(u)$ are continuous ($j = 1, \ldots, m, p = 2, \ldots, m$, for even $m, p = 1, 3, \ldots, m$, for odd m). Then, at these points, the functions $\overset{(j)}{\psi}_{pk}(u)$, $j = 1, \ldots, m$, satisfy the equation*

$$
\Big[\rho(u)\overset{(j)}{\psi}{}'_{pk}(u) + (p^2k^2-1)\rho'(u)\overset{(j)}{\psi}_{pk}(u)\Big]\,\Big|_{\sigma-0}^{\sigma+0} = \overset{(j)}{Q}_{pk}(\sigma), \tag{4.96}
$$

where

$$\overset{(j)}{Q}_{pk}(\sigma)$$
$$= \rho'(\sigma-0)\overset{(j)}{B}_{pk}(\sigma-0) - \rho(\sigma)\overset{(j)}{B}_{pk}{}'(\sigma-0) + ipk\overset{(j)}{C}_{pk}(\sigma-0)$$
$$- \rho'(\sigma+0)\overset{(j)}{B}_{pk}(\sigma+0) + \rho(\sigma)\overset{(j)}{B}_{pk}{}'(\sigma+0) - ipk\overset{(j)}{C}_{pk}(\sigma+0). \tag{4.97}$$

Proof. From (4.90, 2),

$$ipk\overset{(j)}{\chi}_{pk}(u) = \overset{(j)}{B}_{pk}(u) - \overset{(j)}{\psi}_{pk}(u)$$
$$ipk\overset{(j)}{\chi}_{pk}{}'(u) = \overset{(j)}{B}_{pk}{}'(u) - \overset{(j)}{\psi}_{pk}{}'(u),$$

and from (4.90, 3),

$$ipk\overset{(j)}{\varphi}_{pk}(u) = -\rho'(u)\left[ipk\overset{(j)}{\psi}_{pk}(u) - \frac{1}{ipk}\left(\overset{(j)}{B}_{pk}(u) - \overset{(j)}{\psi}_{pk}(u)\right)\right]$$
$$- \frac{\rho(u)}{ipk}\left[\overset{(j)}{B}_{pk}{}'(u) - \overset{(j)}{\psi}_{pk}{}'(u)\right] + \overset{(j)}{C}_{pk}(u).$$

Multiplying the left and the right side with ipk, we get

$$-p^2k^2\overset{(j)}{\varphi}_{pk}(u) = \rho(u)\overset{(j)}{\psi}_{pk}{}'(u) + (p^2k^2-1)\rho'(u)\overset{(j)}{\psi}_{pk}(u)$$
$$+ \rho'(u)\overset{(j)}{B}_{pk}(u) - \rho(u)\overset{(j)}{B}_{pk}{}'(u) + ipk\overset{(j)}{C}_{pk}(u).$$

At the breaking points of the meridian, we have

$$\overset{(j)}{\varphi}_{pk}(\sigma-0) = \overset{(j)}{\varphi}_{pk}(\sigma+0), \tag{4.98}$$

wherefrom,

$$\rho(\sigma+0)\overset{(j)}{\psi}{}'_{pk}(\sigma+0) + (p^2k^2-1)\rho'(\sigma+0)\overset{(j)}{\psi}_{pk}(\sigma+0)$$
$$= \rho(\sigma-0)\overset{(j)}{\psi}{}'_{pk}(\sigma-0) + (p^2k^2-1)\rho'(\sigma-0)\overset{(j)}{\psi}_{pk}(\sigma-0) + \overset{(j)}{Q}_{pk},$$

i.e. we obtained (4.96), where $\overset{(j)}{Q}_{pk}$ is given by (4.97). □

4.5.1 Higher-Order Infinitesimal Bending of Toroid Surfaces of Revolutions Generated by a Polygonal Meridian

We now consider the infinitesimal bending of a toroid generated by a polygonal meridian which does not contain a segment perpendicular to the axis of rotation with the apices $A_l(u_l, \rho_l)$, $l = 1, \dots, n$, $A_{n+1} \equiv A_1$ at the coordinate system $uO\rho$. The equations of the polygon sides are

$$A_l A_{l+1} : \rho_{(l)} = \rho_l + \frac{\rho_{l+1} - \rho_l}{u_{l+1} - u_l}(u - u_l) = k_l u + n_l, \tag{4.99}$$

$l = 1, 2, \dots, n$, $A_{n+1} \equiv A_1$, $\rho_{n+1} \equiv \rho_1$, $u_{n+1} \equiv u_1$, where $\rho_{(l)}$ is the value of ρ on the side $A_l A_{l+1}$:

$$\rho_{(l)}{}' = k_l, \ \rho_{(l)}{}'' = 0, \ l = 1, 2, \dots, n. \tag{4.100}$$

Let us find the field of infinitesimal bending $\overset{(j)}{\mathbf{z}}_k(u, v)$ on the side $A_l A_{l+1}$

$$\begin{aligned}\overset{(j)}{\mathbf{z}}_{k,l}(u, v) = {} & \left[\sum_{(p)} \left(\overset{(j)}{\varphi}_{pk,l}(u) e^{ipkv} + \overset{(j)}{\varphi}_{-pk,l}(u) e^{-ipkv}\right)\right] \mathbf{e} \\ & + \left[\sum_{(p)} \left(\overset{(j)}{\psi}_{pk,l}(u) e^{ipkv} + \overset{(j)}{\psi}_{-pk,l}(u) e^{-ipkv}\right)\right] \mathbf{a}(v) \\ & + \left[\sum_{(p)} \left(\overset{(j)}{\chi}_{pk,l}(u) e^{ipkv} + \overset{(j)}{\chi}_{-pk,l}(u) e^{-ipkv}\right)\right] \mathbf{a}'(v),\end{aligned} \tag{4.101}$$

$l = 1, 2, \dots, n$,

(1) $p = 0, 2, 4, \dots, j$ for even j,
(2) $p = 1, 3, 5, \dots, j$ for odd j.

Starting from (4.91), we get (according to (4.100)):

$$\rho_{(l)} \overset{(j)}{\psi}{}''_{pk,l}(u) = \overset{(j)}{R}_{pk,l}(u), \tag{4.102}$$

where

$$\overset{(j)}{R}_{pk,l}(u) = \rho \overset{(j)}{B}{}''_{pk,l}(u) - p^2k^2 \overset{(j)}{A}_{pk,l}(u) - ipk \overset{(j)}{C}{}'_{pk,l}(u) \tag{4.103}$$

and $\overset{(j)}{A}_{pk,l}(u)$, $\overset{(j)}{B}_{pk,l}(u)$, $\overset{(j)}{C}_{pk,l}(u)$ are given by (4.93)–(4.95) on the side A_lA_{l+1}. Solving (4.102), we have

$$\overset{(j)}{\psi}_{pk,l}(u) = \int du \int \frac{\overset{(j)}{R}_{pk,l}(u)}{\rho_{(l)}(u)} du + \overset{(j)}{M}_{pk,l} u + \overset{(j)}{N}_{pk,l}, \tag{4.104}$$

where $\overset{(j)}{M}_{pk,l}$ and $\overset{(j)}{N}_{pk,l}$ are arbitrary constants. From (4.90, 2) and (4.104), we have

$$\overset{(j)}{\chi}_{pk,l}(u) = \frac{1}{ipk} \left[\overset{(j)}{B}_{pk,l}(u) - \int du \int \frac{\overset{(j)}{R}_{pk,l}(u)}{\rho_{(l)}(u)} du - \overset{(j)}{M}_{pk,l} u - \overset{(j)}{N}_{pk,l} \right]. \tag{4.105}$$

From (4.90, 3), (4.104), and (4.105),

$$\begin{aligned} \overset{(j)}{\varphi}_{pk,l}(u) = {} & k_l \left(\frac{1}{p^2k^2} - 1 \right) \int du \int \frac{\overset{(j)}{R}_{pk,l}(u)}{\rho_{(l)}(u)} du \\ & - \frac{\rho_{(l)}(u)}{p^2k^2} \int \frac{\overset{(j)}{R}_{pk,l}(u)}{\rho_{(l)}(u)} du + \frac{1}{ipk} \overset{(j)}{C}_{pk,l}(u) - \frac{k_l}{p^2k^2} \overset{(j)}{B}_{pk,l}(u) \\ & + \frac{\rho_{(l)}(u)}{p^2k^2} \overset{(j)}{B}_{pk,l}{}'(u) + \left(k_l u + \frac{k_l}{p^2k^2} u - \frac{\rho_{(l)}(u)}{p^2k^2} \right) \overset{(j)}{M}_{pk,l} \\ & + k_l \left(1 + \frac{1}{p^2k^2} \right) \overset{(j)}{N}_{pk,l}, \end{aligned} \tag{4.106}$$

(1) $p = 0, 2, 4, \ldots, j$ for even j,
(2) $p = 1, 3, 5, \ldots, j$ for odd j.

For $p = 0$ from (4.90, 2), we have

$$\overset{(j)}{\psi}_{0,l}(u) = \overset{(j)}{B}_{0,l}(u), \tag{4.107}$$

and from (4.90, 1),

$$\overset{(j)}{\varphi}_{0,l}(u) = \int \overset{(j)}{A}_{0,l}(u)du - k_l \overset{(j)}{B}_{0,l}(u) + \overset{(j)}{M}_{0,l}. \tag{4.108}$$

Using (4.90, 3), (4.107), and (4.108), we get

$$\overset{(j)}{\chi}_{0,l}(u) = \rho_{(l)}(u) \left[\int \frac{\overset{(j)}{C}_{0,l}(u)}{(\rho_{(l)}(u))^2} du + \overset{(j)}{N}_{0,l} \right]. \tag{4.109}$$

The coefficients of the fundamental field $\overset{(j)}{\mathbf{z}}_{k,l}(u, v)$ are given by (4.104)–(4.109). The constants $\overset{(j)}{M}_{pk,l}$, $\overset{(j)}{N}_{pk,l}$ will be found using the fact that the field of infinitesimal bending is to be continuous on the whole surface, including the circles, described by the apices of the meridian. At the apex $A_l(u_l,\ \rho_l)$ from

$$\overset{(j)}{\psi}_{pk,l}(u_{l+1}) = \overset{(j)}{\psi}_{pk,l+1}(u_{l+1}), \quad l = 1, 2, \ldots, n, \ j = 1, 2, \ldots, m, \tag{4.110}$$

and from (4.104), we have

$$u_{l+1}\overset{(j)}{M}_{pk,l} - u_{l+1}\overset{(j)}{M}_{pk,l+1} + \overset{(j)}{N}_{pk,l} - \overset{(j)}{N}_{pk,l+1}$$
$$= \left[\int du \int \frac{\overset{(j)}{R}_{pk,l+1}(u)}{\rho_{(l+1)}(u)} du - \int du \int \frac{\overset{(j)}{R}_{pk,l}(u)}{\rho_{(l)}(u)} du \right] \Bigg|_{u=u_{l+1}}. \tag{4.111}$$

At the mentioned points A_{l+1}, Eq. (4.96) has the form

$$(k_{l+1}u_{l+1} + n_{l+1}) \left[\left(\int \frac{\overset{(j)}{R}_{pk,l+1}(u)}{\rho_{(l+1)}(u)} du \right) \Bigg|_{u=u_{l+1}} + \overset{(j)}{M}_{pk,l+1} \right]$$

$$+(p^2k^2-1)k_{l+1}\left[\left(\int du\int\frac{\overset{(j)}{R}_{pk,l+1}(u)}{\rho_{(l+1)}(u)}du\right)\bigg|_{u=u_{l+1}}\right.$$

$$\left.+\overset{(j)}{M}_{pk,l+1}u_{l+1}+\overset{(j)}{N}_{pk,l+1}\right]$$

$$-\rho_{(l)}(u_{l+1})\left[\left(\int\frac{\overset{(j)}{R}_{pk,l}(u)}{\rho_{(l)}(u)}du\right)\bigg|_{u=u_{l+1}}+\overset{(j)}{M}_{pk,l}\right]$$

$$-(p^2k^2-1)k_l\left[\left(\int du\int\frac{\overset{(j)}{R}_{pk,l}(u)}{\rho_{(l)}(u)}du\right)\bigg|_{u=u_{l+1}}\right.$$

$$\left.+\overset{(j)}{M}_{pk,l}u_{l+1}+\overset{(j)}{N}_{pk,l}\right]$$

$$=\overset{(j)}{Q}_{pk,l}(u_{l+1}),$$

where $\overset{(j)}{Q}_{pk}(u)$ is given by (4.97) $l=1,2,\dots,n,\ j=1,2,\dots,m$. In this way, at the apices of the meridian, we have the system of equations

$$\begin{aligned}&-(n_l+p^2k^2k_lu_{l+1})\overset{(j)}{M}_{pk,l}+(n_{l+1}+p^2k^2k_{l+1}u_{l+1})\overset{(j)}{M}_{pk,l+1}\\&-(p^2k^2-1)k_l\overset{(j)}{N}_{pk,l}+(p^2k^2-1)k_{l+1}\overset{(j)}{N}_{pk,l+1}=\overset{(j)}{V}_{pk}(u_{l+1}),\end{aligned}\tag{4.112}$$

where

$$\begin{aligned}\overset{(j)}{V}_{pk}(u_{l+1})&=k_l\overset{(j)}{B}_{pk,l}-\rho_l\overset{(j)}{B}_{pk,l}{}'+ipk\overset{(j)}{C}_{pk,l}-k_{l+1}\overset{(j)}{B}_{pk,l+1}\\&\quad+\rho_l\overset{(j)}{B}_{pk,l+1}{}'-ipk\overset{(j)}{C}_{pk,l+1}-\rho_l\left(\int\frac{\overset{(j)}{R}_{pk,l+1}(u)}{\rho_{(l+1)}(u)}du\right)\bigg|_{u=u_{l+1}}\end{aligned}$$

$$
\begin{aligned}
&- (p^2k^2-1)k_{l+1}\left(\int du \int \frac{\overset{(j)}{R}_{pk,l+1}(u)}{\rho_{(l+1)}(u)}du\right)\Bigg|_{u=u_{l+1}} \\
&+ \rho_{(l)}\left(\int \frac{\overset{(j)}{R}_{pk,l}(u)}{\rho_{(l)}(u)}du\right)\Bigg|_{u=u_{l+1}} \\
&+ (p^2k^2-1)k_l\left(\int du \int \frac{\overset{(j)}{R}_{pk,l}(u)}{\rho_{(l)}(u)}du\right)\Bigg|_{u=u_{l+1}}.
\end{aligned}
$$

From the above presented, the following theorem holds.

Theorem 4.17. *The sufficient condition for the toroid surface of revolution generated by polygonal, a meridian which does not contain a segment perpendicular to the axis of rotation to be non-rigid of the order* m, i.e. *to have a field of infinitesimal bending* $\overset{(m)}{\mathbf{z}}(u,v)$ *of the order* m*, as an extension of the fields* $\overset{(1)}{\mathbf{z}}(u,v)$, $\overset{(2)}{\mathbf{z}}(u,v),\dots,\overset{(m-1)}{\mathbf{z}}(u,v)$ *is that system of linear equations with respect to unknowns* $\overset{(j)}{M}_{pk,l}$, $\overset{(j)}{N}_{pk,l}$ (4.111) *and* (4.112) *are compatible.*

Remark. We rule out a trivial example of surfaces containing a plane part generated by a side of a polygon perpendicular to the axis of rotation (Cohn-Vossen, 1959; Efimov, 1948).

4.6 Examples of the Infinitesimal Bending of Toroids Generated by a Polygonal Meridian

We are going to analyze the non-rigidity conditions and graphically present a family of toroids which satisfy those conditions through examples. The class of non-rigid toroids discovered by Belov is extended by some new examples of toroids that satisfy the non-rigidity conditions. As an examples of those, the following are given: a toroid with a non-convex quadrangle meridian, a toroid with a convex pentagonal meridian, and a toroid with a nine apices convex meridian:

(1) The first example of non-rigid toroidal surfaces is the surface of a non-convex quadrangle with apices $A(-1,1)$, $B(0,\frac{5}{8})$, $C(1,1)$, and $D(0,\frac{1}{4})$. The quadrangle rotates around the u-axis of the coordinate system $uO\rho$. Revolving AB generates $S_{(1)}$ and revolving BC, CD, and DA, the parts $S_{(2)}S_{(3)}S_{(4)}$ are generated, and also $\mathbf{z}_{(i)}$ denotes the field of infinitesimal bending. Let us denote the deformed toroid as S_ϵ:

$$S_\epsilon : \tilde{\mathbf{r}}(u,v,\epsilon) = \mathbf{r}(u,v) + \epsilon\mathbf{z}(u,v),$$

$$\begin{aligned}
S_{(1)} : \mathbf{r}_{(1)}(u,v) &= \left[-\frac{3}{8}u + \frac{5}{8}\right]\cos(v)\mathbf{i} + \left[-\frac{3}{8}u + \frac{5}{8}\right]\sin(v)\mathbf{j} + u\mathbf{k},\\
&\quad u \in [-1,0], v \in [0,2\pi],\\
S_{(2)} : \mathbf{r}_{(2)}(u,v) &= \left[\frac{3}{8}u + \frac{5}{8}\right]\cos(v)\mathbf{i} + \left[\frac{3}{8}u + \frac{5}{8}\right]\sin(v)\mathbf{j} + u\mathbf{k},\\
&\quad u \in [0,1], v \in [0,2\pi],\\
S_{(3)} : \mathbf{r}_{(3)}(u,v) &= \left[\frac{3}{4}u + \frac{1}{4}\right]\cos(v)\mathbf{i} + \left[\frac{3}{4}u + \frac{1}{4}\right]\sin(v)\mathbf{j} + u\mathbf{k},\\
&\quad u \in [1,0], v \in [0,2\pi],\\
S_{(4)} : \mathbf{r}_{(4)}(u,v) &= \left[-\frac{3}{4}u + \frac{1}{4}\right]\cos(v)\mathbf{i} + \left[-\frac{3}{4}u + \frac{1}{4}\right]\sin(v)\mathbf{j} + u\mathbf{k},\\
&\quad u \in [0,-1], v \in [0,2\pi].
\end{aligned}$$

For the field $\mathbf{z}(u,v)$ of infinitesimal bending of the first order, we have

$$\begin{aligned}
\mathbf{z}_{(1)}(u,v) &= \left[2\left(u + \frac{5}{9}\right)\cos(2v)\cos(v) + \left(u + \frac{5}{9}\right)\sin(2v)\sin(v)\right]\mathbf{i}\\
&\quad + \left[2\left(u + \frac{5}{9}\right)\cos(2v)\sin(v) - \left(u + \frac{5}{9}\right)\sin(2v)\cos(v)\right]\mathbf{j}\\
&\quad + \left[\frac{3}{4}u\cos(2v)\right]\mathbf{k}, u \in [-1,0], v \in [0,2\pi],
\end{aligned}$$

$$\begin{aligned}
\mathbf{z}_{(2)}(u,v) = &\left[2\left(-u+\frac{5}{9}\right)\cos(2v)\cos(v)+\left(-u+\frac{5}{9}\right)\sin(2v)\sin(v)\right]\mathbf{i}\\
&+\left[2\left(-u+\frac{5}{9}\right)\cos(2v)\sin(v)\right.\\
&\left.-\left(-u+\frac{5}{9}\right)\sin(2v)\cos(v)\right]\mathbf{j}\\
&+\left[\frac{3}{4}u\cos(2v)\right]\mathbf{k}, u\in[0,1], v\in[0,2\pi],\\
\mathbf{z}_{(3)}(u,v) = &\left[2\left(-\frac{1}{2}u+\frac{5}{90}\right)\cos(2v)\cos(v)\right.\\
&\left.+\left(-\frac{1}{2}u+\frac{5}{90}\right)\sin(2v)\sin(v)\right]\mathbf{i}\\
&+\left[2\left(-\frac{1}{2}u+\frac{5}{90}\right)\cos(2v)\sin(v)\right.\\
&\left.-\left(-\frac{1}{2}u+\frac{5}{90}\right)\sin(2v)\cos(v)\right]\mathbf{j}\\
&+\left[\frac{3}{4}u\cos(2v)\right]\mathbf{k}, u\in[1,0], v\in[0,2\pi],\\
\mathbf{z}_{(4)}(u,v) = &\left[2\left(\frac{1}{2}u+\frac{5}{90}\right)\cos(2v)\cos(v)\right.\\
&\left.+\left(\frac{1}{2}u+\frac{5}{90}\right)\sin(2v)\sin(v)\right]\mathbf{i}\\
&+\left[2\left(\frac{1}{2}u+\frac{5}{90}\right)\cos(2v)\sin(v)\right.\\
&\left.-\left(\frac{1}{2}u+\frac{5}{90}\right)\sin(2v)\cos(v)\right]\mathbf{j}\\
&+\left[\frac{3}{4}u\cos(2v)\right]\mathbf{k}, u\in[0,-1], v\in[0,2\pi].
\end{aligned}$$

Figure 4.5 shows the surfaces of revolution and the influence of infinitesimal bending. For the purpose of insight view, the rotational parameter v is taken to be less than 2π.

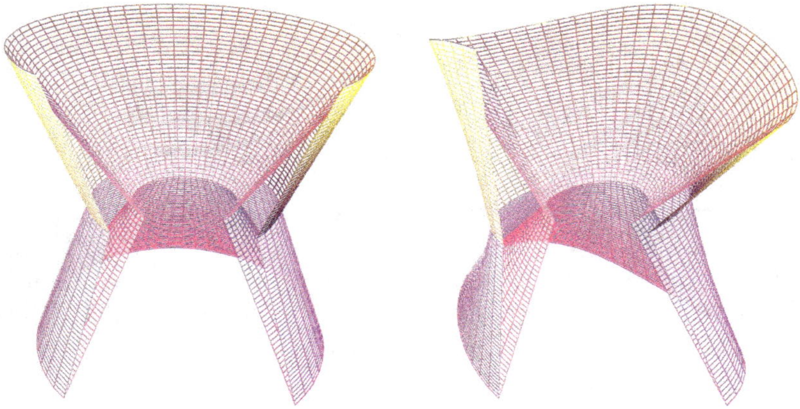

Fig. 4.5. A toroidal surface of a non-convex quadrangle.

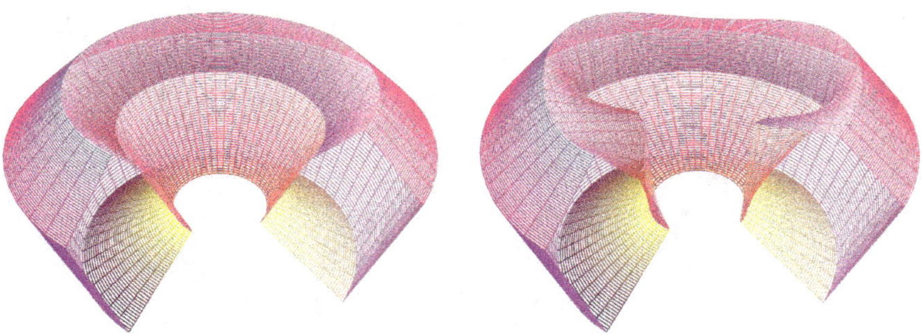

Fig. 4.6. A toroidal surface of a convex pentagon.

Figure 4.5 (left) shows the initial surface from example 1 without any deformations $\epsilon = 0.0$ and $v \in [0, \frac{4}{3}\pi]$.

Figure 4.5 (right) shows the deformed surface from example 1 with deformations $\epsilon = 0.15$ and $v \in [0, \frac{4}{3}\pi]$.

(2) The second example of a non-rigid toroidal surface is a surface with a convex pentagon with apices $A(-1, 2)$, $B(-2, 3)$, $C(0, 4)$, $D(2, 3)$, and $E(1, \frac{119-\sqrt{2641}}{45})$. The polygon rotates around the u-axis of the coordinate system $uO\rho$.

Figure 4.6 (left) shows the initial surface from example 2 without any deformations $\epsilon = 0.0$ and $v \in [0, \frac{3}{2}\pi]$.

Figure 4.6 (right) shows the deformed surface from example 2 with deformations $\epsilon = 0.25$ and $v \in [0, \frac{3}{2}\pi]$.

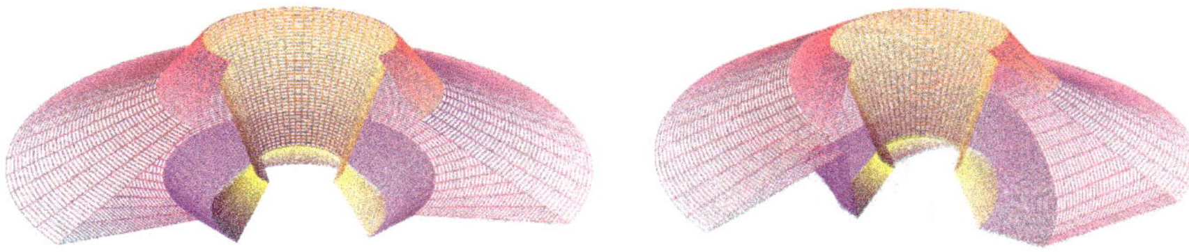

Fig. 4.7. A toroidal surface of a non-convex hexagon.

(3) The third example of a non-rigid toroidal surfaces is a surface with convex polygon with six apices $A(-1,1)$, $B(-2,2)$, $C(-1,3)$, $D(0,6)$, $E(1,3)$, $F(2,2)$. The polygon rotates around the u-axis of the coordinate system $uO\rho$.

Figure 4.7 (left) shows the initial surface from example 3 without any deformations $\epsilon = 0.0$ and $v \in [0, \frac{4}{3}\pi]$.

Figure 4.7 (right) shows the deformed surface from example 3 obtained by rotating parts with deformations $\epsilon = 0.12$ and $v \in [0, 2\pi]$.

Chapter 5

Shape and Energy of Knots: Topological Topics

Louis H. Kauffman[1]

5.1 Knot Dynamics

It is the purpose of this chapter to ask questions about the dynamics of knots under self-repulsion. A knot is arranged in three dimensional space so that it is coated with a mathematical version of electrical charge so that it evolves to geometric positions that minimize its energy under constraints that maintain its total length. Will such evolutions result in the unknotting of loops that are topologically unknotted? What are the forms of extermal positions in such evolutions?

We will give a guided tour of computer experiments that can be performed in the quest for answers to these questions. In the course of our specific computer exploration, we use the knot theory of rational knots and rational tangles [Kauffman and Lambropoulou (2011)] to produce classes of unknots with complex initial configurations that we call *hard unknots* and corresponding complex configurations that are topologically equivalent to simpler knots. We shall see that these hard unknots and complexified knots give examples that do not reduce in the experimental space of the computer program. That is, we find unknotted configurations that will not reduce to simple

[1]This is a special chapter, a bit different from the rest of the book and it is written by Kauffman individually.

circular forms under self-repulsion, and we find complex versions of knots that will not reduce to simpler forms under the self-repulsion. It is clear to us that the phenomena that we have discovered depend very little on the details of the computer program as long as it conforms to our general description of self-repulsion. Thus, we suggest on the basis of our experiments that sufficiently complex examples of hard unknots and sufficiently complex examples of complexified knots will not reduce to global minimal energy states in self-repulsion environments. In the course of the chapter, we will make the character of these examples precise. It is our challenge to other program environments to verify or disprove our assertions.

To make this challenge quite specific, here is a family of examples. Let

$$U_n = N([[n+1]] - [[n]]) = [(n+1) \cdot n]$$

in the notation of Section 5.3. U_n is a diagram for an unknot. The example [11.10] is shown in Figs. 5.17 and 5.18. We conjecture that any algorithm based only on self-repulsion will be unable to unknot U_n for some natural number value for n. In this example, our version of KnotPlot is unable to handle [11.10].

Having made this challenge, we point out that in [Reiter and Schumacher (2020)] refined methods of gradient descent provide ways in which many examples that we do not reach with our methods can be unknotted. And in [Scharein *et al.* (2009)] the BFACF algorithm [Van Rensburg and Rechnitzer (2011)] is used to strong effect. The challenge is given to all of these more powerful algorithms as well.

A knot is represented by a (cyclic) sequence $p_1, p_2, \ldots, p_n$ of points in three dimensional space $\mathcal{R}^3$. In computer graphics, the points can be shown and edges can connect point p_i with point p_{i+1} and of course p_n with p_1. Another computer graphic option is to enclose the resulting piecewise linear curve with tube so that the knot appears as a smooth tube with a chosen diameter. We make use of the computer program KnotPlot [Scharein] created by Robert Scharein, and we thank him for numerous conversations about the use of this program, and for modifying the program to facilitate the use of tangles and rational knots. We shall refer to KnotPlot when speaking of experiments and examples produced in the course of this chapter.

A knot undergoes dynamics of self-repulsion in the computer program by a recursion involving forces of order r^{-5} (r is the distance between a chosen pair of points on the knot) applied to all pairs of non-adjacent points in the sequence of points that define the knot (see above). The recursion is controlled so that the knot-type of the space curve does not change from step to step. We can calculate the Simon Energy [Simon (1994)]. The **Simon Energy** is the sum of $1/|p-q|$ over all pairs of distinct points in the list of points in three dimensional space that describes the knot. The KnotPlot program minimizes this energy by self-repulsion forces of order $1/|p-q|^5$. The Simon Energy corresponds directly to forces of the form $1/|p-q|^2$.

One can examine many aspects of the knot dynamics. Knots and links evolve into special geometric configurations that are not predictable beforehand. One often finds that unknotted curves will unfold and become simplified into curves that are nearly flat circles. In some cases, the force field cannot undo unknots. The final configuration can depend upon the starting configuration and upon degrees of perturbation that are allowed in the process of evolution. The program we use allows, in addition to the repulsion forces, an attraction of nearby points via spring-forces (using an analog of Hooke's Law) [Scharein (1998)]. The balance of the spring and repulsion forces can be used to facilitate experiments. We will give examples of all these phenomena.

5.2 Moving Down Energy Gradients

We begin with a relatively flat representation of the trefoil knot, a torus knot of type $(2,3)$. Generally, we consider torus knots of type (a,b) where a and b are relative prime positive integers. Such a knot winds around a standardly embedded torus in three-space a times around the meridian direction and b times around the longitude direction. It is a fact that a torus knot of type (a,b) is topologically equivalent to a torus knot of type (b,a). We shall write $T(a,b)$ for a torus knot of type (a,b) and so write that $T(a,b)$ and $T(b,a)$ are topologically equivalent. We shall see this equivalence emerge in the dynamics.

Fig. 5.1. Initial Trefoil.

In Fig. 5.1, we see the initial representation of $T(2,3)$. In Fig. 5.2, we see the result of the trefoil after it has reached self-repulsion stability. At this point, the trefoil configuration has calculated Simons energy $E = 169.041560$. The Simons energy is computed by using the fact that the knot we see is represented by a collection of points in 3-dimensional Euclidean space as shown in Fig. 5.3. The Simons energy is the sum of $1/r(p,q)$ ranging over all pairs of points (p,q) in this configuration where $r(p,q)$ is the Euclidean distance between the points p and q. The Simon energy corresponds classically to the force field with exponent 2, but it is convenient to track this energy even when using other force functions.

In Fig. 5.4, we see the initial state for a torus knot of type $T(3,2)$. Topologically, a torus knot of type $T(a,b)$ is equivalent to one of type $T(b,a)$, but geometrically this is quite a different starting position. In one case the curve goes around the torus longitudinally twice and meridianally three times. This is reversed to three and two in the second case. We now examine how the repulsion dynamics worked on the $T(3,2)$. In Fig. 5.5, we see the self-repelled $(3,2)$ torus knot in what appears to be a stable position. Experimentally, this configuration is stable when using the undamped program in KnotPlot. Undamped

Fig. 5.2. Self Repelled Trefoil.

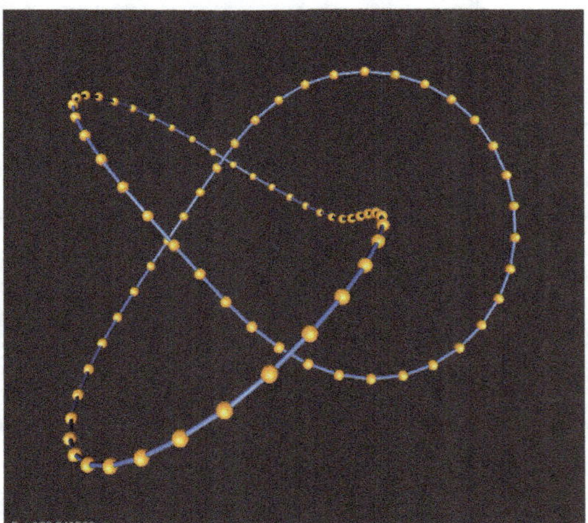

Fig. 5.3. Beaded Trefoil.

means that the springs that are available between successive nodes in the model are not operation (in the sense that the spring constant is high and the springs are hence fully contracted). The energy level of this configuration is 174.59304, considerably higher than the minimum we found for the $(2, 3)$ knot. Now, we have the opportunity to perturb this configuration by using the program in its undamped form. Then, the springs are free to contract and expand, letting the

Fig. 5.4. (3,2) Torus Knot.

Fig. 5.5. Self Repelled (3,2) Torus Knot.

energy move into kinetic energy of motion for the configuration. In Figs. 5.6–5.9, we see that indeed the undamped evolution results in an oscillation that swings arcs of the knot around and lets it fall into the lower energy configuration of the $(2, 3)$ trefoil knot.

Fig. 5.6. Spring Dynamics Destabilizes the $(3, 2)$.

Fig. 5.7. Spring Dynamics Destabilizes the $(3, 2)$.

Figures 5.10–5.15 illustrate the descent of a $(5, 2)$ torus knot. The resulting energy minimum is 221.466258. Note that the internal twist in the knot collapes in the course of this evolution. The same collapse will be observed for $(n, 2)$ torus knots with $n \geq 5$. If one uses the

Fig. 5.8. Toward a New Minimum.

Fig. 5.9. Final Minimum Equivalent to the $(2,3)$ Trefoil Minimum.

undamped descent for a long time, then for the $(5,2)$ the resulting configuration does not change when the undamped evolution is made available. If one runs the damped evolution for a short time, then the

Fig. 5.10. Initial condition of $(5, 2)$ torus knot.

Fig. 5.11. Descending $(5, 2)$ torus knot.

result will be similar to our evolution with the trefoil knot and an exchange of kinetic energy takes the knot to the same configuration as the descent of the $(2, 5)$ torus knot. We illustrate the final descent configuration for a $(2, 5)$ torus knot in Fig. 5.16.

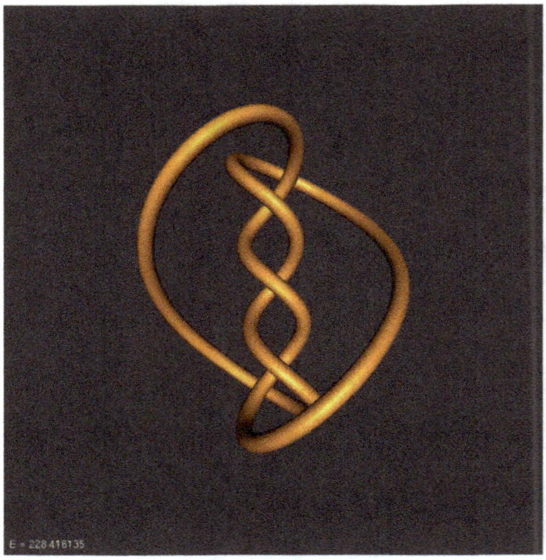

Fig. 5.12. Descending (5, 2) torus knot.

Fig. 5.13. Descending (5, 2) torus knot.

In Fig. 5.17, we illustrate an initial position for an unknot of type [11.10] where we will explain this notation below. Figure 5.18 shows the result of the undamped evolution of this knot. It has not become unknotted and no application of the undamped program will unknot

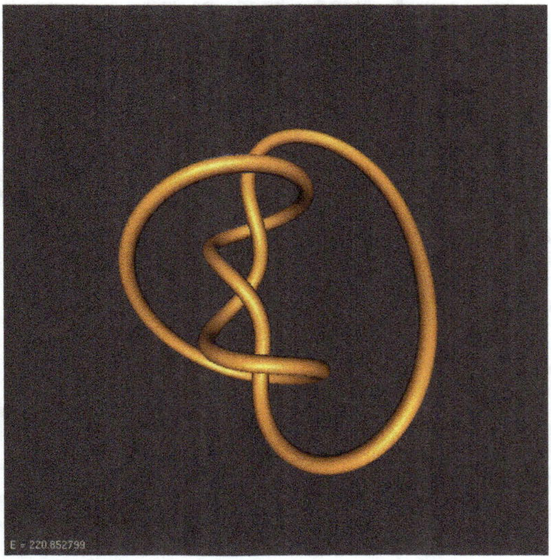

Fig. 5.14. Descending $(5, 2)$ torus knot.

Fig. 5.15. Final Descent of $(5, 2)$ torus knot.

it at any point in the process. This is an experimental example of an unknotted curve that fails to become unknotted in the force field of the Knotplot program. We conjecture that this phenomenon is not specific to this program but rather that some unknots of the form

Fig. 5.16. Minimum for the (2, 5) torus knot.

Fig. 5.17. Starting Position for the unknot [11.10].

$[(n+1).n]$ will not be able to evolve to simple circles in any forcefield of this kind. The reader is advised to experiment with these examples. We explain the corresponding tangle theory in the next section.

Fig. 5.18. End Position for the unknot [11.10] in Local Energy Minimization — A high energy version of the unknot.

Fig. 5.19. End Position for the knot [15.10] in Local Energy Minimization — A high energy version of the figure eight knot.

The knot shown in Fig. 5.19 is obtained by the same techniques as the previous two figures. It is of type [15.10] and is equivalent to a figure eight knot as we will explain in the next section. It will not reduce any further in the KnotPlot environment either via undamped or damped evolution. In fact, as we will see in then next section, we have an infinite collection of figure eight knots in the forms $[(n+5).n]$ and for n greater than 10 we have the empirical result that they all have local energy minima, increasing with the value of n in the

KnotPlot environment. We conjecture that this phenomenon is qualitatively the same in any force field environment for knot repulsion.

5.3 Hard Knots and Collapsing Tangles

Before we begin the details of this section, the reader should view Fig. 5.26. There is illustrated a diagram and a sequence of steps that show that the diagram is unknotted. This is the type of unknot that we have discussed in the previous section. In order to systematically construct configurations of this type, that are complex and yet unknotted, we develop an algebraic language of tangles in the present section. For complete information about all aspects of this discussion, we recommend that the reader consult the paper [Kauffman and Lambropoulou (2011)].

In this section, we give background in terms of a theory of tangles for constructing the unknots and knots in collapsible states. We begin with the definition of a tangle and the notions of combination of tangles (addition, star product, numerator, and denominator) and how this leads to the tangle fraction, the notion of rational tangles and rational knots and how these constructions are related to computer experiments. The notion of a tangle was introduced in 1967 by Conway [Conway (1970)] in his work on enumerating and classifying knots and links.

A **2-tangle** is a proper embedding of two unoriented arcs and a finite number of circles in a 3-ball B^3, so that the four endpoints lie in the boundary of B^3. A **tangle diagram** is a regular projection of the tangle on an equatorial disc of B^3. A **rational tangle** is a special case of a 2-tangle obtained by applying consecutive twists on neighbouring endpoints of two trivial arcs. Such a pair of arcs comprise the [0] or [∞] tangles as shown in Fig. 5.21, depending on their position in the plane. We shall say that the rational tangle is in **twist form** when it is obtained by such successive twists of adjacent strands. For example, see Fig. 5.20 where we show a tangle in the center of the figure that is obtained by first starting with a horizontal 3-twist ([3] in Fig. 5.21 and then twisting two vertical strands negatively twice and then twisting two horizontal strands positively twice. Tangles obtained recursively from the untwisted tangles [0] and [∞] of Fig. 5.21 are called **rational tangles**.

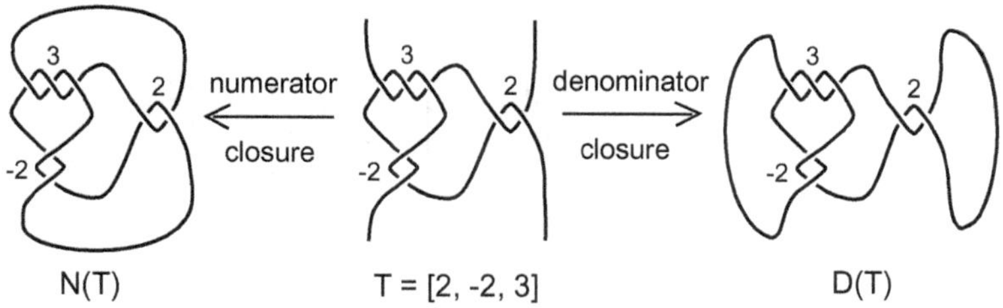

Fig. 5.20. Tangles — Numerators and Denominators.

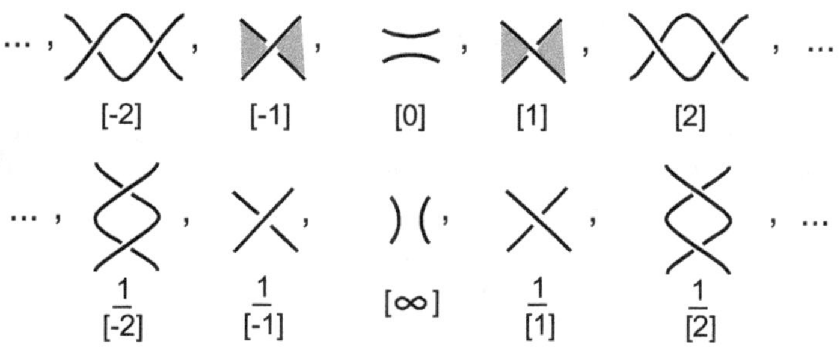

Fig. 5.21. Elementary Rational Tangles.

Note that in Fig. 5.21, we have indicated horizontal twists by bracketed integers $[n]$ and vertical twists by reciprocal integers $1/[n]$, including the convention that $1/[0] = [\infty]$. In Fig. 5.22, we show tangle operations $T + S$, $T \star S$, and T^{rot}. These are called **addition, star product** and **rotation** of tangles. The rotation operation has the effect of turning the tangle by ninety degrees in the plane. Note that twisting horizontal strand n times is the same as forming $T + [n]$ and twisting vertical strands n times is the same as forming $T \star [n]$. Thus, we can describe the tangle in Fig. 5.20 as $([3] \star [-2]) + [2]$.

It turns out that one can associate to any tangle T a fraction $F(T)$ with values in the rational numbers plus the formal value ∞, and that this fraction satisfies the following rules with respect to addition, star product and rotation. Here, $T^\star$ denote the mirror image tangle, obtained by switching all the crossings of T. Once we assume that $F([0]) = 0, F([1]) = 1$, these rules determine the fractions of rational tangles.

(1) $F(S + T) = F(S) + F(T)$
(2) $F(T \star S) = \frac{1}{\frac{1}{F(T)} + \frac{1}{F(S)}}$

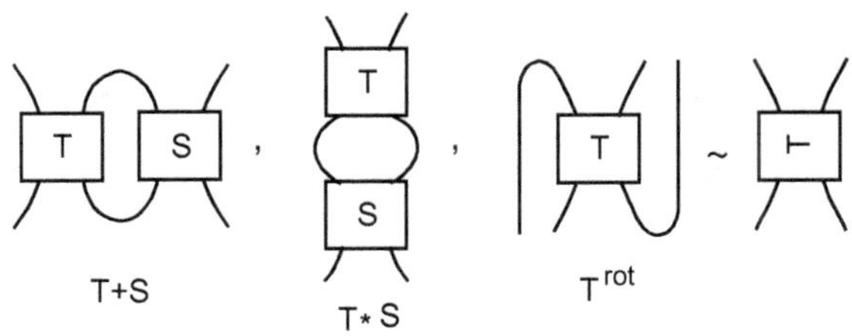

Fig. 5.22. Addition, Star Product and Rotation of Tangles.

(3) $F(T^{rot}) = -1/F(T)$
(4) $F(T^{\star}) = -F(T)$

The second property of the fraction for the star product is a consequence of first property for addition, using the third property that tells us that the fraction of a rotate is the negative reciprocal of the fraction of the original tangle. Note that it follows from these rules that $F([n]) = n$ and that $F(1/[n]) = 1/n$ and that

$$F(T \star (1/[n])) = 1/(n + 1/F(T)).$$

This is the formula for the fraction of a tangle with an extra vertical twist $1/[n]$ at the bottom. From this formula, we see that the fraction of a rational tangle will be represented by a continued fraction. For example,

$$F([a] + ([c] \star (1/[b]))) = a + 1/(b + 1/c).$$

For this reason, it is useful to denote rational tangles by just writing a corresponding continued fraction.

We will illustrate this with an example. In Fig. 5.23, we illustrate these properties by starting with the tangle T from Fig. 5.20 and noting that T is a sum of the form $T = S + [2]$ where $S = [3] \star [-2]$. We could directly use the rules above to compute as follows.

$$\begin{aligned} F(S) &= F([3] \star F(1/[-2]) = 1/(1/F([3]) + F([-2])) \\ &= 1/(1/3 + -2)) = 3/(1-6) = -3/5. \\ F(T) &= F(S + [2]) = F([S]) + 2 = -3/5 + 2 = 7/5. \end{aligned}$$

In Fig. 5.23, we denote $x = F(S)$ and use property 3. above to rotate the tangle and then use property 1. to complete the calculation of

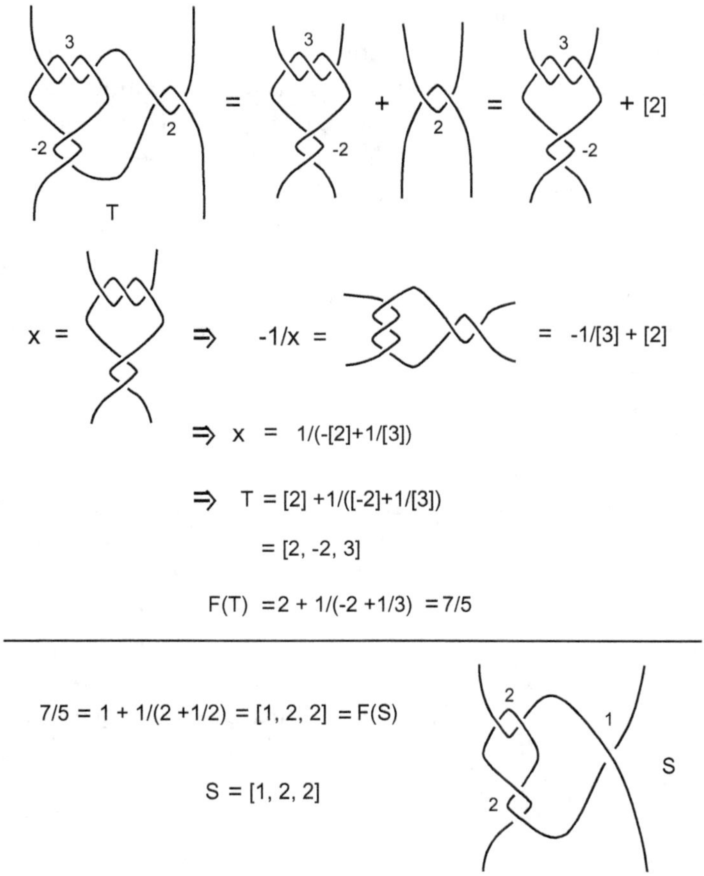

Fig. 5.23. Finding the Fraction of a Rational Tangle.

$F(S)$ without using property 2. In the rest of the figure, we note that $7/5$ has a positive continued fraction expansion as $7/5 = 1 + 2/7 = 1 + 1/(7/2) = 1 + 1/(2 + 1/2) = [1, 2, 2]$. The last symbol, $[1, 2, 2]$ is the symbol for this continued fraction. Thus,

$$[a, b, c] = a + 1/(b + 1/c)$$

and more generally

$$[a_1, a_2, a_3, a_4, \ldots, a_n] = a_1 + 1/(a_2 + 1/(a_3 + 1/(a_4 + \cdots + 1/a_n)))).$$

Thus, we shall from now on use the continued fraction written numerically to indicate the corresponding rational tangle. Thus,

$(1, 2, 2)$ will be our notation for the tangle diagram at the bottom of Fig. 5.23. Note that topologically this tangle is equivalent to $(1, 3, -2)$. These symbols represent the same numerical fraction and they each represent specific diagrams for rational tangles. Figure 5.23 gives the reader enough information to see that these two tangles are topologically equivalent. That is, the reader can take as an exercise to deform the tangle $(1, 3, -2)$ to the tangle $(1, 2, 2)$ by an isotopy that does not move any arcs over the endpoints of the tangle, while leaving the endpoints fixed.

John Horton Conway proved that *two rational tangles are topologically equivalent if and only if they have the same fraction.* See [Conway (1970); Kauffman and Lambropoulou (2003, 2004, 2011)] for statements and proofs of the Conway Theorem. The key to our approach to proving the Conway Theorem, in these cited papers, is the insight that any continued fraction tangle with a mixture of signs such as $(1, 3, -2)$ is not an alternating weave, and can be simplified by standard isotopies (swinging arcs that go over or under consecutive crossings) until it is alternating. The rational form of the tangle is retained under such isotopies, and the final alternating form corresponds to an entirely positive (or entirely negative) sequence such as $(1, 2, 2)$. The only positive or only negative sequence is uniquely obtained from the fraction itself by Euclid's algorithm. We give no more details here, but refer to our papers.

Note that $a + 1/(b + 1/c) = (a, b, c)$. Thus $(0, b, c) = 1/(b + 1/c) = 1/(b, c)$. These sequential symbols are useful for handling rational tangles and continued fractions. Note also that when we draw (a, b, c) the twists occur from right to left. Thus, in Fig. 5.23, we illustrate $(2, -2, 3)$ and the 2 occurs on the right in the diagram.

In Fig. 5.24, we illustrate a useful fact about about tangles. If T is a tangle and we apply rotation to it twice it will be rotated by 180 degrees. It turns out that for rational tangles, one can pick them up, rotate them by 180 degrees and they are topologically unchanged. The figure illustrates one case of the internal topological rotations that accomplish this fact.

The next operation we need is the **numerator construction** $N(T)$ that converts a tangle to a knot of link by connect the top strands to one another and the bottom strands to one another. See Fig. 5.20 for an illustration of the numerator operation and

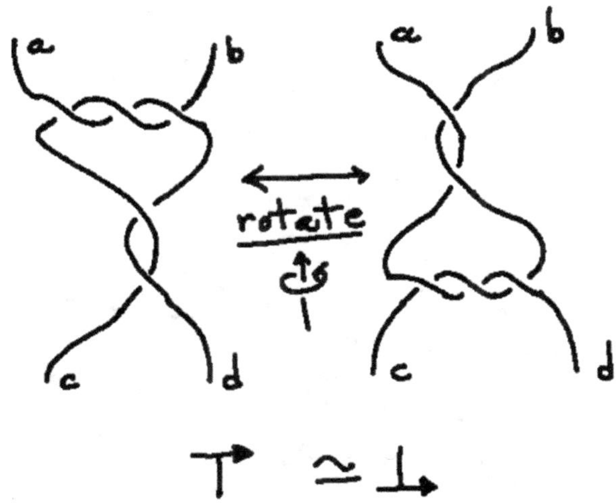

Fig. 5.24. Rotational Isotopy.

its so-called corresponding **denominator** operation. A theorem of Schubert gives a classification of the **rational knots** obtained as numerators of rational tangles. We do not need the full Schubert Theorem here but refer to [Kauffman and Lambropoulou (2003)] for the reader who wants more information about it.

In Fig. 5.25, we illustrate the formula $N(T + [n] + S + [-n]) = N(T + S)$. Actually, this is simply an identity for tangles, but it is often occurring inside a numerator closure. Note that algebraically this is just the cancellation of $[n]$ with $[-n]$ but topologically it can be interpreted as the result of a rotation that involves the intermediate tangle S. Such rotations occur in manipulating rope models and in the motions of curves in the force fields we have considered in the first section of this chapter.

Now, we are in position to look at Fig. 5.26 again. Note that the diagram indicated is $N((1, 2, 3) - (1, 2)) = N([1] + (0, 2, 3) + [-1] + (0, -2))$ and so the $[1]$ and $[-1]$ cancel is in the previous paragraph. Again, one can see this by an internal rotation in the numerator closure. The rest of the figure illustrates how this configuration is unknotted.

We can note that if $T = [a_1, a_2, \ldots, a_n]$ and $T^! = [a_n, a_{n-1}, \ldots, a_1]$, the tangle obtained by reversing all the terms in T,

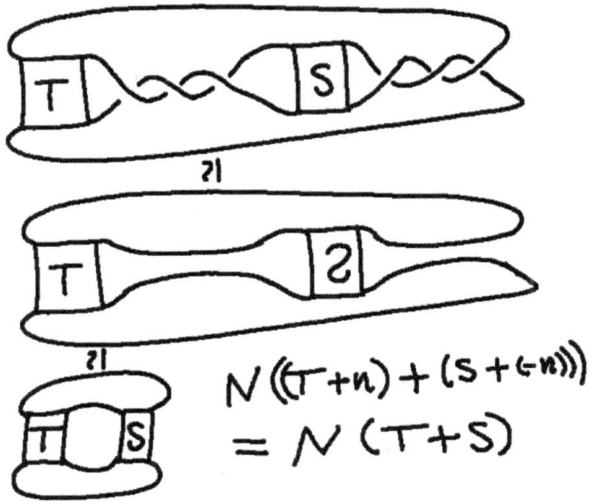

Fig. 5.25. Simplifying a Tangle Sum.

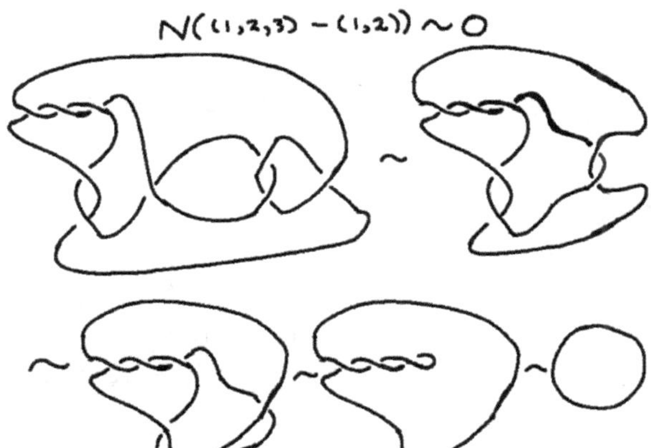

Fig. 5.26. $N((1,2,3)-(1,2))$ is Unknotted.

then $N(T)$ and $N(T^!)$ are topologically equivalent knots. Figure 5.27 will give the reader an appreciation of this fact. In that figure, we illustrate how $N(2,3,4) = N(4,3,2)$ by representing each of these tangles in a partially braided form. The reader can take as an exercise to see that any continued fraction template for a rational tangle can be put into such a braided form.

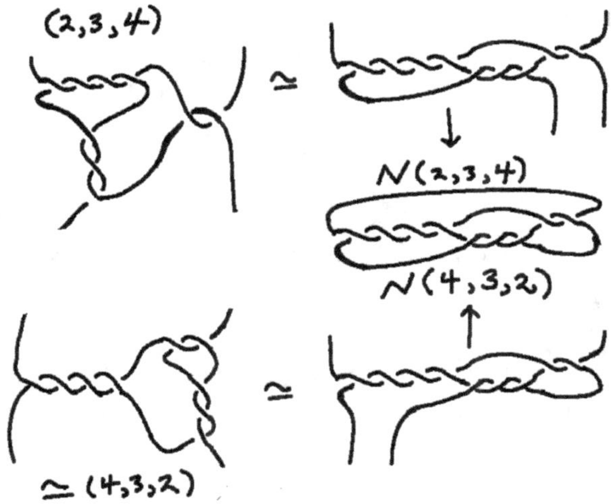

Fig. 5.27. $N(a, b, c) = N(c, b, a)$.

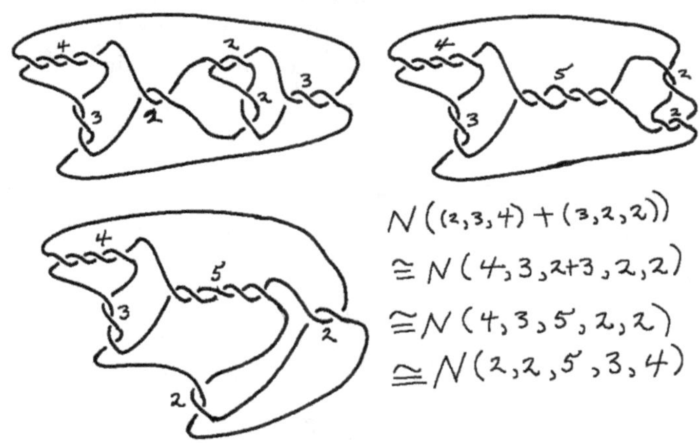

Fig. 5.28. $N((a, b, c) + (d, e, f)) = N(c, b, a + d, e, f)$.

Figure 5.28 illustrates $N(T + S)$ for $T = (2, 3, 4)$ and $S = (3, 2, 2)$. We have the basic fact that the numerator of the sum of two rational tangles T and S is topologically equivalent to the numerator of a rational tangle $T\Box S$ that is constructed from the two of them. Given $T = [a_1, a_2, \ldots, a_n]$ and $S = [b_1, b_2, \ldots, b_{n-1}, b_n]$, let $T\Box S$ be defined by the formula

$$T\Box S = (a_n, a_{n-1}, \ldots, a_2, a_1 + b_1, b_2, \ldots, b_{n-1}, b_n).$$

Then, $N(T+S)$ is topologically equivalent to $N(T\Box S)$. In order to see this topological equivalence the figure shows how to rotate and isotope $N(T+S)$ so that it is clearly equivalent to $N((T\Box S)^!)$. But we know that $N(T^!) = N(T)$ for any T. So this gives the desired result. Thus, in Fig. 5.28, we see that

$$N((2,3,4)+(3,2,2)) = N(4,3,2+3,2,2) = N(4,3,5,2,2).$$

The rotation that we see in $N(T+S)$ in Fig. 5.28 is the key to producing many diagrams of the unknot that can be difficult for the dynamics of the program to undo. The basic result is this: Let $T = [a_1, a_2, \ldots, a_n]$ and $S = [a_1, a_2, \ldots, a_{n-1}]$. Then, $N(T-S)$ is topologically equivalent to the unknot. Figure 5.26 illustrates how this works. The tangle S has just one less term than the tangle T and is obtained from it by removing the last term of T. We call S the *truncate* of T. In the figure, the reader should be able to see that the a_1 from T and the $-a_1$ from $-S$ flank a tangle and allow a rotation that cancels these terms and then brings a_2 and $-a_2$ into canceling position. This continues recursively and allows the knot to unwind and become unknotted. To see this result in tangle terms, the following identity is of use:

$$(\cdots a, b, 0, c, d, \cdots) = (\cdots a, b+c, d, \cdots)$$

where this equality can be read as topological equivalence of tangles, or as identity of continued fractions. For example, in fractions we have

$$\begin{aligned}(a,b,0,c,d) &= a + 1/(b + 1/(0 + 1/(c + 1/d)))) \\ &= a + 1/(b + 1/(1/(c + 1/d)))) \\ &= a + 1/((b+c) + 1/d) \\ &= (a, b+c, d).\end{aligned}$$

The unknotting is then mirrored in a bit of algebra. For example

$$\begin{aligned}&N((a,b,c,d) - (a,b,c)) \\ &= N(d,c,b,a-a,-b,-c) \\ &= N(d,c,b,0,-b,-c) \\ &= N(d,c,b-b,-c) \\ &= N(d,c,0,-c)\end{aligned}$$

$$= N(d, c - c)$$
$$= N(d, 0)$$

and $N(d, 0)$ is unknotted.

By the same token, we can arrange tangle diagrams that collapse to certain knots. For example,

$$\begin{aligned} &N((1,1,1,c,b,a) - (1,1,1)) \\ &\quad = N(a,b,c,1,1,1-1,-1,-1) \\ &\quad = N(a,b,c,1,1,0,-1,-1) \\ &\quad = N(a,b,c,0) \\ &\quad = N(a,b) \end{aligned}$$

This means that we can give the dynamics program tangled versions of knots and unknots and perform experiments to see if the program will undo the unknots and collapse the complexified knots to simpler forms.

5.3.1 *Collapse Examples*

Notation. We make the following definitions. Let $[[n]] = (1, 1, 1, \ldots, 1)$ with n 1's in the continued fraction sequence. Let $[n.m] = N([[n]] - [[m]])$. This is the notation we have used in the first section of the paper. Thus, Figs. 5.17 and 5.18 give images for $[11.10]$, a configuration that we know is unknotted from the above construction. These figures show that the KnotPlot program takes $[11.10]$ to a stable state that is still entangled. The second instance is Fig. 5.19 where we show how $[15.10]$ also has a complex stable state. By our discussion below, we will see that for $n > m$, $[n.m] = N([[n-m-1]])$. Thus, $[15.10] = N([[6]])$ and this is a figure eight knot. The upshot of these constructions is that we can give infinitely many examples of configurations that are either knotted or unknotted but have complex stable states in the force field algorithm of KnotPlot.

To prove that $[n.m] = N([[n - m - 1]])$, consider an example:

$$\begin{aligned} [7.3] &= N((1,1,1,1,1,1,1) + (-1,-1,-1)) \\ &= N(1,1,1,1,1,1,1-1,-1,-1) \end{aligned}$$

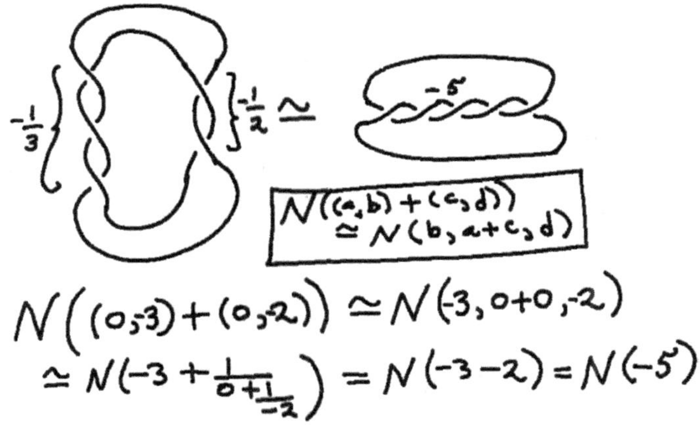

Fig. 5.29. Second example of the numerator of a sum of tangles.

$$= N(1,1,1,1,0) = N(1,1,1)$$
$$= N([[7-3-1]]).$$

Thus, the proof of this fact follows from the work we have already done, plus the use of this compressed notation.

We will also write $([[n]], a, b, c) = (1, 1, 1, \cdots 1, a, b, c)$ so that the $[[n]]$ can indicate n consecutive 1's in the tangle. Then, we have the basic result that

$$K_n(a,b,c,d) = N([[n+1]], a, b, c, d) - ([[n]]))$$

is isotopic to

$$N(a,b,c,d)$$

where (a,b,c,d) can be replaced by any continued fraction tangle. This result follows by the same reasoning as we have outlined above, and it provides a large collection of starting configurations for testing a self-repulsion program.

Our last example illustrates all the principles in this section. The basic knot we work with is the rational knot $N(7,7,7,7)$ illustrated in its initial form in Fig. 5.30 and it its final fully repelled form in Fig. 5.31. In Fig. 5.32, we show the initial form of $K_{11}(7,7,7,7) = N(([[12]], 7, 7, 7, 7) - ([[11]]).)$ In Fig. 5.33, we show the final self-repelled form that occurs from this beginning. We see that in this case $K_{11}(7,7,7,7)$ does not reduce to the minimal form of $N(7,7,7,7)$ in

Fig. 5.30. $N(7, 7, 7, 7)$ – Initial State.

Fig. 5.31. $N(7, 7, 7, 7)$ – Final State.

the KnotPlot environment. It is our conjecture that no environment of this type will fully reduce all of the configurations $K_n(7, 7, 7, 7)$, and we challenge users or designers of other programs to experiment with this family of configurations of the knot $N(7, 7, 7, 7)$. There is,

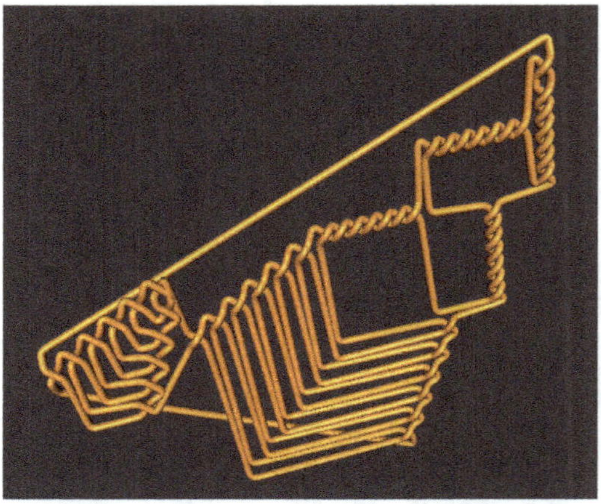

Fig. 5.32. Last Knot – Initial State.

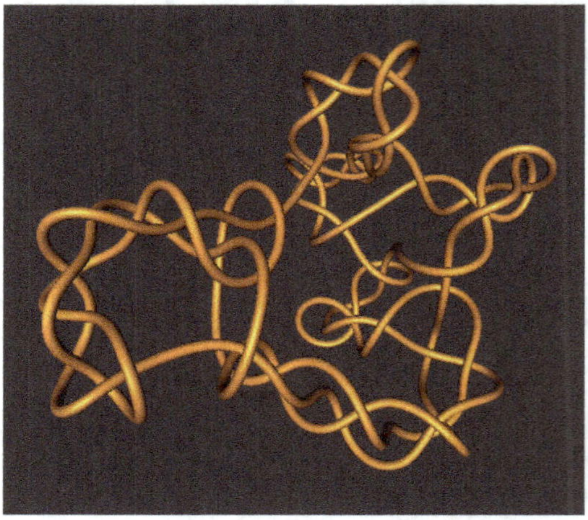

Fig. 5.33. Last Knot – Final State.

of course, nothing special about the choices we have made here. The phenomenon of non-collapse will occur for sufficiently large complexifications of any knot or unknot.

Bibliography

Abdel-Malek, K. Blackmore, D. and Joy, K. (2004). Swept volumes: Foundations, perspectives and applications. *International Journal of Shape Modeling*, 12(2), pp. 87–127.

Abdel-Malek, K. and Othman, S. (1999). Multiple sweeping using the David-Hartenberg representation method. *Computer-Aided Design*, 31, pp. 567–583.

Alexandrov, V. (2009). On the total mean curvature of a nonrigid surface. *Siberian Mathematical Journal*, 50(5), pp. 757–759.

Alexandrov, V. A. (1989). Remarks on Sabitov's conjecture that volume is stationary under an infinitesimal bending of a surface. *Siberian Mathematical Journal*, 30(5), pp. 16–24.

Alexandrov, V. A. (2010). New manifestations of the Darboux's rotation and translation fields of a surface. *New Zealand Journal of Mathematics*, 40, pp. 59–65.

Almgren, F. J. and Rivin, I. (1992). The mean curvature integral is invariant under bending. In I. Rivin *et al.*, (Ed.) *The Epstein Birthday Schrift* (pp. 1–22). University of Warwick.

Baranska, J., Przybyl, S. and Pieranski, P. (2008). Curvature and torsion of the tight closed trefoil knot. *The European Physical Journal B*, 66, pp. 547–556.

Barrett, J. W., Garcke, H. and Nürnberg, R. (2016). Computational parametric Willmore flow with spontaneous curvature and area difference elasticity effects. *SIAM Journal on Numerical Analysis*, 54(3), pp. 1732–1762.

Bauer, M. and Kuwert, E. (2003). Existence of minimizing Willmore surfaces of prescribed genus. *International Mathematics Research Notices*, 2003(10), pp. 553–576.

Belov, K. M. (1968). O beskonechno malyh izgib. toroobraznoi pov. vrashceniya. *Siberian Mathematical Journal*, 9(3), pp. 490–494.

Belova, O., Mikeš, J. and Sherkuziyev, M. (2021). An analytical inflexibility of surfaces attached along a curve to a surface regarding a point and plane. *Results in Mathematics*, 76(56), 13 pp.

Bernard, Y. and Rivière, T. (2014). Energy quantization for Willmore surfaces and applications. *Annals of Mathematics*, 180, pp. 87–136.

Blackmore, D., Leu, M. C. and Shif, F. (1994). Analysis and modeling of deformed swept volume. *Computer-Aided Design*, 26(4), pp. 315–326.

Blatt, S. and Reiter, P. (2014). How nice are critical knots? Regularity theory for knot energies. *Journal of Physics: Conference Series*, 544, 012020.

Borsuk, K. (1947). Sur la courbure totale des courbes fermées. *Annales de la Société Polonaise de Mathématiques*, 20, pp. 251–265.

Broscăteanu, Ş.-C., Mihai, A. and Olteanu, A. (2024). A note on the infinitesimal bending of a rectifying curve. *Symmetry*, 16(10), 1361.

Capovilla, R., Chryssomalakos, C. and Guven, J. (2002). Hamiltonians for curves. *arXiv preprint* arXiv:nlin/0204049.

Carmo, M. P. do. (1976). *Differential Geometry of Curves and Surfaces*, Prentice-Hall.

Chen. B.-Y. (1971a). On a theorem of Fenchel-Borsuk-Willmore-Chern-Lashof. *Mathematische Annalen*, 194, pp. 19–26.

Chen, B.-Y. (1971b). On the total curvature of immersed manifolds. I. An inequality of Fenchel-Borsuk-Willmore. *American Journal of Mathematics*, 93, pp. 148–162.

Chen, B.-Y. (2014). *Total Mean Curvature and Submanifolds of Finite Type*, second edition, Series in Pure Mathematics: Volume 27, World Scientific.

Cohn-Vossen, S. E. (1959). *Nekotorye voprosy differ. geometrii c celom*, J. Fizmatgiz, Moskva.

Connelly, R. (1980). Assumptions and unsolved questions in the theory of bending. In: *Studies in the Metric Theory of Surfaces* (pp. 228–238). (Russian translation), Mir, Moskva.

Conway, J. H. (1970). An enumeration of knots and links and some of their algebraic properties. In J. Leech (Ed.), *Proceedings of the Conference on Computational Problems in Abstract Algebra held at Oxford in 1967*, J. Leech ed. (First edition 1970) (pp. 329–358), Pergamon Press.

Eells, J. (1987). The surfaces of Delaunay, *Mathematical Intelligencer*, 9, pp. 53–57.

Efimov, N. V. (1948). Kachestvennye voprosy teorii deformacii poverhnostei, *Uspekhi Matematicheskikh Nauk*, 3(2), pp. 47–158.

Fáry, M. I. (1949). Sur la Courbure Totale d'une Courbe Gauche Faisant un Nœud, *Bulletin de la Société Mathématique de France*, 77, pp. 128–138.

Fenchel, W. (1929). Uber Krummung und Windung geschlossener Raumkurven, *Mathematische Annalen*, 101, pp. 238–252.

Fernández, L. (2005). *Differential Geometry of Curves and Surfaces*, https://fsw01.bcc.cuny.edu/luis.fernandez01/web/texts/dgcs.pdf.

Fomenko, V. T. (1967). Nekotorye rezultaty teorii beskonechno malyh izgibanii poverhnostei. *Matematicheskii Sbornik*, 72(114), 3, pp. 388–411.

Freedman, M. H., He, Z.-X. and Wang, Z. (1994). M̈bius energy of knots and unknots, *Annals of Mathematics*, 139(1), pp. 1–50.

Gamma, E., Helm, R., Johnston, R. and Vlisides, J. (1995). *Design Patterns: Elements of Reusable Object-Oriented Software*, Addison-Wesley.

Gözütok, Ü., Çoban, H. A., Sağiroglu, Y. (2020). Ruled surfaces obtained by bending of curves. *Turkish Journal of Mathematics*, 139(44), pp. 300–306.

Gray, A. (1998). *Modern Differential Geometry of Curves and Surfaces with Mathematica*, CRC Press, Boca Raton.

Green, A. E. and Zerna, W. (1968). *Theoretical Elasticity* (2nd ed), Oxford University Press.

Gruber, A., Toda, M. and Tran, H. (2019). On the variation of curvature functionals in a space form with application to a generalized willmore energy. *Annals of Global Analysis and Geometry*, 56, pp. 147–165.

Helfrich, W. (1973). Elastic properties of lipid bilayers: Theory and possible experiments. *Zeitschrift für Naturforschung C*, 28(11–12), pp. 693–703.

Henrich, A. and Kauffman, L. H. (2014). Unknotting unknots. *American Mathematical Monthly*, pp. 379–390.

Hinterleitner, I., Mikeš, J. and Stránská, J. (2008). Infinitesimal f-planar transformations. *Journal of Mathematical Sciences*, 52(4), pp. 13–18.

Ivanova-Karatopraklieva, I. (1992). Infinitesimal bending of higher order of rotational surfaces with a planar pole. *Serdica*, 18, pp. 59–78.

Ivanova-Karatopraklieva, I. and Sabitov, I. (1995). Bending of surfaces II. *Journal of Mathematical Sciences*, 74(3), pp. 997–1043.

Jimez, M. I. (2018). Infinitesimal bendings of complete Euclidean hypersurfaces. *Annali di Matematica Pura ed Applicata*, 157, pp. 513–527.

Kauffman, L. K. (1987). State models and the Jones polynomial. *Topology*, 26, pp. 395–407.

Kauffman, L. H. (2001). *Knots and Physics* (3rd ed.), World Scientific Publishing.

Kauffman, L. H. (2013). *Knots and Physics* (4rd ed.), Series on Knots and Everything, 53. World Scientific Publishing Co.

Kauffman, L. H. and Lambropoulou, S. (2003). On the classification of rational knots. *L'Enseignement Mathématiques*, 49, pp. 357–410.

Kauffman, L. H. and Lambropoulou, S. (2004). On the classification of rational tangles. *Advances in Applied Mathematics*, 33(2), pp. 199–237.

Kauffman, L. H. and Lambropoulou, S. (2011). Hard knots and collapsing tangles. In *Introductory Lectures in Knot Theory* (pp. 187–247). World Scientific. (Series on Knots and Everything, Vol. 46).

Kauffman, L. H., Velimirović, Lj. S., Najdanović, M. S. and Rančić, S. (2019). Infinitesimal bending of knots and energy change. *Journal of Knot Theory and Its Ramifications*, 8(11), 1940009.

Maksimović, M., Velimirović, Lj. and Najdanović, M. (2021). Infinitesimal bending of DNA helices. *Turkish Journal of Mathematics*, 45, pp. 520–528.

Maksimović, M. D., Rančić, Najdanović, M. S., Velimirović, Lj. S. and Ljajko, E. S. (2023). On the torsional energy of torus knots under infinitesimal bending. *Analele Stiintifice ale Universitatii Ovidius Constanta, Seria Matematica*, 31(1), pp. 181–197.

Marques, F. C. and Neves, A. (2014). Min-max theory and the Willmore conjecture. *Annals of Mathematics*, 179(2), pp. 683–782.

Milnor, J. W. (1950). On the total curvature of knots. *The Annals of Mathematics*, 52(2), pp. 248–257.

Mikeš J. *et al.* (2019). *Differential Geometry of Special Mappings* (2nd ed.). Palacky University.

Minčić, S. and Velimirović, Lj. (2007). *Differential Geometry-in Serbian*, Faculty of Sciences and Mathematics, University of Niš, Serbia.

Najdanović, M. S. (2014). Infinitesimal bending influence on the volume change. *Applied Mathematics and Computation*, 243, pp. 801–808.

Najdanović, M. S. (2015). Infinitesimal bending influence on the Willmore energy of curves. *Filomat*, 29(10), pp. 2411–2419.

Najdanović, M. S. (2017). On the problem of geodesic mappings and deformations of generalized Riemannian spaces. *Journal of Mathematical Analysis and Applications*, 452, pp. 634–645.

Najdanović, M. S. and Velimirović, Lj. S. (2017a). Second order infinitesimal bending of curves. *Filomat*, 31(13), pp. 4127–4137.

Najdanović, M. S. and Velimirović, Lj. S. (2017b). Willmore energy of curves under second order infinitesimal bending. *Miskolc Mathematical Notes*, 17(2), pp. 979–987.

Najdanović, M. and Velimirović, Lj. (2018). Infinitesimal bending of curves on the ruled surfaces. *University thought — Publication in Natural Sciences*, 8(1), pp. 46–51.

Najdanović, M. S., Rančić, S. R., Kauffman, L. and Velimirović, Lj. S. (2019). The total curvature of knots under second-order infinitesimal bending. *Journal of Knot Theory and Its Ramifications*, 28(1), 1950005.

Najdanović, M., Maksimović, M. and Velimirović, Lj. (2021a). Curves on ruled surfaces under infinitesimal bending. *Bulletin of Natural Sciences Research*, 11(1), pp. 38–43.

Najdanović, M. S., Velimirović, Lj. S. and Rančić, S. R. (2021b). The total torsion of knots under second order infinitesimal bending. *Applicable Analysis and Discrete Mathematics*, 15, pp. 283–294.

Najdanović, M. S., Velimirović, Lj. S. and Vesić, N. O. (2022). Geodesic infinitesimal deformations of generalized Riemannian space. *Mediterranean Journal of Mathematics*, 19, Article 145 (11 pages).

Najdanović, M. S., Rančić, S. R. and Velimirović, Lj. S. (2024). Total torsion and spherical curves bending. *Mediterranean Journal of Mathematics*, 21, Article 74.

O'Hara, J. (1991). Energy of a knot. *Topology*, 30(2), pp. 241–247.

Pansonato, C. C. and Costa, S. I. R. (2008). Total torsion of curves in three-dimensional manifolds. *Geometriae Dedicata*, 136, pp. 111–121. https://doi.org/10.1007/s10711-008-9277-6.

Perlova, N. G. (1969). O beskonechno malyh izgibaniyah 1.,2. i 3.-go poryadkov zam- knutyh rebristyh poverhnostei vrascheniya. *Commentationes Mathematicae Universitatis Carolinae*, 10, pp. 1–35.

Pogorelov, A. V. (1969). Vneshnaya geometriya vypuklyh poverhnostei, Moskva, Nauka.

Pozzetta, M. (2019–2020). *Willmore-type Energies of Curves and Surfaces*, Università degli studi di Pisa, Ph.D. Thesis.

Qin, Y. and Li, S. (2002). Total torsion of closed lines of curvature. *Bulletin of the Australian Mathematical Society*, 65, pp. 73–78.

Rančić, S., Najdanović, M and Velimirović, Lj. (2019). Total normalcy of knots, *Filomatm*, 33, 4, pp. 1259–1266.

Rančić, S., Najdanović, M and Velimirović, Lj. (2023). Surfaces defined by bending of knots. *Filomat*, 37(25), pp. 8635–8640.

Rančić, S. R., Velimirović, Lj. S. and Najdanović, M. S. (2024). On the torsional energy of deformed curves and knots. *Axioms*, 13, Article 661 (15 pages).

Rawdon, E. J. and Simon, J. (2002). Möbius energy of thick knots. *Topology and its Applications*, 125, pp. 97–109.

Reiter, P. and Schumacher, H. (2020). Sobolev gradients for the mobius energy. *arXiv preprint* arXiv:2005.07448.

Rýparová, L. and Mikeš, J. (2019). Infinitesimal rotary transformation. *Filomat*, 33(4), pp. 1153–1157.

Scharein, R. *KnotPlot* (program available from the web), https://knotplot.com.

Scharein, R. (1998). *Interactive Topological Drawing*, PhD Thesis, Department of Computer Science, University of British Columbia.

Scharein, R., Ishihara, K., Arsuaga, J., Diao, Y., Shimokawa, K. and Vazquez, M. (2009). Bounds for the minimum step number of knots in the simple cubic lattice. *Journal of Physics A*, 42(47), 475006.

Schlichtkrull, H. (2011). *Curves and Surfaces-Lecture notes for Geometry 1*, University of Copenhagen.

Shepherd, G., Kruglinski, D. (2003). *Programming with Microsoft Visual C++.NET*, 6th ed. Microsoft Press.

Simon, L. (1993). Existence of surfaces minimizing the Willmore functional. *Communications in Analysis and Geometry*, 1(2), pp. 281–326.

Simon, J. K. (1994). Energy functions for polygonal knots. *Journal of Knot Theory and Its Ramifications*, 3(3), pp. 299–320.

Simon, J. (1996). Energy functions for knots: Beginning to predict physical behavior. In Mesirov J. P., Schulten K., Sumners D. W. (Eds.), *Mathematical Approaches to Biomolecular Structure and Dynamics* (pp. 39–58). The IMA Volumes in Mathematics and its Applications, Vol. 82, Springer, New York, NY.

Spivak, M. (1999). *A Comprehensive Introduction to Differential Geometry* (3rd ed., Vol. 3). Publish or Perish, pp. ix+319.

Struik, D. J. (1961). *Differential Geometry*, Addison-Wesley.

Truesdell, C. (1983). The influence of elasticity on analysis: The classic heritage. *Bulletin of the American Mathematical Society*, 9, pp. 293–310.

Van Rensburg, E. J. and Rechnitzer, A. (2011). BFACF-style algorithms for polygons in the body-centered and face-centered cubic lattices. *Journal of Physics A: Mathematical and Theoretical*, 44(16), 165001.

Vekua, I. (1959). *Obobschennye Analiticheskie Funkcii*, Moskva.

Velimirović, Lj. (2001). Change of geometric magnitudes under infinitesimal bending. *Facta Universitates*, 3(11), pp. 135–148.

Velimirović, Lj. S. (1994). On the infinitesimal rigidity of a class of toroid surfaces of rotation. *Collection of the Scientific Papers of the Faculty of Science Kragujevac*, 16, pp. 123–130.

Velimirović, Lj. S. (1995). On infinitesimal deformations of a toroid rotational surfaces generated by a quadrangular meridian. *Filomat*, 9(2), pp. 197–204.

Velimirović, Lj. S. (1997). Beskonechno malye izgibaniya toroobraznoi poverhnosti vrascheniya s mnogougol'nym meridianom. *Izvestiya VUZ-Matematika* 9, pp. 3–7.

Velimirović, Lj. S. (1998a). A new proof of theorem of Belov. *Publications de l'Institut Mathématique*, 63(77), pp. 102–115.

Velimirović, Lj. S. (1998b). On variation of the volume under infinitesimal bending of a closed rotational surface. *Novi Sad Journal of Mathematics*, 29(3), pp. 377–386.

Velimirović, Lj. S. and Ćirić, M. S. (2011). On the total mean curvature of piecewise smooth surfaces under infinitesimal bending. *Applied Mathematics Letters*. 24, pp. 1515–1519.

Velimirović, Lj. S. and Rančić, R. S. (2007). Rigidity of toroid formed by revolution of parallelogram. *Filomat*, pp. 111–122.

Velimirović, Lj. S., Minčić, S. M. and Stanković, M. S. (2007). Infinitesimal deformations of basic tensor in generalized Riemannian space. *Filomat*, 21(2), pp. 235–242.

Velimirović, Lj. S. and Rančić, R. S. (2009). Notes on infinitesimal bending of a toroid formed by revolution of a polygonal meridian. *Journal for Geometry and Graphics*, 13, pp. 177–186.

Velimirović, Lj. S. and Rančić, R. S. (2010). Higher order infinitesimal bending of a class of toroids. *European Journal of Combinatorics*, 31(4), pp. 1136–1147.

Velimirović, Lj. S., Ćirić, M. S. and Zlatanović, M. Lj. (2010a). Bendings of spherical curves. In *Proceedings of the 25th national and 2st international scientific conference moNGeometrija*, pp. 657–667.

Velimirović, Lj. S., Ćirić, M. S. and Cvetković, M. D. (2010b). Change of the Willmore energy under infinitesimal bending of membranes. *Computers and Mathematics with Applications*, 59(12), pp. 3679–3686.

Velimirović, Lj. S., Ćirić, M. S. and Velimirović, N. (2011). On the Willmore energy of shells under infinitesimal deformations. *Computers and Mathematics with Applications*, 61(11), pp. 3181–3190.

Velimirović, Lj. S., Cvetković, M. D., Ćirić, M. S. and Velimirović, N. (2012). Analysis of Gaudi surfaces at small deformations. *Applied Mathematics and Computation*, 218, pp. 6999–7004.

Velimirović, Lj. S., Cvetković, M. D., Najdanović, M. S. and Velimirović, N. M. (2013). Variation of shape operator under infinitesimal bending of surface. *Applied Mathematics and Computation*, 225, pp. 480–486.

Weiner, J. L. (1978). On a problem of Chen, Willmore, *et al.*, *Indiana University Mathematics Journal*, 27(1), pp. 19–35.

Wente, H. (1986). Counterexample to a conjecture of H. Hopf. *Pacific Journal of Mathematics*, 121, pp. 193–243.

Willmiams, C. J. K. (1987). Use of structural analogy in generation of smooth surfaces for engineering purposes. *Computer-Aided Design*, 19, pp. 310–322.

Willmore, T. J. (1965). Note on embedded surfaces, *An. Şti. Univ. "Al. I. Cuza" Iaşi Secţ. I a Mat. (N.S.) 11B*, pp. 493–496.

Zorin, D. (2005). Curvature-based energy for simulation and variational modeling. In *Proceedings of the International Conference on Shape Modeling and Applications (SMI'05)* (pp. 196–204). IEEE Computer Society.

Index

Vol. 71: *The Geometry of the Universe*
by C. Rourke

Vol. 70: *Geometric Foundations of Design: Old and New*
by J. Kappraff

Vol. 69: *Mathematics of Harmony as a New Interdisciplinary Direction and "Golden" Paradigm of Modern Science*
Volume 3: The "Golden" Paradigm of Modern Science: Prerequisite for the "Golden" Revolution in Mathematics, Computer Science, and Theoretical Natural Sciences
by A. Stakhov

Vol. 68: *Mathematics of Harmony as a New Interdisciplinary Direction and "Golden" Paradigm of Modern Science*
Volume 2: Algorithmic Measurement Theory, Fibonacci and Golden Arithmetic's and Ternary Mirror-Symmetrical Arithmetic
by A. Stakhov

Vol. 67: On Complementarity: A Universal Organizing Principle
by J. S. Avrin

Vol. 66: *Invariants and Pictures: Low-dimensional Topology and Combinatorial Group Theory*
by V. O. Manturov, D. Fedoseev, S. Kim & I. Nikonov

Vol. 65: *Mathematics of Harmony as a New Interdisciplinary Direction and "Golden" Paradigm of Modern Science*
Volume 1: The Golden Section, Fibonacci Numbers, Pascal Triangle, and Platonic Solids
by A. Stakhov

Vol. 64: *Polynomial One-cocycles for Knots and Closed Braids*
by T. Fiedler

Vol. 63: *Board Games: Throughout the History and Multidimensional Spaces*
by J. Kyppö

Vol. 62: *The Mereon Matrix: Everything Connected through (K)nothing*
edited by J. B. McNair, L. Dennis & L. H. Kauffman

Vol. 61: *Numeral Systems with Irrational Bases for Mission-Critical Applications*
by A. Stakhov

Vol. 60: *New Horizons for Second-Order Cybernetics*
edited by A. Riegler, K. H. Müller & S. A. Umpleby

Vol. 59: *Geometry, Language and Strategy: The Dynamics of Decision Processes — Vol. 2*
by G. H. Thomas

Vol. 58: *Representing 3-Manifolds by Filling Dehn Surfaces*
by R. Vigara & A. Lozano-Rojo

Vol. 57: *ADEX Theory: How the ADE Coxeter Graphs Unify Mathematics and Physics*
by S.-P. Sirag

Vol. 56: *New Ideas in Low Dimensional Topology*
edited by L. H. Kauffman & V. O. Manturov

Vol. 55: *Knots, Braids and Möbius Strips*
by J. Avrin

Vol. 54: *Scientific Essays in Honor of H Pierre Noyes on the Occasion of His 90th Birthday*
edited by J. C. Amson & L. H. Kauffman

Vol. 53: *Knots and Physics (Fourth Edition)*
by L. H. Kauffman

Vol. 52: *Algebraic Invariants of Links (Second Edition)*
by J. Hillman

Vol. 51: *Virtual Knots: The State of the Art*
by V. O. Manturov & D. P. Ilyutko

Vol. 50: *Topological Library*
Part 3: Spectral Sequences in Topology
edited by S. P. Novikov & I. A. Taimanov

Vol. 49: *Hopf Algebras*
by D. E. Radford

Vol. 48: *An Excursion in Diagrammatic Algebra: Turning a Sphere from Red to Blue*
by J. S. Carter

Vol. 47: *Introduction to the Anisotropic Geometrodynamics*
by S. Siparov

Vol. 46: *Introductory Lectures on Knot Theory*
edited by L. H. Kauffman, S. Lambropoulou, S. Jablan & J. H. Przytycki

Vol. 45: *Orbiting the Moons of Pluto*
Complex Solutions to the Einstein, Maxwell, Schrödinger and Dirac Equations
by E. A. Rauscher & R. L. Amoroso

Vol. 44: *Topological Library*
Part 2: Characteristic Classes and Smooth Structures on Manifolds
edited by S. P. Novikov & I. A. Taimanov

Vol. 43: *The Holographic Anthropic Multiverse*
by R. L. Amoroso & E. A. Ranscher

Vol. 42: *The Origin of Discrete Particles*
by T. Bastin & C. Kilmister

Vol. 41: *Zero to Infinity: The Fountations of Physics*
by P. Rowlands

Vol. 40: *Intelligence of Low Dimensional Topology 2006*
edited by J. Scott Carter et al.

Vol. 39: *Topological Library*
Part 1: Cobordisms and Their Applications
edited by S. P. Novikov & I. A. Taimanov

Vol. 38: *Current Developments in Mathematical Biology*
edited by K. Mahdavi, R. Culshaw & J. Boucher

Vol. 37: *Geometry, Language, and Strategy*
by G. H. Thomas

Vol. 17: *Hypercomplex Iterations — Distance Estimation and Higher Dimensional Fractals*
by Y. Dang, L. H. Kauffman & D. Sandin

Vol. 16: *Delta — A Paradox Logic*
by N. S. Hellerstein

Vol. 15: *Lectures at KNOTS '96*
by S. Suzuki

Vol. 14: *Diamond — A Paradox Logic*
by N. S. Hellerstein

Vol. 13: *Entropic Spacetime Theory*
by J. Armel

Vol. 12: *Relativistic Reality: A Modern View*
edited by J. D. Edmonds, Jr.

Vol. 11: *History and Science of Knots*
edited by J. C. Turner & P. van de Griend

Vol. 10: *Nonstandard Logics and Nonstandard Metrics in Physics*
by W. M. Honig

Vol. 9: *Combinatorial Physics*
by T. Bastin & C. W. Kilmister

Vol. 8: *Symmetric Bends: How to Join Two Lengths of Cord*
by R. E. Miles

Vol. 7: *Random Knotting and Linking*
edited by K. C. Millett & D. W. Sumners

Vol. 6: *Knots and Applications*
edited by L. H. Kauffman

Vol. 5: *Gems, Computers and Attractors for 3-Manifolds*
by S. Lins

Vol. 4: *Gauge Fields, Knots and Gravity*
by J. Baez & J. P. Muniain

Vol. 3: *Quantum Topology*
edited by L. H. Kauffman & R. A. Baadhio

Vol. 2: *How Surfaces Intersect in Space — An Introduction to Topology (Second Edition)*
by J. S. Carter

Vol. 1: *Knots and Physics (Third Edition)*
by L. H. Kauffman